Welding

Skills, Processes & Practices, Level 2

LARRY JEFFUS   LAWRENCE BOWER

HUGH McPHILLIPS

CENGAGE
Learning·

Australia • Brazil • Japan • Korea • Mexico • Singapore • Spain • United Kingdom • United States

Welding: Skills, Processes & Practices, Level 2
Larry Jeffus, Lawrence Bower and Hugh McPhillips

Publishing Director: Linden Harris

Commissioning Editor: Lucy Mills

Development Editor: Claire Napoli

Production Editor: Beverley Copland

Production Controller: Eyvett Davis

Marketing Manager: Lauren Mottram

Typesetter: S4Carlisle Publishing Services

Cover design: HCT Creative

Text design: Design Deluxe

I would like to take this opportunity to thank my wife Andrea for all her support throughout the adaption of this book. Lucy, Helen and Claire at Cengage Learning for their guidance in helping to bring the detail to life and to all my colleagues up and down the country for their support, not least David Cleghorn and William Roffey for their reviews. I would also like to say a substantial thank you to Andy Ronayne who guided me in my earlier years in teaching and taught me so much about welding and to Roy Davies who as my line manager supported me in all my initiatives. I hope you will find this book to be a support in developing your welding skills and will go on to a rewarding and enjoyable career.

Hugh McPhillips

For product information and technology assistance,
contact **emea.info@cengage.com**.

For permission to use material from this text or product, and for permission queries, email **emea.permissions@cengage.com**.

This work is adapted from *Welding Skills, Processes and Practices for Entry-Level Welders, Books 1, 2 and 3*, 1st edition by *Larry Jeffus and Laurence Bower* published by Cengage Higher Education, a division of Cengage Learning, Inc. ©2010.

British Library Cataloguing-in-Publication Data
A catalogue record for this book is available from the British Library.

ISBN: 978-1-4080-6038-4

Cengage Learning EMEA
Cheriton House, North Way, Andover, Hampshire, SP10 5BE
United Kingdom

Cengage Learning products are represented in Canada by Nelson Education Ltd.

For your lifelong learning solutions, visit **www.cengage.co.uk**

Purchase your next print book, e-book or e-chapter at **www.cengagebrain.com**

Printed in Greece by Bakis
1 2 3 4 5 6 7 8 9 10 – 15 14 13

Contents

6 Thermal Cutting Techniques 119

7 Manual Metal Arc Welding 161

8 Metal Inert Gas Shielded and Flux Cored Welding 209

9 Tungsten Inert Gas Shielded Welding 291

10 Heat Input and Distortion Control Techniques 331

11 Quality Control 346

Mapping Grid

NVQ Unit Title	Chapter 1 Introduction to Welding	Chapter 2 Working Safely, Efficiently & Effectively	Chapter 3 Communicating Technical Information	Chapter 4 Engineering Materials, Measurement & Fitting Skills	Chapter 5 Oxy-fuel Gas Welding, Brazing and Soldering	Chapter 6 Thermal Cutting Techniques	Chapter 7 Manual Metal Arc Welding	Chapter 8 Metal Inert Gas Shielded	Chapter 9 Tungsten Inert Gas Shielded Welding	Chapter 10 Heat Input & Distortion Techniques	Chapter 11 Quality Control
Working Efficiently and Effectively in Engineering											
Complying with Statutory Regulations and Organisational Safety Requirements											
Using and Interpreting Engineering Data and Documentation											
Joining Materials by the Manual Metal Arc Welding Process											
Joining Materials by the Manual MIG/MAG and Other Continuous Wire Processes											
Joining Materials by the Manual TIG and Plasma-arc Welding Processes											
Joining Materials by the Manual Gas Welding Process											
Producing Fillet Welded Joints using a Manual Welding Process											
Joining Materials by Manual Torch Brazing and Soldering											
Cutting Materials using Hand Operated Thermal Cutting Equipment											

Mapping Grid Continued

NVQ Unit Title	Chapter 1 Introduction to Welding	Chapter 2 Working Safely, Efficiently & Effectively	Chapter 3 Communicating Technical Information	Chapter 4 Engineering Materials, Measurement & Fitting Skills	Chapter 5 Oxy-fuel Gas Welding, Brazing and Soldering	Chapter 6 Thermal Cutting Techniques	Chapter 7 Manual Metal Arc Welding	Chapter 8 Metal Inert Gas Shielded	Chapter 9 Tungston Inert Gas Shielded Welding	Chapter 10 Heat Input & Distortion Techniques	Chapter 11 Quality Control
Cutting and Shaping Materials using Gas Cutting Machines											
Cutting Materials using Saws and Abrasive Discs											
Assembling Components using Mechanical Fasteners											
Marking Out Components for Fabrication											
Heat Treating Materials for Fabrication Activities											
VRQ Unit Title											
Engineering Environment Awareness											
Engineering Techniques											
Engineering Principles											
Fabrication and Engineering Principles											
Manual welding techniques											
Producing components from metal plate											
Non-fusion thermal joining methods											
Thermal cutting techniques											

Introduction

Welcome to the world of welding. From the historical data you can see the role that welding developments internationally have had on our present lives and how they continue to expand our horizons. Welding not only provides a career structure but also gives the opportunity for people to express themselves artistically through the use of a range of materials. This gives a diverse experience in the arts from custom-built motorbikes to elaborate works of art. Walk through a theme park and experience a whole range of rides that rely on welding for their integral strength and rigidity and see how much we have come to depend on welding. From commercial to surgical industries, welding continues to play an important part in our lives.

For anyone planning a career or just using welding, there are certain characteristics that he or she will need. You need good eyesight, manual dexterity, hand and eye coordination and an understanding of the principles of welding technology, something which is an ongoing experience.

This book aims to take you through the underlying principles and concepts which make up the four principle welding processes, notably Oxy-Acetylene, Manual Metal Arc, Metal Arc Gas Shielded and Tungsten Arc Gas Shielded welding, as well as cutting techniques.

For those studying for a career, I can say from 40 years of experience you will find it challenging in a number of ways, both rewarding and frustrating. The range of work and the opportunities are enormous with the potential to travel and experience many cultures and work environments. With worldwide shortages in skilled welders the potential for a rewarding career has never been better.

I am often asked what makes a good welder. Some are naturally gifted and some work meticulously to achieve a level of excellence. Essentially what is required is an empathy with the subject and a desire to learn all that you can about the process, materials, effects of heat input and techniques to deliver a quality weld. You also have to be aware of what defects can occur and how to remedy these problems. It takes time to develop these techniques and a considerable amount of practice to achieve a good standard.

OCCUPATIONAL OPPORTUNITIES IN WELDING

Because of the diverse nature of the welding industry, the exact job duties of each skill area will vary. The term 'weldment' will be used throughout this book and this refers to any component that may have more than one weld. The following are general descriptions of the job classifications used in our profession; specific tasks may vary from one location to another.

- Welders perform manual or semi-automatic welding. They are the skilled craftspeople who through their labour produce the welds on a variety of complex products.
- Welding operatives run adaptive control, automatic, mechanized or robotic welding equipment. They need to be fully conversant with all the welding parameters and make adjustments as and when necessary.
- Welder assistants are employed to clean and remove slag from welds, help and move weldments into position for the welder and supply electrodes when requested.
- Welding assemblers/fitters position all the components in the proper places ready for tack welding. These skilled workers must be able to interpret drawings and welding procedures. They must also have a working knowledge of the effects of expansion and contraction of a wide range of materials.

- Welding inspectors are often required to hold a special certification such as one supervised by the Welding Institute in Britain or the American Welding Society. Known as a Certified Welding Inspector, candidates must pass a test covering the welding process, drawing interpretation, weld procedures, weld symbols, metallurgy, and inspection techniques. Eye tests are also required on a regular basis once the technical skills have been demonstrated.

- Welding shop supervisors may or may not weld regularly, depending on the size of the workshop. In addition to their welding skills they must demonstrate good management skills by effectively planning jobs and assigning workers. They must also possess coordination and logistics skills for site work locations.

- Welding trainers have a broad experience of welding processes, materials, metallurgy, codes and standards. They may have considerable experience in a broad range of manufacturing and service industries both at home and overseas. They are usually recognized by professional institutions and are regulated by government or company regulations.

- Welding sales people may be employed by supply manufacturers or equipment manufacturers. Sales personnel require a broad understanding of the welding processes as well as good marketing skills. Good sales people should be able to provide technical information about their products in order to secure sales.

- Self-employed welders are often welders with a high degree of skill and knowledge who prefer to operate their own businesses. These individuals may specialize in one field, such as hardfacing, repair and maintenance or speciality fabricators. They may act as subcontractors of manufactured items and can be as small as a one man operation or as large as a multi-million pound company employing hundreds of workers.

- Welding engineers design, specify and oversee the construction of complex weldments. The welding engineer may work with other engineers in areas such as mechanical, electronic, chemical, or civil engineering in the process of bringing a new building, ship, aircraft, or product into existence. The welding engineer is required to be conversant with all the welding processes concerned and metallurgy of the materials used. They need to have good levels of mathematics, science, communication and design skills and also be able to interpret codes and standards as they apply. Welding engineers are usually university graduates with a chartered engineer status and possess professional certification in welding.

- Welding technicians work as part of the engineering staff and may oversee the actual work for the engineer by providing the engineer with progress reports as well as chemical, physical, non-destructive testing and mechanical test results. Technicians may also require engineers to build prototypes for testing and evaluation.

Large industrial companies employ workers who serve as support for the welders. This group of employees do layouts or make templates for repetitive layouts which are usually made from sheet metal or other suitable materials. These individuals have drawing experience and knowledge of material removal, forming and distortion control procedures.

Employment of welders is expected to increase rapidly due to regeneration projects, renewable energy programmes and increased production of manufactured components such as rolling stock, transport, air conditioning and maintenance programmes. Another factor which will impact on job opportunities is the large numbers of welders that are reaching retirement age.

Foreword

Welding: Skills, Processes & Practices, Level 2 is an up-to-date, comprehensive and well informed publication and fills a gap for a textbook suitable for learners studying on programmes of both theoretical and practical based learning containing fabrication and welding. It is of particular value in supporting apprentice developmental learning at Level 2 and provides an excellent source of relevant and well explained theory around welding processes that can be easily understood by learners, and most importantly mapped to the assessment criteria of both the mandatory and specific welding units of EAL qualifications.

The excellent text is supported by clear and high definition technical illustrations that greatly support and enhance learning. Each chapter is concluded with an excellent summary that ensures learning is fully understood and embedded.

I am confident in endorsing this book as an important contribution to the welders' toolbox. Every learner on a welding programme should have access to one!

Allan Macdonald AWeldI
Product Specialist
EAL (Excellence, Achievement & Learning Limited)

CREDITS

Although every effort has been made to contact copyright holders prior to publication, this has not always been possible. If notified, the publisher will undertake to rectify any errors or omissions at the earliest opportunity.

Images, Figures and Tables:

The publisher would like to thank the following sources for permission to reproduce their copyright protected images, figures and tables:

American Torch Tip – pp302c; **Arcon Welding, L.L.C.** – pp170tr; **BP p.l.c. 2003** – pp62cl; **Brett V. Hahn** – pp233bm; **CMOS X-ray** – pp375cr, pp375br; **Controls Corporation of America** – pp301bl, pp301bc; **Dynatorque** – pp233tr; **E.I. Du Pont de Nemours & Co., Inc.** – pp376; **ESAB Welding & Cutting Products** – pp123bl, pp123br, pp125, pp126tl, pp150bl, pp151tr, pp153br, pp154, pp172tl, pp194tl, pp211t, pp221bl, pp221br, pp222tm, pp223tr, pp225tr, pp227, pp302t, pp302br, pp304cl; **Hornell, Inc.** – pp24c, pp26bl; **HSE** – pp19, pp20; **Kedman Co., Huntsman Product Division** – pp23; **Larry Jeffus** – pp34b(a), pp35cm, pp35cr, pp83cl, pp83bl, pp84tl, pp84cl, pp84tr, pp87, pp114bl, pp126bl, pp126br, pp129, pp131tl, pp132l, pp132tr, pp132br, pp139tl, pp139tr, pp139bl, pp139br, pp140bl, pp140br, pp145tl, pp148bm, pp152, pp157b, pp175tl, pp175tr, pp176c, pp181tr, pp184b, pp186t, pp188t, pp190tl, pp198tr, pp232tl, pp233c, pp233br, pp246, pp252, pp257t, pp258, pp267c, pp270, pp285tr, pp285ml, pp304cr, pp309bl, pp309br, pp310t, pp310c, pp310b, pp311br, pp312tl, pp312tr, pp312cl, pp312cr, pp313tl, pp315cr, pp315bl, pp316tr, pp316b, pp318tl, 99318tr, pp319cl, pp319cr, pp322bl, pp322br, pp323b, pp325tl, pp325tr, pp325c, pp326bl, pp326br, pp328tl, pp328tr, pp329, pp357br; **Lincoln Electric Company** – pp168bl, pp170c, pp222tl, pp222cm, pp223tl, pp225tl, pp233tl, pp234, pp261tl, pp261b, pp264bl, pp292bl, pp292br, pp311bl; **Magnaflux Corporation** – pp379br; **Mine Safety Appliances Company** – pp26br, pp230bl, pp230br; **Nathan Portlock Allan Photography/Media Select International for Cengage Learning** – pp124cl, pp124cr, pp124bl, pp124br, pp177cl, pp177cr, pp177bl, pp179cm, pp219cl, pp219cr, pp231tr, pp231cl, pp231cr, pp232cr, pp235cr, pp235bm, pp313tr, pp313cr, pp314tl, pp314tr, pp315ml, pp316bm; **NASA** – pp62bl, pp62br; **Newage Testing Instruments, Inc.** – pp364tl, pp364tr; **Prince & Izant Co.** – pp113tl, pp114br; **Thermadyne Holding Corporation** – pp145tr; **Tinius Olsen Testing Machine Co, Inc.** – pp361b, pp365tr; **TWECO, a thermadyne company** – pp172c, pp226tr; **Victory Equipment, a Thermadyne Company** – pp122tm, pp122cm, pp123cl, pp123cr, pp128, pp136br; **Woodworker's Supply Inc.** – pp83cr, pp83br

Photo Credits

Photographed by Nathan Portlock Allan Photography & Media Select International

About the Authors

LARRY JEFFUS is a dedicated teacher and author with over twenty years experience in the classroom and several Delmar Cengage Learning welding publications to his credit. He has been nominated by several colleges for the Innovator of the Year award for setting up non-traditional technical training programs. He was also selected as the Outstanding Post-Secondary Technical Educator in the State of Texas by the Texas Technical Society. Now retired from teaching, he remains very active in the welding community, especially in the field of education.

LAWRENCE BOWER is a welding instructor at Blackhawk Technical College, an AWS SENSE School, in Janesville, Wisconsin. Mr. Bower is an AWS-certified Welding Inspector and Welding Educator. In helping to create *Welding: Skills, Processes and Practices, Level 2*, he has brought to bear an excellent mix of training experience and manufacturing know-how from his work in industry, including fourteen years at United Airlines, and six years in the US Navy as an aerospace welder.

HUGH McPHILLIPS IEng, IncMWeldI, AWS.

Hugh has been involved for over 35 years in the Fabrication and Welding Industry as a lecturer and now consultant. He has worked closely with the major awarding bodies such as TWI, EAL, City & Guilds, and ECITB. Hugh has been Technical Director for a Skills2Learn Virtual Reality Fabrication & Welding Programme and is working on a distance learning programme. He is Chair of the Association of Fabrication and Welding Trainers in Education, a national body covering lecturers and private traing providers. He is also a National Judge for WorldSkillsUK Construction Metalwork.

About the Book

Web link boxes suggest websites for further research and understanding of the topic. **WWW** ⊙

Tip boxes share the authors' experience and provide positive suggestions to improve knowledge and skills. **TIP** ⊙

Health and safety boxes draw your attention to related health and safety information essential for each technical skill. **HEALTH & SAFETY** ⊙

Functional skills icons indicate sections, activities and questions which address Functional skills for Maths, English and ICT.

Workshop task boxes suggest practical activities to test and observe the results of a particular process.

Review questions are provided at the end of all core chapters. You can use the questions to prepare for oral and written assessments and help test your own knowledge throughout. Seek guidance from your supervisor if there are areas you are unsure of.

Unit reference boxes are provided at the beginning of all core chapters to indicate the NVQ and VRQ units covered in each chapter.

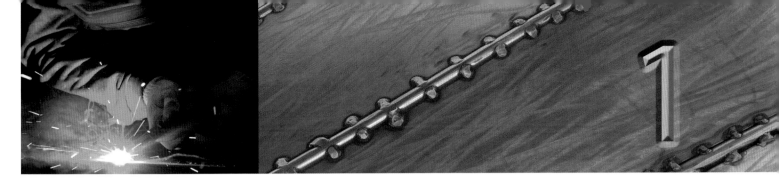

Introduction to Welding

LEARNING OBJECTIVES

After completing this chapter, you should be able to:

- describe how O/A, MMA, MIG, MAGS, FCAW, TIG and TAGS processes work
- list four factors that must be considered before a welding process is selected
- discuss three events in the history of welding
- describe the purpose of a welding procedure specification (WPS)
- define the terms *weld, forge welding, autogenous welding, fusion welding* and *certification*.

FUNCTIONAL SKILLS

- Preparing reports and records.
- Following verbal instructions to complete work assignments.
- Following written details to complete work assignments.
- Using IT to find information.

UNIT REFERENCES

NVQ:

Heat Treating Materials for Fabrication Activities.

VRQ:

Engineering Environment Awareness.

KEY TERMS

autogenous weld a weld in which all of the weld metal has come from the parent material only and no filler material is added.

automated operation operations are performed repetitively by a robot or other machine that is programmed flexibly to do a variety of processes.

automatic operation operations are performed repetitively by a machine that has been programmed to do an entire operation without the intervention of the operator.

brazing a process that uses heat from a fuel-gas flame or electrical induction with a low melting point filler material to flow between a joint by capillary attraction.

certification approval a widely respected document certifying that a welder has passed a performance qualification test at an accredited test facility.

confined spaces a space with limited or restricted means for entry or exit, and it is not designed for continuous employee occupancy. Confined spaces include, but are not limited to, underground vaults, tanks, storage bins, manholes, pits, silos, process vessels and pipelines.

defect an imperfection that is unable to meet minimum acceptance standards or specifications.

dexterity the ability to manipulate one or more objects.

electrical resistance a ratio of the degree to which an object opposes an electric current through it, measured in Ohms.

flux cored arc welding (FCAW) an arc welding process that uses an arc between a continuous filler metal electrode and the weld pool. The process is used with shielding gas from a flux contained within the tubular electrode, with or without additional shielding from an externally supplied gas, and without the application of pressure.

forge welding a solid state welding process that produces a weld by heating the workpieces to welding temperature and applying blows sufficient to cause permanent deformation at the hot surfaces.

hard facing the surface deposition of a wear resistant compound on an inferior carrier material.

imperfection an interruption of the typical structure of a material, such as a lack of bonding in its mechanical, metallurgical or physical characteristics. An imperfection is not necessarily a defect.

kerf the width of a saw, oxy-fuel or plasma cut.

machine operation welding operations are performed automatically under the observation and control of the operator.

manual metal arc is an arc welding process in which an arc is created between a covered electrode and the weld metal. The process is used with shielding from the decomposition of the electrode covering, without the application of pressure, and with filler metal from the electrode.

manual operation the entire welding process is manipulated by the welding operator.

metallurgy the science of materials and structures formed.

non-destructive testing (NDT) testing that does not destroy the test object; also called non-destructive examination (NDE) and non-destructive inspection (NDI).

prototype a one-off product designed to prove specification and become the model for future production.

semi-automatic operation during the welding process, the filler metal is added automatically, and all other manipulation is performed manually by the operator.

slag a non-metallic product resulting from the mutual breakdown of flux and non-metallic impurities in some welding processes.

spatter a fine deposition of filler material as the result of too high a current range or too long an arc length.

standard a technical standard is an established norm or requirement. It is usually a formal document that establishes uniform engineering or technical criteria, methods, processes and practices.

tungsten inert gas shielded (TIG) and tungsten active gas shielded (TAGS) are arc welding processes that use an arc between a tungsten electrode (non-consumable) and the weld pool. The process is used with shielding gas and without the application of pressure.

weld test a welding performance test to a specific code or standard.

weld a localized joining of metals or non-metals produced either by heating the materials to suitable temperatures, with or without the application of pressure, or by the application of pressure alone and with or without the use of the filler material.

welder certification a widely respected document certifying that a welder has passed a performance qualification test at an accredited test facility.

welder performance qualification the demonstration of a welder's or welding operator's ability to produce welds meeting prescribed standards.

welding procedure qualification record (WPQR) a record of welding variables used to produce an acceptable test weldment and the results of tests conducted on the weldment to qualify a welding procedure specification.

welding procedure specification (WPS) a document providing in detail the required variables for specific application to assure repeatability by properly trained welders and welding operators.

welding symbol a graphical representation of a weld.

INTRODUCTION

Welding can trace its history back from ancient times, through the Industrial Revolution and up to the modern day. The blacksmith was the forerunner of today's welder, using **forge welding** to join two pieces of hot metal by hammering them together. The World Wars saw an increasing demand for a greater range of materials and the technology to develop new methods of joining. From the timeline on page 3 (Figure 1.1) you can see how rapidly this development took place and the significance it has on our day-to-day activities.

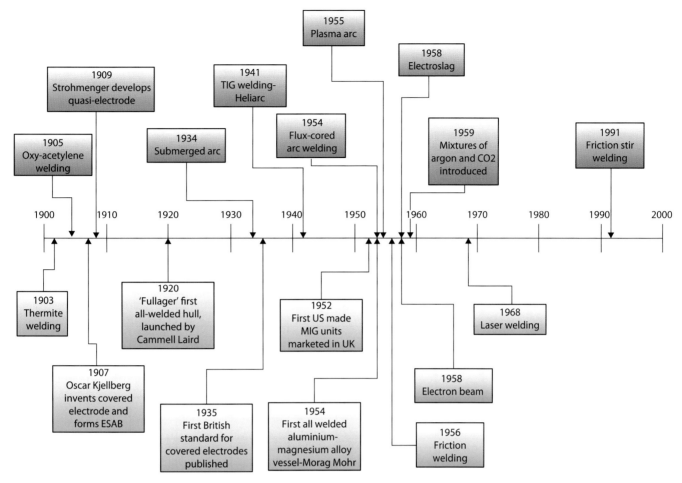

FIGURE 1.1 Welding timeline

HEAT GENERATION

The timeline above shows that many of the developments in welding have been devised around a series of heat generating systems: thermo-chemical, electric arc generation, electrical resistance, mechanical processes, and radiation of heat and light.

- *Thermo-chemical* energy-combustion is a reaction where oxygen and fuel gas such as acetylene, hydrogen, or MAPP react to produce heat and bring about the conditions for welding to take place. Oxy-acetylene welding was invented at the turn of the twentieth Century and was used extensively in both World Wars. Today it is regarded by many as a process which requires great dexterity, especially when used in heating and ventilation and refrigeration pipe work. Gases have also been developed to support other heat generating systems.

- *Electric arc energy* processes such as TIG and MIG are the result of the flow of electrons across a gap created when an arc is struck between an electrical source and a conducting material which releases heat in excess of 4000°C. Tungsten inert gas shielded welding (TIG) and metal inert gas shielded welding (MIG) are two of the most commonly recognized processes which use the combination of an electric arc and gas shield technology to produce an efficient method of joining materials.

- *Electric resistance* can be used to induce heat or offer resistance to the flow of an electrical current (resistance welding) in order to release heat and is the basis of a wide range of modern welding processes.
- *Mechanical processes* have been used extensively to generate heat to join a wide range of materials. Typical processes that use this principle include friction, explosive, ultrasonic and friction stir welding.
- *Thermit welding* is a welding process where heat is generated by combining metal oxides to produce an endothermic reaction which generates the heat and makes a weld deposit in complicated situations, such as railway lines, possible.
- *Radiation of heat and light* have been used successfully to join certain exotic materials such as titanium by the use of Lasers or electron beam.

Welding Process Classification									
Source of Heat									
No heat or heat by conduction	Mechanical	Thermo-chemical		Electric resistance		Electric arc		Radiation	
		Flames, Plasma	Endothermic reaction	Induction	Direct	Consumable electrode	Non consumable	Electro magnetic	Particle
Cold pressure	Explosive	Plasma	Thermit	High frequency Induction	Electro-slag	Metal inert Gas shielded	Tungsten inert gas Shielded	Laser	Electron beam
Hot pressure	Friction	Atomic Hydrogen		Induction butt	Flash butt	CO2 metal arc	Carbon arc		
Thermo-compression bonding	Ultrasonic	Oxy-fuel Gas			High frequency resistance	Covered electrode MMA			
	Friction Stir	Forge			Projection	Submerged arc			
		Pressure butt			Spot. Seam and resistance butt	Stud			
						Spark Discharge			
						Percussion			

FIGURE 1.2 **Welding process classification**

WELDING DEFINED

Before we go any further it is important to clarify what we mean by a weld. Welding can be defined in simple terms as: 'the joining of two pieces of material together to form one piece by heating them to a temperature high enough to cause softening or melting to join the materials'.

Welding can take place where the weld is in a liquid state or in a solid state, with or without the application of pressure. It is important to note that the word 'material' is used because today's welds can be made from a growing list of materials, including plastics, glass, fabrics and ceramics.

An autogenous weld is one in which all of the weld metal has come from the parent material only and no filler material is added.

From commercial applications to surgical industries, welding continues to play an important part in our lives.

Welded joints are a critical component of structures   Spiral staircase

FIGURE 1.3 **The uses of modern welding techniques**

PRE-REQUISITES FOR WELDING

If you are planning a career in welding, or just want to gain welding skills, there are certain characteristics that you will need. You will need good eyesight, manual dexterity, hand and eye co-ordination and an understanding of the principles of welding technology, something which will develop as you gather more experience.

This book takes you through the underlying principles and concepts which make up the five key welding processes:

1 oxy-acetylene welding

2 manual metal arc welding

3 metal arc gas shielded welding

4 flux cored arc welding

5 tungsten arc gas shielded welding

Welding can be both a rewarding and challenging career. The range of work and opportunities is enormous with the potential to travel and experience many cultures and work environments. What makes a good welder? Some welders are naturally gifted and some work meticulously to achieve a level of excellence. Essentially what is required is a desire to learn all that you can about the processes, the materials, the effects of heat input and techniques to deliver a quality weld. You also have to be aware of what defects can occur and how to remedy these problems. It takes time to develop these techniques and a considerable amount of practice to achieve a good standard.

OCCUPATIONAL OPPORTUNITIES IN WELDING

Because of the diverse nature of the welding industry, the exact job duties of each skill area vary. The following are general descriptions of the job classifications used in this profession; specific tasks may change from one location to another.

● *Welders* perform manual or semi-automatic welding. They are the skilled craftspeople who produce the welds on a variety of complex products.

- *Welding operatives* run adaptive control, automatic, mechanized or robotic welding equipment. They need to be fully conversant with all the welding parameters and make adjustments as and when necessary.
- *Welders mates* are employed to clean and remove slag from welds, help and move weldments into position for the welder and supply electrodes when requested.
- *Welding assemblers/fitters* position all the components in the proper places ready for tack welding. These skilled workers must be able to interpret drawings and welding procedures and have a working knowledge of the effects of expansion and contraction of a wide range of materials.
- *Welding inspectors* are often required to hold a special certification such as one supervised by The Welding Institute (TWI) in Britain or the American Welding Society (AWS). Candidates must pass a test covering the welding process, drawing interpretation and symbols, weld procedures, metallurgy and inspection techniques. Vision screening is also required on a regular basis.
- *Welding shop supervisors* may or may not weld on a regular basis, depending on the size of the workshop. In addition to their welding expertise they must demonstrate good management skills by effectively planning jobs and assigning workers. They also need co-ordination and logistics skills for site work locations.
- *Welding trainers* have a broad experience of welding processes, materials, metallurgy, codes and standards, often gained in a wide range of manufacturing and service industries both at home and overseas. They are usually recognized by professional institutions and are regulated by government or company regulations.
- *Welding sales people* may be employed by supply manufacturers or equipment manufacturers. They require a broad understanding of the welding processes as well as good marketing skills. Good sales people should be able to provide technical information about their products in order to secure sales.
- *Self-employed welders* are often welders with a high degree of skill and knowledge who prefer to operate their own businesses. These individuals may specialize in one field, such as hard facing, repair and maintenance or be speciality fabricators. They may act as subcontractors of manufactured items.
- *Welding engineers* design, specify and oversee the construction of complex weldments. The welding engineer may work with other engineers in areas such as mechanical, electronic, chemical or civil engineering in the process of constructing a new building, ship, aircraft or product. The welding engineer must be conversant with all the welding processes concerned and the metallurgy of the materials used. He or she needs to have good levels of mathematics, science, communication and design skills and must also be able to interpret codes and standards as they apply. Welding engineers are usually university graduates with a chartered engineer status and possess professional certification in welding.
- *Welding technicians* work as part of the engineering staff and may oversee the actual work for the engineer by providing progress reports as well as chemical, physical, non-destructive testing and mechanical test results. Technicians may also need engineers to build prototypes for testing and evaluation.

WELDING PROCESSES

1. Oxy-acetylene

- This is a versatile process which can be used for welding, brazing and soldering.
- It uses comparatively inexpensive equipment.
- Mobility allows for use on site.

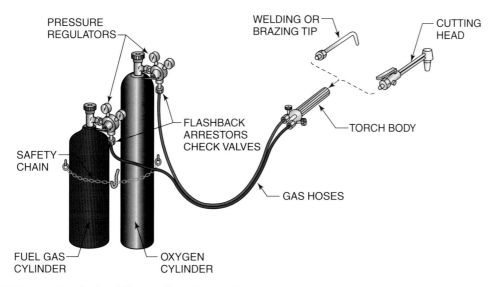

FIGURE 1.4 Oxy-fuel welding and cutting equipment

2. Manual Metal Arc (MMA)

- Wide range of metals can be joined using this process.
- It has a higher deposition rate than oxy-acetylene.
- It is capable of being used in fairly constricted access situations, as electrodes can be manipulated to shape.
- New lightweight inverters have extended the range applications, especially on site.

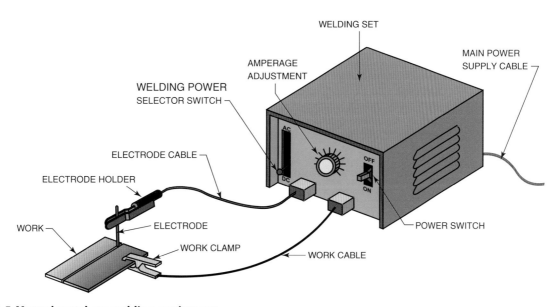

FIGURE 1.5 Manual metal arc welding equipment

3. Tungsten Inert Gas Shielded (TIG)

- It provides high quality weld deposits on a wide range of metals including exotic metals such as titanium.
- It requires little or no post-weld finishing.
- The process can be automated.
- By making the gas 'active' (composite gas such as argon/hydrogen – tungsten active gas shielded TAGS) you can improve the penetration characteristics.

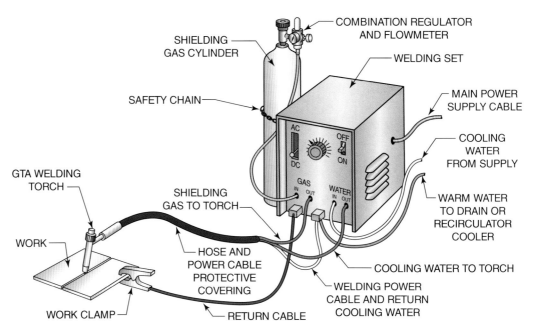

FIGURE 1.6 TIG fill welding equipment

4. Metal Inert Gas Shielded (MIG/MAGS)

- It offers extremely fast deposition rates (continuous feed wire electrode).
- It can be used for thick and thin applications.
- Post-weld finishing is generally reduced (reduced spatter).
- The process can be automated.
- MIG = Metal Inert gas: Used for aluminium with an inert gas (argon or argon helium mix).
- MAGS = Metal Active Gas Shielded: Used for carbon steels. Uses small percentage of Carbon Dioxide (CO_2) (5–15%) mixed with an inert gas (usually argon).

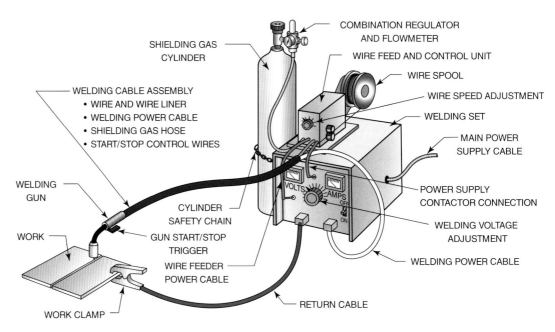

FIGURE 1.7 Metal inert gas welding

5. Flux Cored Arc Welding (FCAW)

TIP ◉

The introduction of new, low-cost equipment and improved availability of filler metals has resulted in MIG/MAGS and FCAW becoming the most commonly used welding processes.

- This is a variant of the MIG/MAGS process in which the wire contains a self-shielding flux and alloying elements in the core.
- It can be used on standard MIG welding sets.
- It can also be used with additional gas coverage.
- Deposition rates are high.
- Post-weld cleaning (removal of slag coating) is required.
- It is less prone to draughty conditions.

FACTORS AFFECTING THE SELECTION OF A WELDING PROCESS

The selection of the joining process for a particular job depends upon many factors:

- *Availability of equipment.* What are the types, capacity, condition of equipment available to make the welds and what are the weight considerations on site?
- *Repetitiveness of the operation.* How many welds will be required to complete the job, and are they all the same?
- *Quality requirements.* Is this weld going to be used on a piece of furniture, to repair a piece of equipment or to join a pipeline?
- *Location of work.* Will the welding be done in a workshop or on a remote job site?
- *Materials to be joined.* Are the parts made out of a standard stock or some exotic alloy?
- *Size of the parts to be joined.* Are the parts small, large, or of different sizes, and can they be moved or must they be welded in place?
- *Skill or experience of workers.* Do the welders have the ability to do the job?
- *Access requirements.* This may well determine which process is the most suitable, especially if working at height.
- *Safety considerations.* Working in confined spaces with inert shielding gases or certain types of electrodes with the potential to reduce the welders breathing zone.
- *Cost of material*
- *Code or specification requirements.* Often the selection of the process is dictated by a code, or standards.

The welder must decide on the welding process and also select the method of applying it. The following methods are used to perform welding, cutting or brazing operations:

- *Manual operation.* The welder is required to manipulate the entire process.
- *Semiautomatic operation.* The filler metal is added automatically, and all other manipulation is done manually by the welder.
- *Machine operation.* Operations are done mechanically under the observation and correction of a welding operator.
- *Automatic operation.* Operations are performed repetitively by a machine that has been programmed to do an entire operation without the intervention of the operator.
- *Automated operation.* Operations are performed repetitively by a robot or other machine that is programmed flexibly to do a variety of processes.

TRAINING FOR WELDING OCCUPATIONS

Generally, years of training are required to learn the basics of a welding process. To become a skilled welder, both college or training centre and on-the-job experience are required. Because of the diverse nature of the welding industry, no single list of skills can be given. However, there are specific skills that are required of most entry-level welders. The ability to interpret instructions, read workshop drawings and retrieve measurements from them are skills that can be acquired by good communication with designers, colleagues and responsible personnel.

In addition to welding skills, an entry-level welder must possess workplace skills, a theoretical knowledge of welding, welding symbols, metal properties, electricity and good communication skills in order to be a valuable member of any team. Welding is like a community, we all respect other people's skills and often learn a lot by working with colleagues who have experienced different working procedures.

Awarding Organisations develop qualifications, and the training can be provided by a formal apprenticeship following:

- company training schemes;
- private training providers;
- professional institutions (The Welding Institute).

All of these providers conform to qualifications based on the National Occupational Standards.

Skill and technical knowledge requirements are higher in some industries, such as atomic energy, aerospace and pressure vessel construction, where high standards for welders must be met to ensure that weldments will withstand the critical forces that they will be subjected to in use.

Before being assigned a job where the service requirements of the weld are critical, welders usually must pass a certification test given by an employer or awarding body. Some authorities also require welders to obtain a license for certain types of site construction. After a welder, or welding operator, has received a certification approval or qualification by passing a standardized test, he or she is approved to produce only the welds covered within the range of approval of the test. Certifications are usually good for a maximum of six months unless a welder is doing code-quality welds routinely and has his certification endorsed by the company.

> **TIP** ⊙
>
> The Welding Institute have international status and allow qualifications achieved within the UK to be recognized overseas.

QUALIFIED AND CERTIFIED WELDERS

Welder qualification and welder certification are often misunderstood. Being certified does not mean that a welder can weld anything, nor that every weld the welder makes is acceptable. It means that the welder has demonstrated the skills and knowledge necessary to; make acceptable to the standard they are approved against on a *specified alloy* and in *one or more positions.*

Individual codes control test requirements. Within these codes, changes in any of a number of essential variables can result in the need to recertify:

- Welders can be certified in each welding *process,* such as O/A, MMA, MIG/MAGS, TIG/ TAGS, FCAW, and resistance spot welding (RSW). A separate test is required for each process.

- The type of *material* (e.g. steel, aluminum, stainless steel, titanium) being welded will require a change in certification. Even a change in the alloy within a base metal type can require a change.

- Each certification is valid for a specific range of *thickness* of parent material used in the test. For example, if a 6mm plain carbon steel plate is used in the test, then under some codes the welder would be qualified to make welds in a plate thickness range from 3mm to 12mm.

- Changes in the classification of the *filler metal* can require recertification.

- If the process requires a *shielding gas,* then changes in gas type or mixture can affect the procedure certification.

- In most cases, a weld test taken in a flat *position* limits certification to flat and possibly horizontal welding. A test taken in the vertical position, however, usually allows the welder to work in the flat, horizontal, and vertical positions, depending on the code requirements.

> **TIP** ⊙
>
> Welders who have passed this type of testing are often referred to as *qualified welders* or approved welders.

Any welder qualification or certification process must include the specific welding skill level. The detailed information for a welding test is often given as part of a welding procedure specification (WPS) or similar set of welding specifications or schedules. These standards inform everyone about which skills are required, enabling the welder to prepare for the welding test and to demonstrate welding skills to the company.

PROFESSIONAL ASSOCIATION

The Welding Institute is the professional body for welders within the UK. It has different membership grades, and students studying a formal welding qualification have free membership in the first year with discounted rates while they are training. The benefits of belonging to this organization are recognition within the welding community at large and access to a considerable amount of knowledge of new techniques, course availability, job opportunities and of course world recognition of certification in welder coding, non-destructive testing, and European welder directives. Being part of such an organization gives a student a broader perspective of the world of welding and the opportunity to network, which could open up a spectrum of job opportunities.

SKILLS COMPETITIONS

Each year WorldSkills sponsors a SkillWeld and Constructional Metalwork Competition for students on welding and fabrication courses. Students are selected by the local college, training provider or sponsor to compete within their local centre, and from there go on to compete at regional finals. The national competition brings together those candidates with the highest scores and with the potential to represent the UK at the WorldSkills Competition, which is held every two years. Like the Olympic Games, this is an opportunity for young people from around the world to demonstrate their skills on an international stage. Many of the students involved have gone on to successful careers in welding and fabrication management.

WWW ▸
For further information visit **www.twi.co.uk.**

WORKSHOP TASKS AND PRACTICES

This textbook contains tasks and practices that are intended to help you develop your welding knowledge and skills. The tasks are designed to allow you to see what effect changes in the process settings, operation or techniques have on the type of weld produced.

When you complete a task, you should observe and possibly take notes on how the change affected the weld. It will often be necessary for you to make changes in your equipment settings or technique to ensure that you are making an acceptable weld. By watching what happens when you make the changes in the welding shop, you will be better prepared to decide on changes required to make good welds on the job. The practices are designed to build your welding skills.

Each practice gives the evaluation criteria or acceptable limits for the weld.

TIP ▸
All welds have some *imperfections*, but if they are within the acceptable limits, they are not called *defects* but *imperfections*.

Layout

The tasks and practices will require that you read and interpret simple drawings and sketches including welding symbols. You will fill out a bill of materials that will be required to fabricate the weldment. The material specifications must be given in SI units. Once the bill of materials is complete, you must lay out on appropriate metal stock the individual parts that are to be cut out. The parts must be laid out to within a tolerance of + or – 1.6mm with an angular tolerance of +5° or –5°. Be sure to leave an appropriate amount of space between parts for the kerf (cut) if the parts are not to be sheared.

Signpost Functional Skills

Written Procedures

Each weldment drawing includes a written list of notes that must be followed. You must also follow guidelines for each weldment.

Written Records

You will be asked to fill out reports and other records as needed. Written records must be complete, neat and legible. They must be turned in with the completed weldment and will be considered in the overall evaluation of your skills. Similar records are required by most large welding companies to determine the productivity of welders and to ensure that each job is charged correctly for time and materials.

Verbal Instructions

In any working shop, verbal instructions are given from time to time. These instructions are as important as written ones. In some cases critical information such as safety concerns are given verbally. Your safety and the safety of others could depend on your ability to remember and follow verbal instructions.

Units of Measurement

Metric units are the norm within the UK and Europe and it is important that you make yourself familiar with the units of measurement, pressure and temperature as these all impact on welding to some degree.

SUMMARY

Welding is a very diverse trade. Almost every manufactured product utilizes a welding or joining process in its production. Products that are produced by welding range from small objects, to larger structures, such as buildings, ships and transport. Your knowledge and understanding of the various processes and their applications will provide you with employable skills that can result in a rich and rewarding career.

ACTIVITIES AND REVIEW

1 Identify three pre-requisites for being a welder.

2 Describe three forms of heat generation used in welding.

3 State what the abbreviation MAGS stands for and what advantages can be gained in this process.

4 Using the job descriptions of the various welding roles, plot a career path and line of responsibility that leads to the Welding Engineer.

5 List four considerations to be taken into account when selecting an appropriate welding process for a project.

6 Give a brief description for the following:
  a Forge weld.
  b Autogenous weld.
  c Fusion weld.

7 Identify the advantages of belonging to a professional institute.

8 List three advantages of using oxy-acetylene welding.

9 State how good work habits could create occupational opportunities for you.

10 List four criteria that would be listed on a weld procedure specification (WPS).

11 Identify how flux cored wire arc welding (FCAW) differs from the other arc welding processes.

12 Name four industries in which welding is an integral part of production.

13 List three advantages for each of the following welding processes:
  a MMA.
  b TIG.
  c MIG.

14 State the advantages of working as part of a team.

15 Name the ideal process for high deposition rates on thin-gauge metal.

16 In terms of reading and creating working drawings state what are SI units for measurement.

17 Define what makes an imperfection into a defect.

18 Using the internet identify the SI units for:
  a Pressure.
  b Temperature.
  c Mass.

19 Identify the welding process which requires the welder to manipulate the whole process.

20 Name the person responsible for overseeing the following:
  a Giving advice on consumables and equipment.
  b Overseeing weld testing and weld procedures.
  c Determining welding process, materials, and assembly procedures.

Working Safely, Efficiently and Effectively

LEARNING OBJECTIVES

After completing this chapter, the student should be able to:

- describe the duties of the employer and employee under the Health and Safety at Work Act and other current legislation
- describe the specific regulations and safe working practices and procedures that apply to your work activities and identify the relevant sources of information
- describe your responsibilities for dealing with hazards and minimizing risks in your workplace
- describe the first aid facilities that exist in your work area and within your organization in general, and the procedures to be followed in the case of accidents involving injury
- describe workplace policies and procedures for emergencies, including fire-fighting and evacuation procedures
- describe the personal protective equipment (PPE) and protective clothing that is available for your area of work and the need to observe personal protection and hygiene procedures at all times
- explain how to act responsibly within the work environment
- describe the methods of manually handling and moving loads
- correctly explain when to act on your own initiative and when to seek help and advice from others and explain to whom you should report in the event of problems that you cannot resolve.

FUNCTIONAL SKILLS

- Preparing a risk assessment form.
- Extracting information from a chart.
- Interpretating legislation required for specific circumstances.

UNIT REFERENCES

NVQ:

Working Efficiently and Effectively in Engineering. Complying with Statutory Regulations and Organizational Safety Requirements.

VRQ:

Engineering Environmental Awareness.

KEY TERMS

acetone a fragrant (garlic smelling) liquid chemical used in acetylene cylinders. The cylinder is filled with a porous material (kapok or prepared charcoal) and acetone is then absorbed by this material to stabilize the gas. Acetylene is then added and absorbed by the acetone, which can absorb up to twenty five times its own volume of the gas.

earmuffs a type of hearing protection that covers the entire ear.

earplugs a type of hearing protection that is fitted into the ear.

electric shock an electric shock can occur upon contact of a human's body with any source of voltage high enough to cause sufficient current through the body.

explosimeter a piece of equipment which can indicate the parts per million of explosive compounds present within a structure or air.

flash glasses eye protection specifically designed to filter out UV light.

forced ventilation to remove excessive fumes, ozone or smoke from a welding area, a ventilation system may be required to supplement natural ventilation.

french chalk available in rectangular form commonly used for marking metal.

full face shield protective equipment designed to cover the entire face.

goggles special eye protection designed to seal around each eye.

hot work permit a document required to be completed before beginning hot work operations in areas not specifically designated for welding or cutting.

housekeeping performance of duties to keep a welding shop or job site clean and free of hazards.

infrared a form of electromagnetic radiation whose wavelength is longer than that of visible light, but shorter than that of microwaves. Infrared radiation is heat that can be felt at a distance.

mandatory the law to conform with statutory regulations.

material safety data sheet (MSDS) hazards and properties form containing data regarding the properties of a particular substance.

natural ventilation the process of supplying and removing air through an indoor space by natural means.

oxy-fuel gas welding a group of welding processes that produces fusion of the work pieces by heating them with an oxy-fuel gas flame. The processes are used with or without filler metal.

safety glasses eye protection worn on the face to protect the eyes from impact, sparks or dust.

tack weld a weld made to hold the parts of a weldment in proper alignment until the final welds are made.

ultraviolet light is a form of electromagnetic radiation with a wavelength shorter than that of visible light, but longer than X-rays.

valve protection cap a protective cover which fits on a compressed gas cylinder.

ventilation the intentional movement of air from outside a building to the inside.

visible the visible spectrum (sometimes called the optical spectrum) is the portion of the electromagnetic spectrum that is visible to (can be detected by) the human eye.

volt a unit of electrical pressure.

warning label a form of hazard communication that attaches directly to an object.

water table a special table designed for plasma arc cutting operations, where the torch head is submerged under water in order to reduce smoke and noise.

welding helmet equipment designed to protect the welder's face and head from radiation, sparks, spatter and fumes associated with welding and cutting operations.

weldment a component in which one or more joints are joined by a range of welding processes.

INTRODUCTION

This chapter covers the underpinning knowledge required to work safely, efficiently and effectively in a welding environment. The information included in this text is intended as a guide only; there is no substitute for caution and common sense.

Welding, like other industrial jobs, has potential hazards. Learning to work safely is as important as learning to be a skilled welder.

Most large welding manufacturers have **mandatory** health and safety classes that must be successfully completed before beginning work. These classes may cover company-specific regulations, government regulations, health and safety regulations and European directives. Companies provide this training to protect you, others, and the business from injury and losses in production resulting from accidents. Violation of company safety policies and practices may result in suspension or dismissal.

LEGISLATION

You need to have a good understanding of all legislation relating to welding. It is important that you obtain and read all relevant publications from the Health and Safety Executive to keep up to date with any changes following European directives.

Some of the key pieces of legislation which apply to welding are:

- Health and Safety at Work Act (HASAWA);
- Reporting of Injuries Diseases and Dangerous Occurrences Regulations (RIDDOR);
- Noise at Work Regulations;
- Electricity at Work Regulations (EAWR);
- Manual Handling Operations Regulations;
- Health and Safety (Display Screen Equipment) Regulations;
- Workplace (Health, Safety and Welfare) Regulations;
- Provision and Use of Work Equipment Regulations (PUWER);
- Lifting Operations and Lifting Equipment Regulations (LOLER);
- Management of Health and Safety at Work Regulations;
- Control of Substances Hazardous to Health (COSHH) Regulations;
- Personal Protective Equipment (PPE) Regulations.

The Health and Safety at Work Act (HASAWA) provides a legal framework to encourage high standards of health and safety at work. It requires minimum standards of health, safety and welfare to be upheld in each area of the workplace.

Employers' legal responsibilities are:

- to provide a safe working environment with adequate welfare facilities;
- to provide and maintain safe plant and safe systems of work;
- to provide information, instruction and training to ensure safe working practices;
- to provide all personal protective equipment (PPE) required to carry out these processes.

Employees' legal responsibilities are:

- to take reasonable care for their health and safety and others who may be affected by them;
- to cooperate with their employers on all matters with regards to health and safety;
- to wear all PPE as provided;
- to not tamper with anything provided in the interests of health, safety or welfare.

WWW >

Visit the Health and Safety Executive website for all general health and safety information and also specific information related to welding:
www.hse.gov.uk
www.hse.gov.uk/welding.

HAZARDS AND RISK

A *hazard* is anything with the *potential* to cause harm. Walk around the workplace and try to identify what could cause harm. Seek the views of your co-workers as they may have noticed things which are not immediately obvious to you. The use of material safety data sheets (MSDS) and equipment checklists can also help to spot the hazards and put any risks into perspective.

Hazards may be associated with the following:

- machinery;
- electricity (damaged plugs, insulation, incorrect fuse or no earth connection – electric shock potential);
- slippery or uneven surfaces;
- handling and transporting;
- contaminants and irritants;
- dust and fumes;
- material ejection;
- fire;
- working at height;
- environment;

HEALTH & SAFETY >

Accident and ill-health records can also be useful indicators when identifying hazards.

- moving parts;
- pressured and stored energy systems (e.g. gas canisters);
- toxic or volatile materials;
- unshielded processes.

A *risk* is the *likelihood* that a hazard will cause harm. It can depend on a number of different factors. For example, the risk of a person slipping on a wet floor depends on the amount of water on the floor, the smoothness of the floor's surface, the type of shoe sole the person is wearing, the number of people walking over the area and the size of the area covered in water.

Risk is measured on a scale of 1–5:

1 Very unlikely to happen, causing harm

2 Unlikely to happen, causing harm

3 Possible to happen, causing harm

4 Likely to happen, causing harm

5 Very likely to happen, causing harm

Linked to this is a scale for *severity*:

1 Minor injury

2 Major injury

3 Loss of limb

4 Death of an individual

5 Multiple deaths

From the above, a formula can be used to calculate risk:

$$\text{Risk} = \text{Likelihood rating} \times \text{Severity rating}$$

Risk Assessment

HEALTH & SAFETY ⊙

All employers, including self-employed persons, are legally required to assess the risk within the workplace.

In recent years, risk assessment has assumed a very high profile. This is due to an increased knowledge of health and safety issues and the rising use of litigation against companies and individuals following incidents.

As part of any learning programme it is important to be able to carry out your own risk assessment. This will help you appreciate what risk assessment is all about. It will also ensure that you have considered all the potential hazards and risks and have put in place procedures and good working practices to promote a safe working environment for yourself and those around you.

Risks associated with the welding environment include the tools, materials and equipment you use, oil or chemical spills, accidental breakages of tools or equipment that are not reported properly and workers who do not follow the correct working practices and procedures.

To conduct a risk assessment you must identify if a hazard exists and then evaluate the level of risk related to it. Is the hazard covered by existing precautions? If not, you may need to introduce new precautions to reduce the risk.

Anyone who works within the area may be at risk including:

- fellow workers, trainees and labourers;
- people working in a supporting capacity such as cleaners, office staff and deliveries;
- visitors to the workplace;
- on site members of the public who may stray into a work area.

Once you have evaluated a risk, rate it as high, medium or low. If you find something that needs to be addressed, draw up an action plan giving priority to hazards which present a high risk or could affect most people.

It may be possible to remove the hazard by using an alternative technique or process. If not, the risk must be controlled, possibly by preventing access to the hazard by guarding or cordoning off the area.

Drawing up a *code of practice* is a good way to limit the risk and issue appropriate PPE. Provide welfare facilities for removal of contamination and first aid to comply with all legislation. Codes of practice have been drawn up over a period of time and have been 'proven' to ensure that this is the safest method of carrying out an activity.

Make your report available to everyone concerned, and make sure that the precautions in place are 'reasonably practicable' and that the remaining risk is low. Systematically review your assessments, especially if new equipment, processes, materials or procedures are implemented which could lead to additional hazards. This review should indicate that the precautions adopted are effective.

HEALTH & SAFETY ◉

Failure to comply with codes of practice could result in prosecution if someone is injured as a direct result.

COSHH

COSHH stands for the Control of Substances Hazardous to Health. COSHH requires employers to control exposure to hazardous substances to protect employees and others who may be exposed from work activities. Hazardous substances are anything that can harm your health when you work with them if they are not properly controlled e.g. by using adequate **ventilation.** They can include:

- substances used directly in work activities such as acids, cleaning agents, paints or glues;
- substances generated during work activities such as fumes from welding, brazing or soldering;
- naturally occurring substances such as blood, bacteria or assorted grain dusts.

For the vast majority of commercial chemicals, the presence (or lack) of a **warning label** will indicate whether COSHH is relevant. An everyday example of this is that washing up liquid does not have a warning label while bleach does, telling us that COSHH applies to bleach.

There are several ways to tell whether a substance is hazardous:

- Manufacturers, importers and suppliers have a 'duty of care' to inform you.
- Consult warning labels on containers.
- Read Material Safety Data Sheets (MSDS) (also known as Safety Data Sheets and Product Safety Data sheets).
- Consult Health and Safety Executive Publication EH40.

Hazardous substances are kept on a list and there are regulations regarding their transportation which require data sheets known as Chemical Hazard Information and Packaging for Supply regulations (CHIP) and hazard chemical symbols for clear identification. EH40 lists substances which have occupational exposure limits assigned to them, based on the concentration of the substance in the air. The entry routes for these substances can be by:

TIP ◉

MSDS contain 16 specific sections on all aspects of safety.

- inhalation (breathed in);
- ingestion (swallowed);
- absorption (entry through the skin).

These substances are therefore controlled by minimum Occupational Exposure Limit (OEL) and Maximum Exposure Limit (MEL) exposure times within a working day and maximum PPE to control the hazard.

MATERIAL SAFETY DATA SHEETS (MSDSs)

All manufacturers of potentially hazardous materials must provide to the users of their products detailed information regarding possible hazards resulting from the use of their products. These **material safety data sheets** (sometimes referred to as COSHH sheets dependant upon

HEALTH & SAFETY ◐

COSHH regulations apply within the UK and may exist in a slightly different format in other countries, therefore it is important to become fully conversant with the regulations of the country you are operating in.

their nature) are often known by their abbreviation, MSDSs. They must be provided to anyone using the product or anyone working in an area where the products are in use. Often companies will post these sheets on a bulletin board or put them in a convenient place near the work area. All employees who handle or work in areas with hazardous materials should be familiar with the hazards and precautions for dealing with these products. MSDSs are an important part of an overall hazard communication program. Other types of hazard communications are product labelling (as laid down in the Control of Substances Hazardous to Health (COSHH) Regulations, special danger symbols (see Figure 2.1a on page 19), flashing lights, and audible signals.

WASTE MATERIAL DISPOSAL

There is an increasing concern for the environment and an awareness of the need to improve management of 'finite' resources. One of the most effective ways to manage a company's commitment and ensure action is taken is to implement an environmental management system (EMS) such as BS EN ISO 14000 series. Other directives which interact with ISO 14000 are the Clean Air Act and Pollution Prevention and Control Act. All of these regulations place a 'duty of care' on all those involved in the management of waste, including collecting, disposing or treating controlled waste which is subject to licensing.

Failure to comply may result in prosecution. There are many benefits to using a system like this:

● potential cost savings;

● providing a competitive edge when tendering for contracts, especially government contracts;

● minimizing the potential for environmental incidents and reducing or eliminating their impact.

Welding shops generate a lot of waste materials. Much of the waste is scrap metal. All scrap metal, including electrode stubs, can easily be recycled. Recycling metal is good for the environment and can be a source of revenue for the welding shop.

Other forms of waste, such as burned flux, cleaning solvents and dust collected in shop air-filtration systems, may be considered hazardous materials and may come under COSHH regulations. The material manufacturer or an environmental consultant will be able to advise whether any waste material is considered hazardous. Throwing hazardous waste material into the bin, pouring it on the ground, or dumping it down the drain is illegal.

BURNS

Burns are among the most common and painful injuries that occur in the welding workshop. The chance of infection is high and it is important that all burns receive proper medical treatment. Burns can be caused by ultraviolet light rays as well as by contact with hot welding material. Both types can be avoided if proper clothing and personal protective equipment (PPE) are worn. There are three types of burns you should be aware of:

● *First degree burns* the surface of the skin is reddish in colour, tender and painful, but the burn does not involve any broken skin.

● *Second-degree burns* the surface of the skin is severely damaged, resulting in the formation of blisters and possible breaks in the skin.

● *Third-degree burns* the surface of the skin and possibly the tissue below the skin appear white or charred. Initially, there may be little pain because nerve endings have been destroyed. Do not remove any clothes that are stuck to the burn.

 [TOXIC symbol] 1	TOXIC	☐	Can cause poisoning by: ingestion, inhalation, absorption through tissues
 [ENVIRONMENTAL symbol] 2	ENVIRONMENTAL	☐	The use of, or disposal of any equipment or substance that could have a detrimental effect on the environment
 [CORROSIVE symbol] 3	CORROSIVE	☐	Can cause corrosion to: skin, eyes, repertory system
 [IRRITANT symbol] 4	IRRITANT	☐	Can cause harmful effect although not actually toxic or corrosive
 [OXIDIZING symbol] 5	OXIDIZING	☐	Many oxidising substances (nitric acid, peroxides, and chlorates) provide oxygen supply to potential explosion or fire
 [BIO CHEMICAL symbol] 6	BIO CHEMICAL	☐	Manufactured chemicals which can cause injury or death to anyone who comes into contact with them
 [RADIOACTIVE symbol] 7	RADIOACTIVE	☐	Radio active substances are used in some non destructive testing processes (ironising emissions and X-Rays)
 [EXPLOSIVE symbol] 8	EXPLOSIVE	☐	Apart from the obvious explosives (e.g. TNT) many substances will explode given the right conditions e.g. flour, coal dust, glasses etc.
 [FLAMMABLE symbol] 9	FLAMMABLE	☐	Apply to many organic chemicals with low flash points: petroleum products, acetone, etc.

FIGURE 2.1a COSHH labeling chart

	Example of hazard statement	Example of precautionary statement
	Heating may cause an explosion	Keep away from heat/sparks/open flames/hot surfaces – no smoking
	Heating may cause a fire	Keep only in original container
	May intensify fire; oxidizer	Take any precaution to avoid mixing with combustibles
	Causes serious eye damage	Wear eye protection
	Toxic if swallowed	Do not eat, drink or smoke when using this product
	Toxic to the aquatic life, with long-lasting effects	Avoid release to the environment
	Reflects serious longer-term health hazards such as carcinogenicity and respiratory sensitization e.g. may cause allergy or asthma symptoms or breathing difficulties if inhaled	In case of inadequate ventilation, wear respiratory protection
	Refers to less serious health hazards such as skin irritancy/sensitization and replaces the CHIP ✗ symbol e.g. may cause an allergic skin reaction	Contaminated work clothing should not be allowed out of the workplace
	Used when the containers hold gas under pressure e.g. may explode when heated	None

FIGURE 2.1b **Warning labels**

HEALTH & SAFETY ▸

Ensure that a RIDDOR form has been completed.

Treatment

With all burns the first line of treatment is to reduce the area of the burn by passing cold water (not ice water – this could cause a potential trauma if used) over the affected area for a good period of time. If the burn has broken the skin, cover with lint free cloth or cling film to prevent bacteria forming, and dispatch to hospital for medical supervision.

Burns Caused by Light

The three types of light are ultraviolet, infrared and visible. Ultraviolet and infrared light are not visible to the unaided human eye, and they are types of light that can cause burns. During welding, one or more of the three types of light may be present. Arc welding produces all three types, but gas welding produces visible and infrared light only. The light from the welding process can be reflected from walls, ceilings, floors or any other large surface. This reflected light is as dangerous as the direct welding light. To reduce the danger from reflected light the welding area should, if possible, be painted with a flat, dark-coloured or black paint. Matt black will reduce the reflected light by absorbing more of it than any other colour. When the welding is to be done on site, in a large shop, or other area that cannot be painted, weld curtains can be placed to absorb the welding light, Figure 2.2. These special portable welding curtains may be either transparent or opaque. Transparent welding curtains are made of a special high-temperature, flame-resistant plastic that will prevent the harmful light from passing through.

Ultraviolet Light

Ultraviolet light waves are the most dangerous and can cause first-degree and second-degree burns to the eyes and exposed skin. Because ultraviolet light cannot be seen or felt, the welder must stay protected when in the area of any of the arc welding processes. The closer a welder is to the arc and the higher the current, the quicker a burn can occur. The ultraviolet light is so intense during some welding processes that it can cause a condition known as 'arc eye', which may come on ten to twelve hours after exposure. The symptoms are nausea, headache, and a feeling of hot sand in the eyes. This will normally pass within twenty four hours with no after effects. The skin can be burned within minutes and you should always protect yourself from radiation. Ultraviolet light can pass through loosely woven, thin or light-coloured clothing, and arc welding helmets that are damaged or poorly maintained.

Infrared Light

Infrared light is the light wave that is felt as heat. Although infrared light can cause burns, a person will immediately feel this type of light, making it easy to avoid burns. When you are welding and you feel infrared light, you are probably being exposed to ultraviolet light at the same time and you should take protective action.

FIGURE 2.2 Portable welding curtain

Visible Light

Visible light is the light that we see. It is produced in varying quantities and colours during welding. Too much visible light may cause temporary night blindness (poor eyesight under low light levels). Too little visible light may cause eye strain, but visible light is not hazardous.

RIDDOR

RIDDOR stands for the Reporting of Injuries, Diseases, and Dangerous Occurrences Regulations and these forms are all about reporting accidents such as:

- deaths and major injuries;
- incapacity to work for more than three days;
- specified diseases;
- dangerous occurrences of near misses.

Any of the factors listed on the previous page *must* be reported to the Health and Safety Executive (HSE) using the prescribed form. It is essential that accurate records are kept of any incident. These regulations cover employers, employees, self-employed persons, trainees and other people injured on an employer's premises such as visitors. Examples of *major* injuries include: any fracture, other than to the fingers, thumbs or toes, amputation of any limb, dislocation of the shoulder, hip, knee or spine, loss of sight or penetrating eye injury.

Examples of *serious* conditions include; certain eye conditions, electric shock requiring attention. Unconsciousness through lack of oxygen, poisoning and acute illness due to exposure to certain materials.

Examples of reportable *diseases* include; certain poisonings, skin diseases such as occupational dermatitis, lung diseases such as occupational asthma and infections such as hepatitis. Other conditions such as hand-arm vibration (HAV) system and repetitive strain injury (RSI) are also reportable.

Finally, examples of reportable *occurrences* include; structural collapses such as buildings or scaffoldings, fires and explosions, release of gases or other dangerous substances, failure of breathing equipment while in use, incidents with dangerous substances in transit and contact with, or arcing of overhead cables.

In any workshop there should be an accident report book for minor injuries such as a cut to the finger. Cuts should always be covered in the working environment to avoid exposure to substances and bacteria.

All establishments need to make provision for first aid. In a small workshop, there should be a designated person. Within a large organization there should be nominated first aiders who have been trained to a minimum standard to deal with predominately minor injuries. Ideally there will be a designated location with access to clean running water; this is especially important with site locations. This location should be supplied with enough first aid boxes to cover the number of personnel on site or in an organization. The first aiders should have access to an emergency line to call for an ambulance should one be required.

HEALTH & SAFETY ❯

All incidents should be recorded in the accident book or on a RIDDOR form.

WWW ❯

RIDDOR forms are available to download from **www. riddor.gov.uk**.

PERSONAL PROTECTIVE EQUIPMENT

Personal protective equipment (PPE) should always be seen as the last form of defence. If it is designated for use with any equipment or process it is essential that it is worn at all times when using the equipment. It is essential that the PPE is checked for functionality and conforms to the specifications before it is used. For example, some categories of safety glasses are specifically stated for grinding purposes as they have impact resistance, and only those should be worn.

Eye protection must be worn in the workshop at all times. Eye protection can be safety glasses, with side shields, goggles, or a full face shield, see Figure 2.3. To give better protection when working in brightly lit areas or outdoors, some welders wear flash glasses, which are special, lightly tinted, safety glasses. These provide protection from both flying debris and reflected light.

Suitable eye protection is important because eye damage caused by excessive exposure to arc light is not noticed at the time. Welders must take care to select filters or goggles that are suitable for the process being used, see Table 2.1. Using the correct shade of lens is also important, because both extremes of too light or too dark can cause eye strain, especially if using oxy-acetylene with fluxes as these give off more glare. New welders often select too dark a lens, assuming it will give them better protection, but this results in eye strain. Any approved arc welding lenses will filter out the harmful ultraviolet light. Select a lens that lets you see comfortably.

SIDE SHIELDS

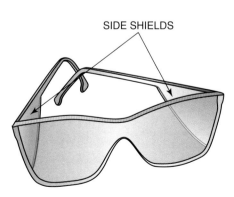

FIGURE 2.3 Safety glasses with side shields

TABLE 2.1   Huntsman® Selector Chart

1

Goggles, flexible fitting, regular ventilation

2

Goggles, flexible fitting, hooded ventilation

3

Goggles, cushioned fitting, rigid body

4

Spectacles

5

Spectacles, eyecup type eyeshields

6

Spectacles, semi-flat-fold sideshields

7

Welding goggles, eyecup type, tinted lenses

7A
Chipping goggles, eyecup type, tinted lenses

8

Welding goggles, coverspec type, tinted lenses

8A
Chipping goggles, coverspec type, clear safety lenses

9

Welding goggles, coverspec type, tinted plate lens

10

Face shield, plastic or mesh window (see caution note)

11

Welding helmet

Non-sideshield spectacles are available for limited hazard use requiring only frontal protection.

Applications

Operation	Hazards	Protectors
Acetylene-Burning Acetylene-Cutting Acetylene-Welding	Sparks, Harmful Rays, Molten Metal, Flying Particles	7,8,9
Chemical Handling	Splash, Acid Burns, Fumes	2 (for severe exposure add 10)
Chipping	Flying Particles	1,2,4,5,6,7A,8A
Electric (Arc) Welding	Sparks, Intense Rays, Molten Metal	11 (in combination with 4,5,6 in tinted lenses advisable)
Furnace Operations	Glare, Heat, Molten Metal	7,8,9 (for severe exposure add 10)
Grinding-Light	Flying Particles	1,3,5,6 (for severe exposure add 10)
Grinding-Heavy	Flying Particles	1,3,7A,8A (for severe exposure add 10)
Laboratory	Chemical Splash, Glass Breakage	2 (10 when in combination with 5,6)
Machining	Flying Particles	1,3,5,6 (for severe exposure add 10)
Molten Metals	Heat, Glare, Sparks, Splash	7,8 (10 in combination with 5,6 in tinted lenses)
Spot Welding	Flying Particles, Sparks	1,3,4,5,6 (tinted lenses advisable, for severe exposure add 10)

CAUTION:
Face shields alone do not provide adequate protection. Plastic lenses are advised for protection against molten metal splash.
Contact lenses, of themselves, do not provide eye protection in the industrial sense and shall not be worn in a hazardous environment without appropriate covering safety eyewear.

Source: Courtesy of Kedman Co., Huntsman Product Division

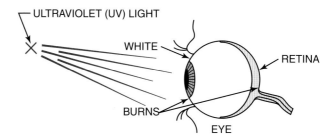

FIGURE 2.4 Eye damage from ultraviolet light

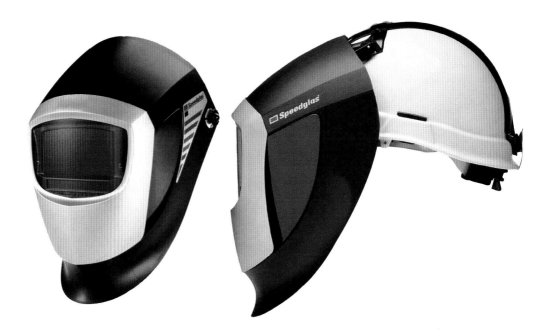

FIGURE 2.5 Typical arc welding helmets

Welding Helmets

Even with quality welding helmets, like those shown in Figure 2.5, the welder must check for potential problems that may occur from accidents or daily use. Small, undetectable leaks of ultraviolet light can cause a welder's eyes to itch or feel sore after a day of welding. To prevent these leaks, make sure the lens gasket is installed correctly, Figure 2.6. The outer and inner clear lenses must be plastic. As shown in Figure 2.7, the lens can be checked for cracks by twisting it between your fingers. Worn or cracked spots on a helmet must be repaired. Tape can be used as a temporary repair until the helmet can be replaced or permanently repaired. Approved safety glasses with side shields should always be worn under your welding hood, even for small jobs or tack welds.

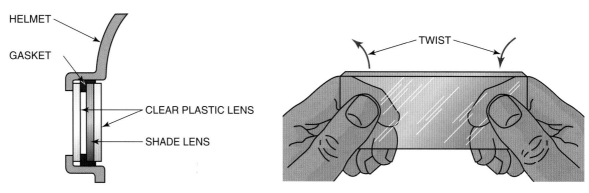

FIGURE 2.6 Gasket placement

FIGURE 2.7 Shade lens

Safety Glasses

Safety glasses with side shields are adequate for general use, but if heavy grinding, chipping or overhead work is being done, goggles or a full face shield should be worn in addition to safety glasses, Figure 2.8. Safety glasses are best for general protection. They must always be worn under an arc welding helmet and at all times in the shop or on the work site.

Ear Protection

The Noise at Work regulations stipulates certain decibel levels at which ear protection should be worn. The welding environment can be very noisy. The sound level is at times high enough to cause pain and some loss of hearing if the welder's ears are unprotected. In the UK noise levels are measured from 80dB and it is recommended that for anything above this you should use ear defenders. Hot sparks can also drop into an open ear, causing severe burns. Ear protection is available in several forms. Earmuffs cover the outer ear completely, Figure 2.9. Earplugs fit into the ear canal, Figure 2.10. Both protect a person's hearing, but only the earmuffs protect the outer ear from burns.

FIGURE 2.8 Full face shield

FIGURE 2.9 Earmuffs

FIGURE 2.10 Earplugs

HEALTH & SAFETY

Damage to your hearing caused by high sound levels may not be detected until later in life, and the resulting loss in hearing is permanent. Your hearing will not improve with time, and each exposure to high levels of sound will cause further damage.

TIP

If the noise level is such that you cannot clearly hear someone 3 metres away, you should wear ear defenders.

RESPIRATORY PROTECTION

All welding and cutting processes produce undesirable by-products, such as harmful dusts, fumes, gases, smoke, airborne contaminants and vapours. For your own safety and the safety of others, your primary objective will be to prevent these contaminants from forming and collecting in the shop atmosphere. Preventative measures include water tables for cutting, general and local ventilation, thorough cleaning of surface contaminants (away from sources of heat) before starting work, and confinement of the operation to outdoor or open spaces.

Production of welding by-products cannot be avoided. They are created when the temperature of metals and fluxes is raised above the temperatures at which they vaporize or decompose. Most are recondensed in the weld. However, some do escape into the atmosphere, producing the haze that occurs in improperly ventilated welding shops. Some fluxes used in welding electrodes, produce fumes that can irritate the welder's nose, throat and lungs; typical of these are the fluxes used on stainless electrodes.

When welders must work in an area where effective general controls to remove air-borne welding by-products are not feasible, respirators should be provided by employers when they are necessary to protect the welders' health. The respirators supplied must be suitable for the purpose intended and the workshop should establish and implement a written respiratory protection program with worksite-specific procedures. Guidelines for developing a respiratory protection program are available from the Health and Safety Executive website.

WWW

www.hse.gov.uk.

Equipment

All respiratory protection equipment used in a welding shop should conform to European Standards (CE). The following are a few different types of respirator:

- *Air-purifying respirators* have an air-purifying filter, cartridge or canister that removes specific air contaminants.
- *Atmosphere-supplying respirators* supply breathing air from a source independent of the ambient atmosphere. They include supplied-air respirators (SARs) or airline respirators, where the source of breathing air is not carried by the user and self-contained breathing apparatus (SCBA) units where the breathing-air source is carried by the user.
- *Demand respirators* are atmosphere-supplying respirators that admit breathing air to the face piece only when a negative pressure is created inside the face piece by inhalation.
- *Positive-pressure respirators* are respirators in which the pressure inside the respiratory inlet covering exceeds the ambient air pressure outside the respirator.
- *Powered air-purifying respirators (PAPRs)* are air-purifying respirators that use a blower to force the ambient air through air-purifying elements to the inlet covering, Figure 2.11.

Respiratory protection equipment used in many welding applications is of the filtering face piece (dust mask) type, Figure 2.12. These masks use the negative pressure as you inhale to draw air through a filter. In areas of severe contamination, you can use a hood-type respirator that covers your head and neck and may even cover portions of your shoulders and torso.

Fume Sources

Some materials used as paints, coating or plating on metals to prevent rust or corrosion can cause respiratory problems. Other potentially hazardous materials might be used as alloys in metals.

Before it is welded or cut, any metal that has been painted or has any grease, oil or chemicals on its surface must be thoroughly cleaned. This can be done by grinding, sand blasting, or applying an approved solvent. It may not be possible to clean metals that are plated or alloyed before welding or cutting begins.

FRESH AIR

FIGURE 2.11 Breathing equipment

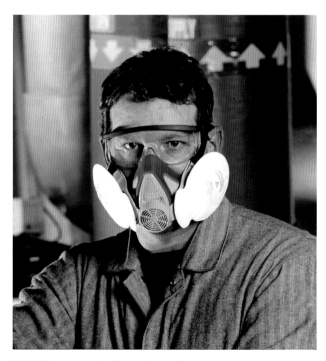

FIGURE 2.12 Typical respirator for contaminated environments

Most paints containing lead have been removed from the market. But some industries still use these lead-based paints, as in marine or shipping applications. Solder often contains lead alloys. The welding and cutting of lead-bearing alloys or metals whose surfaces have been painted with lead-based paint can generate lead oxide fumes. Inhalation and ingestion of lead oxide fumes and other lead compounds will cause lead poisoning. Symptoms include a metallic taste in the mouth, loss of appetite, nausea, abdominal cramps and insomnia. In time, anemia and a general weakness, chiefly in the muscles of the wrists, develop.

Cadmium and *zinc* are plating materials used to prevent iron or steel from rusting. Cadmium is often used on bolts, nuts, hinges and other hardware, and it gives the surface a yellowish-gold appearance. Exposure to high concentrations of cadmium fumes can produce severe lung irritation. Long-term exposure to low levels of cadmium in air can result in emphysema (a disease affecting the lungs' ability to absorb oxygen) and kidney damages.

Zinc, often in the form of galvanizing, may be found on pipes, sheet metal, bolts, nuts and other types of hardware. Zinc plating that is thin may appear as a shiny, metallic patchwork or crystal pattern; thicker, hot-dipped zinc appears rough and may look dull. Zinc is used in large quantities in the manufacture of brass and is found in brazing rods. Inhalation of zinc oxide fumes can occur when welding or cutting on these materials and can cause metal fume fever, whose symptoms are very similar to those of common influenza.

Some concern has been expressed about the possibility of lung cancer being caused by some of the chromium compounds that are produced in the welding of stainless steels.

Chromium hexavalent (CrVI) compounds, often called 'hexavalent chromium', 'Hex Chrome' or 'Chrome 6', exist in several forms. Industrial uses include chromate pigments in dyes, paints, inks and plastics; chromates added as anticorrosive agents to paints, primers and other surface coatings; and chromic acid electroplated onto metal parts to provide a decorative or protective coating. Hexavalent chromium can also be formed when performing 'hot work' such as welding on stainless steel or melting chromium metal. In these situations, the chromium is not originally hexavalent, but the extreme temperatures result in oxidation that can convert the chromium to a hexavalent state. Employers working with this material must establish a safety program that measures Hex Chrome levels and provides adequate respiratory protection for workers. Further respirator information and guidance can be found on the HSE website.

Rather than take chances, welders should recognize that fumes of any type, regardless of their source, should not be inhaled. The best way to avoid problems is to provide adequate ventilation. If this is not possible, breathing protection should be used. Protective devices for use in poorly ventilated areas are shown in Figures 2.11 and 2.12.

Vapour Sources

Potentially dangerous gases also can be present in a welding shop. Proper ventilation or respirators are necessary when welding in confined spaces, regardless of the welding process being used.

Ozone is a gas that is produced by the ultraviolet radiation in the air in the vicinity of arc welding and cutting operations. Ozone is irritating to all mucous membranes, with excessive exposure producing pulmonary oedema, or fluid on the lung, making it difficult to breathe. Severe cases of pulmonary oedema may require immediate care. Other effects include headache, chest pain and dryness in the respiratory tract.

Phosgene is formed when ultraviolet radiation decomposes chlorinated hydrocarbons, which can come from solvents such as those used for degreasing metals and from refrigerants from air-conditioning systems. They decompose in the arc to produce a potentially dangerous chlorine acid compound which reacts with the moisture in the lungs to produce hydrogen chloride, which in turn destroys lung tissue. For this reason, any use of chlorinated solvents should be well away from welding operations in which ultraviolet radiation or intense heat is generated. Any welding or cutting on refrigeration or air-conditioning piping must be done only after the refrigerant has been completely removed in accordance with Environmental Protection Agency (EPA) regulations.

Nitric oxide is a sharp sweet-smelling gas at room temperature, Nitrogen oxides are released to the air from the exhaust of motor vehicles, the burning of coal, oil, or natural gas and during processes such as arc welding, They are also produced commercially by reacting nitric acid with

WWW ⊙

Download the Respiratory Protective Equipment at work guide: www.hse.gov.uk/ pubns/priced/hsg53.pdf.

HEALTH & SAFETY ⊙

Extreme care must be taken to avoid the fumes produced when welding is done on dirty or used metal. Any chemicals that are on the metal will become mixed with the welding fumes, a combination that can be extremely hazardous. All metal must be cleaned before welding to avoid this potential problem.

metals or cellulose. Low levels of nitrous oxide in the air can irritate your eyes, nose, throat and lungs, possibly causing you to cough and experience shortness of breath, tiredness and nausea. Exposure to low levels can also result in fluid build-up in the lungs 1–2 days after exposure. Breathing high levels of nitrous oxide can cause rapid burning, spasms and swelling of tissue in the throat and upper respiratory tract, and a build-up of fluids in your lungs.

Carbon monoxide is produced from the breakdown of carbon dioxide shielding in arc welding or cellulosic electrode coatings. Carbon monoxide has no smell and is readily absorbed into the bloodstream, causing headaches, dizziness, or muscular weakness. High concentrations may result in unconsciousness and death.

Care must be taken to avoid the infiltration of any fumes or gases, including argon or carbon dioxide, into a confined working space, such as when welding in tanks. The collection of some fumes and gases can go unnoticed by the welders. Concentrated fumes or gases can cause a fire or explosion if they are flammable, asphyxiation if they replace the oxygen in the air, or death if they are toxic.

VENTILATION

The welding area should be well ventilated. Excessive fumes, nitrous oxide, carbon monoxide, ozone or smoke may collect in the welding area; ventilation should be provided for their removal. Natural ventilation is best, but forced ventilation may be required. Areas that have 283m^3 or more per welder or that have ceilings 4.9m high or higher, may not require forced ventilation unless fumes or smoke begin to collect.

Small workshops with large numbers of welders require forced ventilation. This can be general or localized using fixed or flexible exhaust hoods, Figure 2.14. General room ventilation must be at a specified rate per person welding. Localized exhaust hoods must provide a sufficient air velocity to pull the welding fumes away from the welder. Local government regulations may require that welding fumes be treated to remove hazardous components before they are released into the atmosphere.

Any system of ventilation should draw the fumes or smoke away before they rise past the level of the welder's face.

Forced ventilation is always required when welding on metals that contain zinc, lead, beryllium, cadmium, mercury, copper, austenitic manganese or other materials that give off dangerous fumes.

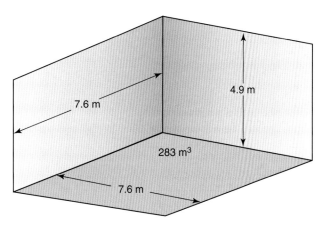

FIGURE 2.13 Forced ventilation

FIGURE 2.14 An exhaust hood

CONFINED SPACES

Work in confined spaces requires special precautions. Owners, contractors and workers all need to be familiar with written confined-space working procedures. Often work is supervized by a specially trained person. Asphyxiation (lack of breathing air) causes unconsciousness and even death without warning. Confined-space atmospheres that are oxygen enriched will greatly intensify combustion, which can cause rapid, severe and often fatal burns.

Confined spaces must also be tested for toxic or flammable gases, dusts and vapours, and for excessive or inadequate oxygen levels before entering and during operations. These same precautions apply to areas such as tank bottoms, pits, low areas and spaces near floors when heavier-than-air gases and vapours are present, and to high areas such as tank tops and near ceilings when lighter-than-air gases are present. Gases such as carbon dioxide, argon and propane are heavier than air; gases such as natural gas and helium are lighter than air. Anyone working in a confined space should be observed at all times. If possible, a continuous monitoring system with audible alarms should be used. Further precautions can be found on the HSE website.

WORK CLOTHING

Fire retardant overalls should be worn especially when welding. Further to this, it is important to choose work clothing that will minimize the possibility of getting burned because of the high temperature and amount of hot sparks, metal and slag produced during welding, cutting or brazing.

Work clothing must also stop ultraviolet light from passing through it, which means the material should be dark, thick and tightly woven. The best choice is 100 per cent wool, but it is difficult to find. Another good choice is 100 per cent cotton clothing, the most popular fabric used.

You must avoid wearing synthetic materials, including nylon, rayon and polyester. They can easily melt or catch fire. Some synthetics produce a hot sticky residue that can make burns more severe. Others may produce poisonous gases.

All clothing must be free of frayed edges and holes. It should be relatively tight-fitting to prevent excessive folds or wrinkles that might trap sparks.

SPECIAL PROTECTIVE CLOTHING

Overalls should be worn by everyone in the workshop. In addition to this clothing, extra protection is needed for each person who is in direct contact with hot materials. Leather is often the best material to use, as it is lightweight and flexible, resists burning and is readily available. Synthetic insulating materials such as chrome leather are also available. Ready-to-wear leather protection includes capes, jackets, aprons, sleeves, gloves, caps, pants, knee pads and spats.

Hand Protection

All-leather, gauntlet-type gloves should be worn when doing any welding. Gauntlet gloves that have a cloth liner for insulation are best for hot work. Non-insulated gloves will give greater flexibility for fine work. Some leather gloves are available with a canvas gauntlet top; they should be used for light work only.

When a great deal of manual dexterity is required for gas tungsten arc welding, brazing, soldering, oxy-fuel gas welding and other delicate processes, soft leather gloves may be used.

Body Protection

Full-leather jackets and capes will protect a welder's shoulders, arms and chest. A jacket, unlike a cape, protects a welder's back and complete chest. A cape is open and much cooler but offers less protection. The cape can be used with a bib apron to provide some additional protection while leaving the back cooler. Either the full jacket or the cape with a bib apron should be worn for any out-of-position work.

TIP ❯

The use of a rope secured around the welder can be used for monitoring and can also be used to remove an unconscious worker without putting others at risk by going into the environment.

HEALTH & SAFETY ❯

If you have been injured while using a product, you should, if possible, take the material's MSDS with you when you seek medical treatment.

HEALTH & SAFETY ❯

Do not have cigarettes or any loose paper in your pockets because sparks can travel anywhere especially into top and back pockets.

HEALTH & SAFETY ❯

There is no safe place to carry butane lighters or matches while welding or cutting. They can catch fire or explode if subjected to welding heat or sparks. Matches can erupt into a ball of fire. Butane lighters and matches must always be removed from the welder's pockets and placed a safe distance away before any work is started.

Waist and Lap Protection

Bib aprons or full aprons will protect a welder's lap. This is especially important if you squat or sit while working and when you bend over or lean against a table.

Arm Protection

For some vertical welding, a full or half sleeve can protect a person's arm. The sleeves work best if the work level is not above the welder's chest. Work levels higher than this usually require a jacket or cape to keep sparks off the shoulders.

Leg and Foot Protection

When heavy cutting or welding is being done and a large number of sparks are falling, leather spats and leggings should be used to protect the welder's legs and feet. Leggings can be strapped to the legs, leaving the back open. Spats will prevent sparks from burning through the front of lace-up boots.

HANDLING AND STORING CYLINDERS

Oxygen and fuel-gas cylinders or other flammable materials must be stored separately. The two storage areas must be separated by 3m or by a wall 2.4m high with at least a half-hour burn rating. The purpose of this is to prevent the heat of a small fire from causing the compressed gas cylinder safety valve to release. If oxygen were released, a small fire would become a raging inferno. Inert-gas cylinders may be stored separately or with oxygen cylinders.

Empty cylinders must be stored separately or with full cylinders of the same gas. All cylinders must be stored vertically and have the protective caps screwed on firmly.

Securing Gas Cylinders

Cylinders must be secured with a chain or other device so that they cannot be knocked over accidentally. Cylinders attached to a manifold or stored in a special room used only for cylinder storage should be chained.

Storage Areas

Cylinder storage areas must be located away from halls, stairwells and exits so that they will not block an escape route in an emergency. Ideally gas cylinder storage areas should be outside protected from the elements and with through ventilation and secured at all times, with only nominated personnel access. Storage areas should also be located away from heat, radiators, furnaces and welding sparks. Stored fuel-gas cylinders should be separated from any flammable material by a minimum distance of 6.1m or a wall 1.5m high. The location of storage areas should be such that unauthorized people cannot tamper with the cylinders. A warning sign that reads 'Danger – No Smoking, Matches or Open Lights', or with similar wording, must be posted in the storage area. A separate room used to store acetylene must have good ventilation and should have a warning sign posted on the door.

Cylinders with Valve Protection Caps

Cylinders equipped with a valve protection cap must have the cap in place unless the cylinder is in use. The cap prevents the valve from being broken off if the cylinder is knocked over. If the valve of a full high-pressure cylinder (argon, oxygen, carbon dioxide, mixed gases) is broken off, the cylinder can travel around the shop like a missile if it has not been secured properly.

- Never lift a cylinder by the safety cap or the valve. The valve can easily break off or be damaged.
- When cylinders are moved, the valve protection cap must be on, especially if the cylinders are mounted on a truck or trailer for site work.
- Cylinders must never be dropped or handled roughly.
- Use warm (not boiling) water to loosen cylinder valves that are frozen.
- Any cylinder that leaks, has a bad valve, or has damaged threads must be identified and reported to the supplier. French chalk can be used to write the problem on the cylinder.
- If the leak cannot be stopped by closing the cylinder valve and it is not alight, then the cylinder should be moved to a vacant lot or an open area and a warning sign posted.

Acetylene cylinders that have been lying on their sides must stand upright for a minimum one hour or more before they are used. The acetylene is dissolved in acetone, and the acetone is absorbed in a mass. The mass does not allow the liquid to settle back away from the valve very quickly, Figure 2.15. If the cylinder has been in a horizontal position, using it too soon after it is placed in a vertical position may draw acetone out of the cylinder, and this makes the gas in the cylinder unstable and potentially explosive. Acetone lowers the flame temperature and can damage regulator or torch valve settings.

Oxygen cylinders are pressurized to 230 bar and therefore need careful handling and storage., They should be secured upright at all times and at a distance of 3m from combustible gases such as acetylene and propane. Oxygen is used to intensify the heat generation and should *never* be used to 'sweeten' the atmosphere in a confined space as this can bring about instantaneous ignition of materials especially in the presence of grease that may be on overalls.

Argon and *Argon mixtures* are used as inert or active shielding gases with tungsten inert gas shielded or with metal arc gas shielded processes. They are heavier than air and will displace the normal breathing atmotsphere, resulting in asphyxiation of the operator if not controlled. Therefore local extraction is to be recommended at **ALL** times when using these gases.

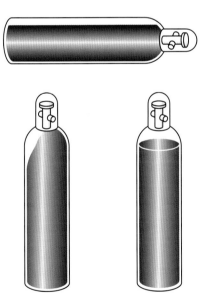

FIGURE 2.15 **Acetone settling**

FIRE PROTECTION

Fire is a constant danger to the welder. Welding is considered to be 'hot work' by the Fire Service Authorities and when welding on site, the welder may be required to obtain a hot work permit from the company. Most permits provide a checklist of items that must be inspected before hot work begins. Floors must be swept and floor drains must be checked with an **explosimeter** tool, which detects and measures concentrations of combustible gases or vapours in the air. Walls, floor openings, and any ductwork in the area must also be inspected.

Hot work permits (Figure 2.16) usually require at least two signatures, one from the area supervisor and one from the fire watcher. In particularly hazardous situations, the company fire marshal should be notified before and after hot work begins.

Even with a permit, the welder can be held liable for any damage resulting from a fire caused by his or her welding. Highly combustible materials should be 10m or more away from any welding. When it is necessary to weld within 10m of combustible materials, when sparks can reach materials further than 10m away, or when anything more than a minor fire might start, a fire watch is required.

HOT WORK PERMIT

(Sample)

Instructions
1. Evaluate if the hot work can be avoided or completed in a safer way.
2. Follow precautions listed to the right.
3. Complete permit and display in area where work is being done.

Hot work done by:
Name of Employee ————————————
Name of Contractor:_____
Job Date _____ Job No. _____
Location (Building & Floor)
———————————————————
Type of Job:
———————————————————
Name of Person Performing Work:
———————————————————
I verify that the above location has been examined, that precautions on the checklist have been taken to prevent fire, and that permission is authorized to perform the work.
Signed by:
Safety Supervisor _____
(signature)
Area Supervisor:_____
(signature)
Fire Watch:_____
(signature)
Permit Expires_____ _____
(Date)   (Time)

Recommended Precautions Checklist

☐ Available sprinklers, hose streams and extinguishers are in service and good repair.
☐ Hot work equipment is in good repair.

Requirements within 11m. of Work
☐ Combustible floors wet down, covered with damp sand or other shields.
☐ Floors swept clean of combustibles.
☐ Explosive atmosphere in area eliminated.
☐ Flammable liquids, dust, lint, and oily deposits removed.
☐ All wall and floor openings covered to prevent sparks from passing thru.
☐ Ducts and conveyor systems that might carry sparks to distant combustibles are protected or shut down.
☐ Fire-resistant covers suspended beneath work.
☐ Other combustible materials removed or covered with fire-resistant covers.

Work on Walls and Ceilings
☐ Combustibles on the other side of the wall moved away.
☐ Construction is noncombustible and without combustible coverings.

Work on Enclosed Equipment
☐ Equipment cleaned of all combustibles.
☐ Equipment purged of all flammable vapours.

Fire Watch and Work Area Monitoring
☐ Fire watch should be provided during, and for at least 30 minutes after work is completed.
☐ Fire watch trained on facility alarms and equipped with fire extinguishers.
☐ Fire watch may be required above, below and in adjacent areas.
☐ Other precautions taken_____

FIGURE 2.16 Hot work permit

Fire Watch

A fire watch can be provided by any competent person who knows how to sound the alarm and use a fire extinguisher. The fire extinguisher must be the type required to put out a fire on the type of combustible materials near the welding. Combustible materials that cannot be removed from the welding area should be soaked with water or covered with sand or noncombustible insulating blankets, whichever is available.

Fire Extinguishers

Fire is classified into categories – Class A, B, C, D and E defined by the types of flammable materials. There is also class F for oils and fats expected to be found in kitchens. Different fire extinguishers are designed to be used on different types of fire. Some fire extinguishers can be used on more than one type. However, using the wrong type of fire extinguisher can be dangerous, either causing the fire to spread, electric shock or an explosion.

A fire extinguisher works by breaking the 'fire triangle' of heat, fuel and oxygen. Most extinguishers both cool the fire and remove the oxygen. They use a variety of materials to extinguish the fire. The majority of fire extinguishers found in welding shops use carbon dioxide or powder. There are a variety of different extinguishers with different uses:

TIP

BS.EN3 is the standard that governs the colouring and labelling of fire extinguishers.

- *Water fire extinguishers* are used on flammable solids (Class A), such as paper, wood and cloth. The cylinder body is red with a red label. Water is a fast efficient method of extinguishing fires by providing a rapid cooling effect.
- *Foam fire extinguishers* are used on flammable liquids (Class B), such as oil, gas and paint thinner and can also be used on Class A fires. The cylinder is red and has a cream label. This type of extinguisher sprays foam over the fire which starves the fire of oxygen and prevents re-ignition.
- *Powder fire extinguishers* are used for electrical fires e.g. fires involving motors, fuse boxes and welding machines. They can also be used on Class A and B fires. The cylinder is red and has a blue label. This type of extinguisher sprays dry powder over the fire which starves the fire of oxygen and prevents re-ignition.
- CO_2 *fire extinguishers* can be used on flammable liquids (Class B) or electrical fires (Class E). The cylinder is red and has a black label. This type of extinguisher delivers CO_2 gas under pressure which displaces air from the fire and prevents re-ignition. Care must be taken if used in confined spaces.
- *Special metal powder fire extinguishers* are used on flammable metals, such as zinc, magnesium and titanium. These extinguishers should only be used after specific training.

Fire blankets should be available in all locations dealing with sources of heat, to wrap and smoother anyone's clothes that may have caught fire.

- Should read suitable for welding process being used and for the type of combustible materials located nearby.
- The extinguishers should be placed so that they can be easily removed without reaching over combustible material and should also be low enough to be easily lifted off the mounting, see Figure 2.21 on page 34.
- The location of fire extinguishers should be marked with red paint and signs, high enough so that their location can be seen from a distance over people and equipment.
- The location should also be marked near the floor so that they can be found even if a room is full of smoke.

In the Event of a Fire Occurring

If a fire breaks out you must warn fellow workers by breaking the nearest fire alarm, which may give a ringing tone or klaxon warning, and informing whoever is in charge to contact the fire brigade. *Always* report to the designated assembly point so that your name can be checked off as being safe, and *do not* enter the building until the fire brigade or fire marshall tells you it is safe to do so.

EQUIPMENT MAINTENANCE

A routine schedule for planned maintenance of equipment will aid in detecting potential problems such as leaking coolant, loose wires, poor earth leads, frayed insulation or split hoses. Fixing small problems promptly can prevent the loss of valuable time later.

FIGURE 2.17 Type A fire extinguisher symbol

FIGURE 2.18 Type B fire extinguisher symbol

FIGURE 2.19 Type C fire extinguisher symbol

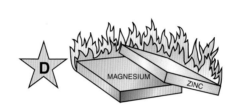

FIGURE 2.20 Type D fire extinguisher symbol

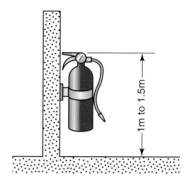

FIGURE 2.21 Mounting fire extinguishers

Any maintenance beyond routine external maintenance should be referred to a trained service technician. In most areas, it is against the law for anyone but a licensed electrician to work on arc welders and for anyone but a factory-trained repair technician to work on regulators. Electric shock and exploding regulators can cause serious injury or death.

TIP

Always check your equipment prior to use.

Hoses

Hoses must be used only for the gas or liquid for which they were designed. Blue hoses are to be used only for oxygen and red hoses are to be used only for acetylene or other fuel gases. Avoid using unnecessarily long lengths of hose. Never use oil, grease or other pipe-fitting compounds on any joints. Hoses should also be kept out of the direct line of sparks. Any leaking or bad joints in gas hoses must be repaired or replaced by a competent person.

WORK AREA HOUSEKEEPING

The work area should be kept clean and tidy. Collections of steel, welding electrode stubs, wire, hoses and cables are difficult to work around and easy to trip over. An electrode caddy can be used to hold the electrodes and stubs, Figure 2.22. Hooks can be made to hold hoses and cables and scrap steel should be thrown into scrap bins.

If a piece of hot metal is going to be left unattended, write the word 'hot' on it before leaving. Do the same to warn people of hot tables, vices, firebricks and tools.

FIGURE 2.22 Electrode caddy

PUSH IN THIS DIRECTION ONLY.

FIGURE 2.23 Adjustable wrench

HAND TOOLS

Hand tools are used by the welder to assemble and disassemble parts for welding and to perform routine equipment maintenance.

The adjustable wrench is the most popular tool used by the welder. It should be adjusted tightly on the nut and pushed so that most of the force is on the fixed jaw, Figure 2.23. When working on a tight bolt or nut, the wrench should be pushed with the palm of an open hand or pulled to prevent injuring the hand. If a nut or bolt is too tight to be loosened with a wrench, obtain a longer wrench. An extension bar should not be used. Also remember the following safety points:

● The fewer points a box end wrench or socket has, the stronger it is and the less likely it is to slip or damage the nut or bolt, Figure 2.24.

● Striking a hammer directly against a hard surface such as another hammer face or anvil may result in chips flying off and causing injury.

● The mushroomed heads of chisels, punches and the faces of hammers should be ground off, to prevent injuring your hand or stop particles flying into your eye, Figure 2.25.

● A handle should always be placed on the tang of a file to avoid injuring your hand. A file can be kept free of filings by rubbing a piece of chalk on it before it is used, Figure 2.26.

TIP ❯

Always use the correct tool for the job. Do not try to force a tool to do a job it was not designed to do.

FIGURE 2.24 **Wrench slippage**

CORRECTLY GROUND

MUSHROOMED

FIGURE 2.25 **Punches and chisels**

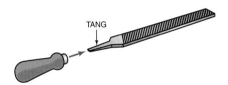
TANG

FIGURE 2.26 **File**

ELECTRICAL SAFETY

All electrical equipment comes under the Electricity at Work Act and the Provision and Use of Work Equipment Regulations (PUWER) and must conform to the guidelines they set out. Electric shock can cause injury and death unless proper precautions are taken. Most welding and cutting operations involve electrical equipment in addition to the arc welding power supply. Grinders, electric motors on automatic cutting machines and drills are examples. Most electrical equipment in a welding shop is powered by AC sources with input voltages ranging from 110, 240 to 415 volts. However, fatalities have occurred where people were working with equipment operating at less than 80 volts. Most electric shocks in the welding industry are a result of accidental contact with bare or poorly insulated conductors. Electrical resistance is lowered in the presence of water or moisture, so welders must take special precautions when working under damp or wet conditions, including perspiration.

The work piece being welded and the frame or chassis of all electrically powered machines must be connected to a good electrical earth terminal. The work lead from the welding power supply does not operate as an earth but supplies power to the point of the weld contact. A separate lead, the welding return lead, acts as an earth to the work piece and power source.

Electrical connections must be tight. Terminals for welding leads and power cables must be shielded from accidental contact by personnel or by metal objects. Cables must be used within their current-carrying and duty-cycle capacities, or they will overheat and break down the insulation rapidly. Cable connectors for lengthening leads must be fully insulated.

HEALTH & SAFETY ❯

Cables must be checked periodically to be sure that they have not become frayed, which could cause fire or electric shock; if they have, they must be replaced immediately.

Welders should not allow the metal parts of electrodes or electrode holders to touch their skin or wet coverings on their bodies. Dry gloves in good condition must always be worn. Rubber-soled shoes are advisable. Precautions must be taken against accidental contact with bare conducting surfaces when the welder is required to work in cramped kneeling, sitting or lying positions. Insulated mats or dry wooden boards are desirable protection in conditions where moisture could be present.

Welding circuits must be turned off when the work station is left unattended. When working on the welder, welding leads, electrode holder, torches, wire feeder, guns or other parts, the main power supply must be turned off and locked or tagged to prevent electrocution. Since the electrode holder is energized when changing coated electrodes, the welder must wear dry gloves.

ELECTRICAL SAFETY SYSTEMS

For protection from electric shock, standard portable power tools are built with either external earthing or double insulation.

A tool with external earthing has a wire that runs from the housing through the power lead to an earth terminal on the power plug. When this third terminal is connected to an earth by the electrical outlet, the earthing wire will carry any current that leaks past the electrical insulation of the tool away from the user and into the ground.

A double-insulated tool does not require earthing because it has an extra layer of electrical insulation that eliminates the need for an earthed outlet. Double-insulated tools are always labelled as such on their nameplate or case.

VOLTAGE WARNINGS

Before connecting a tool to a power supply, be sure the voltage supplied is the same as that specified on the nameplate of the tool. A power source with a voltage greater than that specified for the tool can lead to serious injury to the user as well as damage to the tool. Using a voltage lower than the rating on the nameplate is harmful to the motor.

In the UK any appliance which plugs into a mains supply must be Portable Appliance Tested (PAT) for earth continuity and insulation.

EXTENSION LEADS

If the power source is some distance from the work area or if the portable tool is equipped with a short power lead, an extension lead must be used. When extension leads are used on portable power tools, the conductors must be large enough to prevent an excessive drop in voltage between the power tool and the supply. This drop in voltage occurs because of electrical resistance in the wire and can cause loss of power, overheating and possible motor damage.

Current specifications require outdoor supplies to be protected with residual current devices. These safety devices are often referred to as RCDs.

When using extension leads, keep in mind the following safety tips:

- Always connect the lead of a portable electric power tool into the extension lead before connecting it to the outlet. Always unplug the extension lead from the supply before unplugging the lead of the portable power tool.
- Extension leads should be long enough to make connections without being pulled taut, which creates unnecessary strain or wear, but they should not be excessively long.
- Be sure that the extension lead does not come into contact with sharp objects or hot surfaces. The lead should not be allowed to kink, nor should it be dipped in or splattered with oil, grease or chemicals.
- Before using a lead, inspect it for loose or exposed wires and damaged insulation. If a lead is damaged, replace it. This also applies to the tool's power lead.

- Extension leads should be checked frequently while in use to detect any unusual heating. Any cable that feels more than slightly warm to a bare hand should be checked immediately for overloading.
- See that the extension lead is positioned so that no one trips or stumbles over it.
- To prevent the accidental separation of a tool lead from an extension lead during operation, make a connection or use a lead connector.
- Extension leads that go through dirt and mud must be cleaned before storing.
- Never use an extension lead on a reel, without pulling all the cable out, as this can create a back electromotive force (EMF), causing heating of the cable and reducing the effectiveness of the cable insulation.

TIP

For further information see BS EN 60309-2 1999 Plugs, Sockets, Outlets and Couplers for Industrial Purposes.

MANUAL HANDLING

The Manual Handling Regulations place responsibilities on employers to identify manual handling operations, and to reduce the risks associated with them by supplying information, training and equipment to support the employee. It is the employee's responsibility to take care in all situations and pay due care and attention for others. Employees must follow any instructions or procedures issued and co-operate with their employer on all health and safety matters. It is also their duty to report any identified hazardous handling activity.

The regulations cover the movement and support of any load by physical effort including; lifting, carrying, putting down, pushing and pulling.

Proper lifting, moving, and handling of large, heavy welded assemblies are important to the safety of workers and the weldment. Improper work habits can cause serious personal injury and damage to equipment and materials.

Lifting

When you are lifting a heavy object, the weight of the object should be distributed evenly between both hands, and your legs should be used to lift, not your back, always keep a straight back when lifting, see Figure 2.27. Do not try to lift a large or bulky object without help if it is heavier than 25kg.

FIGURE 2.27 Correct lifting

LIFTING OPERATIONS AND LIFTING EQUIPMENT REGULATIONS (LOLER)

These regulations cover any equipment used to lift loads, such as cranes, lifting tackle, chains, straps and motorized fork lifts, etc. Under these regulations all equipment *must* be monitored regularly, and any repairs recorded in a logbook for inspection by Health and Safety. Inspection is recorded at designated periods and *no* equipment can be used that does not pass.

Hoists or Cranes

The capacity of hoists or cranes should be checked (safe working limit/load) before trying to lift a load. They can be accidentally overloaded with welded assemblies. Keep any load as close to the ground as possible while it is being moved. Pushing a load on a crane is better than pulling. It is advisable to stand to one side of ropes, chains, and cables that are being used to move or lift a load. If they break and snap back, they will miss you.

TIP

If it is necessary to pull a load, use a rope.

WORKING AT HEIGHTS

'Working at heights' means working at any height from which you may injure yourself. If you need to work at heights for any period of time then you *must* work from a platform and *not* from ladders.

Platforms can be fixed scaffolding which must be put up by a certified scaffolder, who must also make any subsequent adjustments. There must be a kick-board all around the scaffold and a hand-rail with intermediary rails or screens to prevent someone falling off, or equipment dropping. Ladders must be secured and extend 1m past the landing stage for easy access. Alternatively, increasing use is made of mobile platforms which can be rolled or driven into position.

When working at heights, the immediate area around the installation should be cordoned off with safety signs to prevent individuals entering the danger area. All scaffolds should have '**keep nets**' on the underside of the platform extending outwards for a given distance to catch any tools, materials or debris.

Ladder Safety

Ladders come under the Provision and Safe Use of Work Equipment Regulations. Improper use of ladders is often a factor in falls and even short step stools can pose a potential hazard. Never approach a climb assuming that because it is not high, it cannot be that dangerous. All ladder use poses a danger to your safety. Keep the area around the base of the ladder clear so that if you do fall, it will not be into debris or equipment.

Types of Ladders

Both step ladders and straight ladders are used extensively in welding fabrication. Straight ladders may be single-section or extension-type. Most ladders used in welding are made from aluminum, or fibreglass. All ladders used in welding should be listed under lifting operations and load equipment regulations (LOLER).

Ladder Inspection

Over time, ladders can become worn or damaged and should be inspected each time they are used. Look for loose or damaged steps, rungs, rails, braces and safety feet. Check to see that all hardware is tight, including hinges, locks, nuts, bolts, screws and rivets. Never use a defective ladder.

Rules for Ladder Use

Step ladders must be locked in the full opened position with the spreaders. Straight or extension ladders must be used at the proper angle; either too steep or too flat is dangerous, Figure 2.28.

The following are general safety and usage rules for ladders:

● Follow all recommended practices for safe use and storage.
● Do not exceed the manufacturer's recommended maximum weight limit for the ladder.
● Before setting up a ladder, make certain that it will be erected on a level, solid surface.
● When using extending ladders, always ensure that the two sections overlap each other by three rungs to ensure rigidity.
● Never use a ladder in a wet or muddy area where water or mud will be tracked up the ladder's steps or rungs. Only climb or descend ladders when you are wearing clean, dry shoes.
● Wear well-fitted shoes or boots.
● Tie the ladder securely in place.
● Climb and descend the ladder cautiously, using both hands at all times.
● Do not carry tools and supplies in your hand as you climb or descend a ladder. Use a rope to raise or lower the items once you are safely in place.
● Never use metal ladders around live electrical wires, fibreglass ladders are available for this type of work.
● Never use a ladder that is too short for the job so you have to reach or stand on the top step.

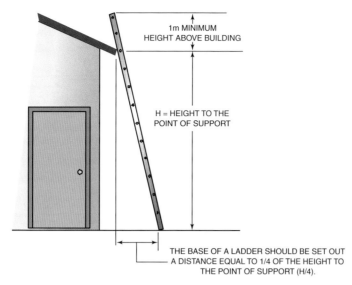

1m MINIMUM
HEIGHT ABOVE BUILDING

H = HEIGHT TO THE
POINT OF SUPPORT

THE BASE OF A LADDER SHOULD BE SET OUT
A DISTANCE EQUAL TO 1/4 OF THE HEIGHT TO
THE POINT OF SUPPORT (H/4).

FIGURE 2.28 Ladder safety

WORKING EFFICIENTLY AND EFFECTIVELY

Using your Initiative

At some time you will work on your own and will need to make some informed choices with regards to your personal safety and the safety of those around you. If you are not sure of anything, ask for advice from supervisors or fellow workers. Always remember it is better to be safe than sorry.

Attitudes

It is important that you are fully aware of your work environment and comply with all statutory regulations, abide by all safety signs, and behave in a responsible manner with due regard to others. Never try to play pranks in an engineering environment. The consequences could be much more serious than you imagine, and could include costly damage, prosecution, lifelong disability or even death.

Think about how you ask for advice and help from your fellow workers, and how you respond when you are asked for help yourself. In the industry you will come across a wide range of personalities and it will sometimes take all your tact to deal with people. But if you remember one guiding principle – 'treat others as you would like to be treated' you will not go far wrong.

Always be respectful of other people's personal property, especially their tools. Never use or take somebody's tools without asking them first. Try to be aware of people's personal beliefs and cultural backgrounds. Think about how your behaviour and the things you say affect those around you. What you may think is a joke may be upsetting to someone else. Always try to adopt a positive attitude towards the people you are working with. Your personal attitudes and your ability to work as a team member are just as important as the quality of your work.

Make sure you are ready for any tasks that need carrying out, and show that you are someone who can be relied on. No matter how tired you are or how inconvenient, trivial and unnecessary a request may seem to you, always try to be cheerful, helpful and efficient.

TIP

Good time keeping is essential in the work environment and in training; it shows you have a positive attitude towards your work.

Personal Hygiene

If you are to keep yourself safe in an engineering environment, personal hygiene is of utmost importance. Some of the substances you will come across are extremely hazardous. Always wash your hands when you have used fluxes, acids, alkalis or solvents. These can all be ingested if you eat food without washing your hands first. Cutting fluids used with metal cutting saws can impregnate your overalls and clothes and make you susceptible to some forms of skin cancer, so change your overalls regularly and keep yourself safe.

SUMMARY

Health and safety is of utmost importance to the industry. A sizable amount of money is spent on the protection of welders. Usually manufacturers have a safety department with one individual in charge of plant safety, and in small workshops this responsibility falls upon the workshop manager or to a delegated person. The safety officer's job is to make sure that all welders comply with safety rules during production. The proper clothing, shoes and eye protection to be worn are emphasized. Any worker who does not follow established safety rules is subject to dismissal.

If an accident does occur, it is important that appropriate and immediate first aid steps be taken. All welding shops should have established plans for the action to take in case of accidents. You should take time to learn the proper procedure for accident response and reporting before you need to respond in an emergency. After the situation has been properly taken care of, you should fill out a RIDDOR report.

Equipment must be periodically checked to be sure that it is safe and in proper working condition in accordance with the electricity at work act and the Provision and Use of Work Equipment Regulations (PUWER). Maintenance workers are employed to see that the equipment is in proper working condition at all times.

ACTIVITIES AND REVIEW

1   Produce a risk assessment for using oxy-acetylene equipment in a confined space.

2   You have decided to open a welding distribution business, selling and hiring welding goods, identify four pieces of legislation that would apply to this business.

3   State four sources of fume within a welding environment.

4   List the spectrum of light radiation associated with welding.

5   Name three instances in which a RIDDOR form would need to be completed.

6   Give a brief description of the following:
    a   Arc eye.
    b   Metal fume fever.

7   Identify, according to Table 2.1, what eye and/or face protection should be used for each of the following?
    a   Acetylene welding.
    b   Chipping.
    c   Electric arc welding.
    d   Spot welding.

8   Name a common metal for which a welder would encounter hexavalent chromium fumes while welding or grinding that metal.

9   List two types of respirators and describe how they work.

10 State what the following abbreviations stand for:
    a   PUWER
    b   HASAWA
    c   LOLER
    d   COSHH
    e   MSDS

11 Why is it unsafe to carry butane lighters or matches in your pockets while welding?

12 Describe an acceptable storage area for a cylinder of fuel gas.

13 Name four types of fire extinguishers and what type of burning material they are used to extinguish.

14 Give a brief description of what a hot permit is.

15 Name two advantages of recycling scrap metal.

16 State why it is important that acetylene cylinders are not stored horizontally, and what procedures should be adopted if you have to transport to site in this condition and prior to use.

17 Define what is meant by 'duty of care' with regards to waste disposal.

18 Give a brief description of how to safely lift a heavy object.

19 State why it is important to keep the welding area clean and tidy.

20 Give a brief description of the procedures to be followed in the following circumstances:
    a   The lead of the extension lead is warm.
    b   You are going to be working outdoors.
    c   You notice that the insulation on the welding lead has been damaged.

21 State what the definition of working at height is.

22 Identify which fabrics are the best choice for general work clothing in the welding workshop.

23 What is the key to preventing accidents in a welding workshop?

24 Name four metals that can give off dangerous fumes/vapours during welding and require forced ventilation.

25 List what special protective items can be worn to provide extra protection for a welder's:

a Hands.
b Arms.
c Body.
d Feet.

Communicating Technical Information

LEARNING OBJECTIVES

After completing this chapter, you should be able to:

- list five basic factors related to joint design
- identify the major parts of a welding symbol
- list the component parts of a V-butt weld preparation prior to welding
- list three ways in which the edge preparation can be carried out
- list the five major types of joints
- identify butt joints on pipe and plate in the flat, horizontal, vertical and overhead positions
- identify fillet welds on pipe and plate in the flat, horizontal, vertical and overhead positions
- make a simple sketch in first angle orthographic projection of a workbench from the information supplied and produce a cutting list.

FUNCTIONAL SKILLS

- Interpreting basic elements of a drawing or sketch.
- Interpreting welding symbol information.

UNIT REFERENCES

NVQ:

Working Efficiently and Effectively in Engineering.
Complying with Statutory Regulations and Organizational Safety Requirements.
Using and Interpreting Engineering Data and Documentation.

VRQ:

Engineering Environment Awareness.
Engineering Techniques.
Engineering Principles.
Fabrication and Engineering Principles.

KEY TERMS

american welding society (AWS) a multi-faceted, non-profit organization with a goal to advance the science, technology and application of welding and related joining disciplines.

ASME an American welding standard used as described in text. ASME is the American Society of Mechanical Engineers.

fillet weld a weld of approximately triangular cross section joining two surfaces approximately at right angles to each other in a lap joint, T-joint, or corner joint.

frustrum a cone with its top cut off parallel to the base of the cone.

fusion the flowing together or deposition into one body of the materials being welded.

groove an opening or a channel in the surface of a part or between two components, that provides space to contain a weld.

joint configuration the type of joint produced by the joint dimensions. The details set out in prints or plans to note the shape required.

joint type a weld joint classification based on the five basic arrangements of the component parts such as a butt joint, corner joint, edge joint, lap joint and T-joint.

measuring the process of estimating the magnitude of some attribute of an object, such as its length or weight, relative to some standard (unit of measurement), such as a metre or a kilogram.

peening using the ball part of a ball-peen hammer to counteract contraction on weld deposits in certain materials.

plasma arc cutting (PAC) an arc cutting process that uses a constricted arc and removes the molten metal with a high-velocity jet of ionized gas issuing from the constricting orifice.

plug weld a joint commonly used in motor vehicle bodywork in which one of the components to be joined has a preprepared section removed to allow access to the bottom panel. The weld is completed by filling the section and the bottom panel together.

projection drawings are sketches, drawings or computer models presented in one of several view types, such as isometric or orthographic projections.

shear strength as applied to a soldered or brazed joint, it is the ability of the joint to withstand a force applied parallel to the joint.

tolerances the allowable deviation in accuracy or precision between the measurement specified and the part as laid out or produced.

weld joint the junction of members or the edges of members that are to be joined or have been joined by welding.

welding position the relationship between the weld pool, joint, joint members and the welding heat source during welding.

welding symbol a graphical representation of a weld.

INTRODUCTION

Joint design affects the quality and cost of the completed weld and requires special attention and skill. The factors influencing the choice of weld include:

- the welding style to be used;
- whether the joint is a single weld or is to be mass produced;
- access for welding, access to one or both sides of the joint;
- material thickness;
- residual stress and control distortion.

Joint selection involves compromise. For example, compromises may be made between strength and cost, or between equipment available and welder skill. Your ability to choose the best design will increase as you gain more experience of welding. Even with experience, trial welds are necessary before selecting the final joint configuration and welding parameters.

This chapter will give you some understanding of joint design and weld symbols, the universal language of welders.

WELD JOINT DESIGN

The purpose of a weld joint is to join parts together so that the stresses are distributed. The forces causing stresses in welded joints are tensile, compression, bending, torsion and shear (Figure 3.1). The ability of a joint to withstand these forces depends upon both the joint design and the weld integrity. Some joints can withstand certain types of forces better than others.

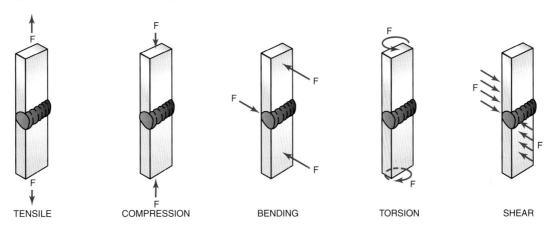

| TENSILE | COMPRESSION | BENDING | TORSION | SHEAR |

FIGURE 3.1 **Forces on a weld**

The basic parts of a weld joint design that can be changed include:

- *Joint type.* The type of joint is chosen by analyzing the way that the joint members come together, Figure 3.2.

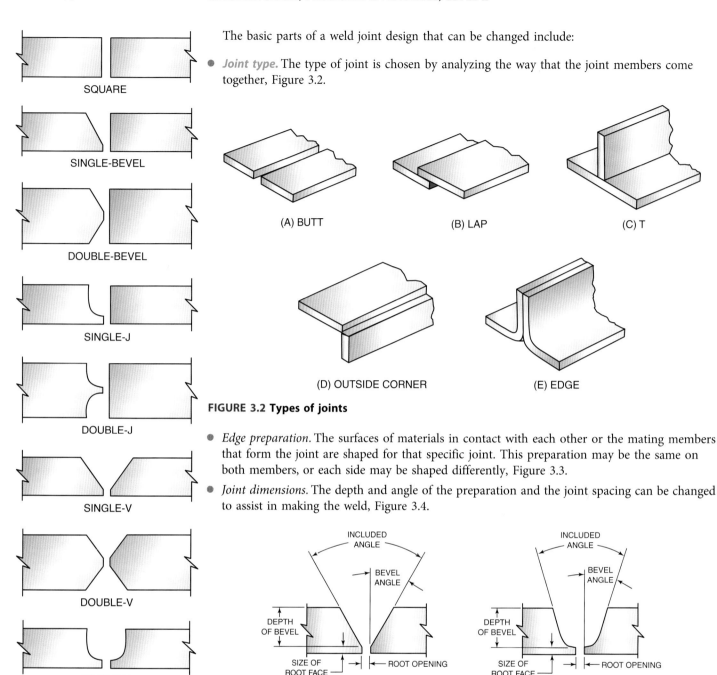

SQUARE

SINGLE-BEVEL

DOUBLE-BEVEL

SINGLE-J

DOUBLE-J

SINGLE-V

DOUBLE-V

SINGLE-U

DOUBLE-U

FIGURE 3.3 Edge preparation

(A) BUTT   (B) LAP   (C) T

(D) OUTSIDE CORNER   (E) EDGE

FIGURE 3.2 Types of joints

- *Edge preparation.* The surfaces of materials in contact with each other or the mating members that form the joint are shaped for that specific joint. This preparation may be the same on both members, or each side may be shaped differently, Figure 3.3.

- *Joint dimensions.* The depth and angle of the preparation and the joint spacing can be changed to assist in making the weld, Figure 3.4.

INCLUDED ANGLE   BEVEL ANGLE   DEPTH OF BEVEL   SIZE OF ROOT FACE   ROOT OPENING   V-GROOVE

INCLUDED ANGLE   BEVEL ANGLE   DEPTH OF BEVEL   SIZE OF ROOT FACE   ROOT OPENING   U-GROOVE

FIGURE 3.4 Included angle

Welding Process

The welding process to be used has a major effect on the selection of the joint design. Each welding process has characteristics that affect its performance. Some can be used easily in any position; others may be restricted to particular positions. The rate of travel, penetration, deposition rate and heat input also affect the welds used on some joint designs.

Parent Material

Some metals present specific problems in terms of thermal expansion, crack sensitivity or distortion, and the joint selected must control these problems. For example, magnesium is very susceptible to post-weld stresses, and the U-butt preparation works best for thick sections.

Plate Welding Positions

The ideal welding position for most joints is the flat position, because it allows for larger molten weld pools to be controlled. A large weld pool usually means that the joint can be completed more quickly. However, it is not always possible, that all welds can be made in the flat position. Special joint designs may be used for certain types of out-of-position welding. For example, the single bevel joint is often the best choice for horizontal welding.

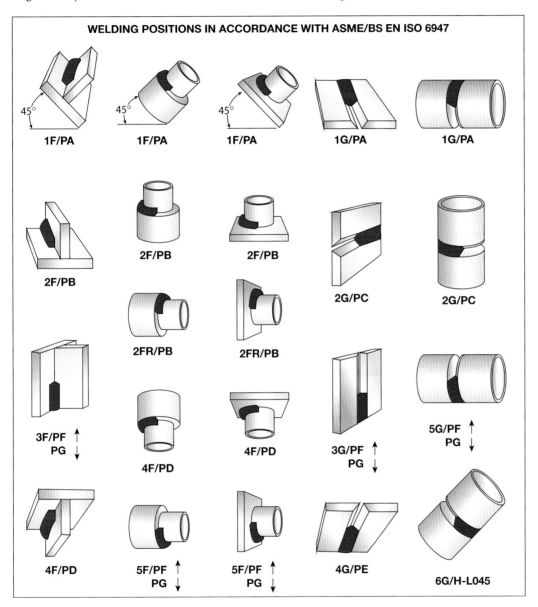

WELDING POSITIONS IN ACCORDANCE WITH ASME/BS EN ISO 6947

FIGURE 3.5 **Typical welding positions**

Welding positions in accordance with BS EN ISO 6947 or the American Society of Mechanical Engineers (ASME) has divided plate welding into four basic positions for butts and fillet welds as follows:

- *Flat PA or 1F (IG).* Welding is performed from the upper side of the joint, and the face of the weld is approximately horizontal.
- *Horizontal/Vertical PB or 2F.* Welding is performed so that one plate is in the horizontal axis while the other plate is in the vertical position.
- *Horizontal PC or 2F (2G).* The axis of the weld is approximately horizontal, but the type of weld dictates the complete definition. For a fillet weld, welding is performed on the upper side

of an approximately vertical surface. For a butt weld, the face of the weld lies in an approximately vertical plane.

- *Vertical PF/PG or 3F (3G)*. The axis of the weld is approximately vertical, PF being vertical up and PG being vertical down.
- *Overhead PD/PE or 4F (4G)*. Welding is performed from the underside of the joint.

Pipe welding positions in accordance with BS EN ISO 6947 or ASME has identified five basic positions for pipe welding:

1 *Horizontal rolled PA (1G)*. The pipe is rolled either continuously or intermittently so that the weld can be performed within 0° to 15° of the top of the pipe.

2 *Horizontal fixed PF or PG (5G)*. The pipe is parallel to the horizon, and the weld is made vertically around the pipe. PF the weld is made from the bottom of the pipe upwards, while PG the weld is made from the top down to bottom of pipe.

3 *Vertical PC (2G)*. The pipe is vertical to the horizon, and the weld is made horizontally around the pipe.

4 *Inclined H-LO45 (6G)*. The pipe is fixed at a 45° inclined angle, and the weld is made around the pipe.

5 *Inclined with a restriction ring 6GR*. The pipe is fixed at a 45° inclined angle, and a restricting ring is placed around the pipe below the weld preparation. This configuration is normally associated with structural applications such as oil rig or offshore wind farm jackets.

Metal Thickness

Metal thickness affects joint design. On thin sections, it is often possible to make full-penetration welds using a square butt joint. But with thicker plates or pipe the edge must be prepared with a weld preparation on one or both sides. The edge may be shaped with a bevel, V-butt, J-butt or U-butt. The choice of shape depends on the type of metal, its thickness and whether it is made before or after assembly. When welding on thick plate or pipe, it is often impossible to get 100 per cent penetration without weld preparation. This may be cut on either one of the plates or pipes or both. On some plates it may be necessary to prepare both sides of the joint, to help control weld distortion (Figure 3.6). The weld preparation may be ground, flame-cut, or machined on the edge of the plate before assembly.

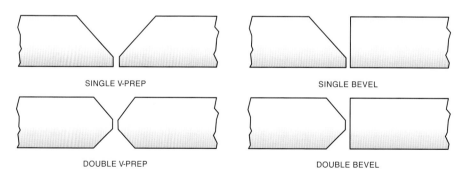

SINGLE V-PREP   SINGLE BEVEL

DOUBLE V-PREP   DOUBLE BEVEL

FIGURE 3.6 V-prep and bevel joint types

For most welding processes, plates that are thicker than 10mm may be prepared on both sides of the joint, dependent upon joint design, position, code and application. Plates in the flat position are usually prepared on only one side unless they can be repositioned or must be welded on both sides. T-joints in thick plates are easier to weld and show less distortion if they are prepared on both sides.

Sometimes plates are either V-butt welded or just welded on one side and then back-gouged and welded, Figure 3.8. Back-gouging involves cutting a groove in the back side of a joint that has been welded. It can ensure 100 per cent joint fusion at the root and remove impurities in the root pass.

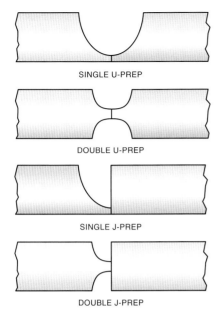

SINGLE U-PREP

DOUBLE U-PREP

SINGLE J-PREP

DOUBLE J-PREP

FIGURE 3.7 U-prep and J-prep joint types

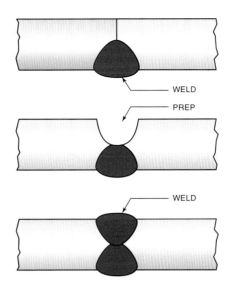

WELD

PREP

WELD

FIGURE 3.8 Back gouging a weld joint to ensure 100% joint penetration

Code or Standards Requirements

The type, depth, angle and location of the weld preparation are usually determined by a code or standard that has been qualified for the specific job. The British, European, International Organization for Standards (ISO), American Welding Society (AWS), ASME Boiler and Pressure Vessel (BPV) Section IX Standards are among the agencies that issue these codes and specifications.

The joint design for a particular set of specifications often must be *pre-qualified*. This means that it has been tested and found to be reliable for the weldments for specific applications. The joint design can be modified, but the cost of having the new design accepted under the standard is often prohibitive.

Welder Skill

The skills of the welder can limit the choice of joint design. A joint must be designed in such a way that the welder can reliably reproduce it. Some joints have been designed without adequate room for the welder to see the molten weld pool or to get the electrode or torch into the joint.

Acceptable Cost

It is often possible to reduce costs, and still meet the weldment's strength requirements, by changing the design. Reducing the weld preparation angle can also help by decreasing the welding filler metal required as well as the time required to fill the weld joint, Figure 3.9.

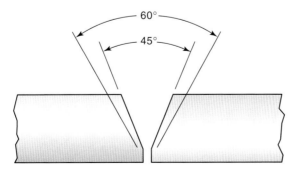

60°

45°

FIGURE 3.9 V-butt preparation

MECHANICAL DRAWINGS

Drawings are used as a means of communicating information between people involved in the manufacture of engineering components and assemblies. Initially, free hand sketches are used to convey ideas. From these ideas formalized drawings are produced which give specific measurements and data and take into account such things as method of manufacture, material selection, environmental conditions and tolerances.

A group of drawings, known as a 'set of drawings', should contain enough information to enable the welder to produce the weldment. It may contain pages showing different aspects of the project to aid in its fabrication. The pages may include: title page, pictorial view either in isometric or oblique projection, assembly drawings, single details drawing and exploded view, Figure 3.10.

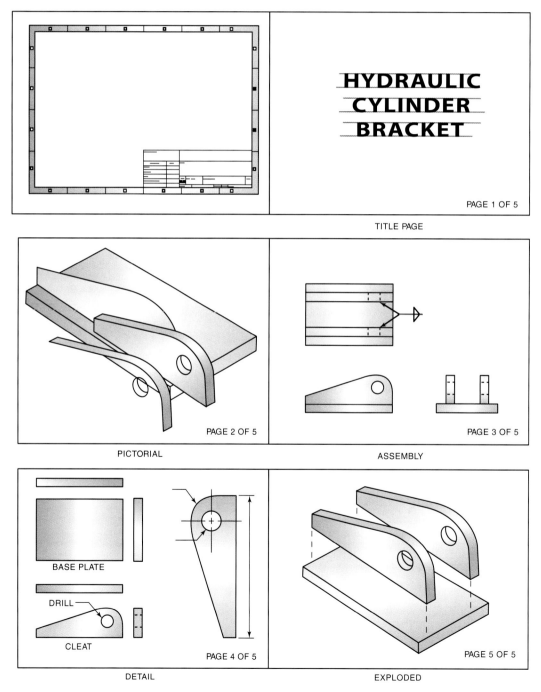

PAGE 1 OF 5

TITLE PAGE

PAGE 2 OF 5

PICTORIAL

PAGE 3 OF 5

ASSEMBLY

BASE PLATE

DRILL

CLEAT

PAGE 4 OF 5

DETAIL

PAGE 5 OF 5

EXPLODED

FIGURE 3.10 Drawings that can make up a set of drawings

As well as graphic representations of the weldment, a set of drawings may also contain information in the title box and bill of materials.

- *The title box*, which appears in one corner of the drawing, should contain the name of the part, the company name, the scale of the drawing, the date of the drawing, the name of the person who made the drawing, the drawing number, the number of drawings in the set and tolerances.

- *A bill of materials* is a list of the various items that will be needed to build the weldment, Table 3.1.

TABLE 3.1   Bill of Materials

Part	Number Required	Type of Material	Size (SI Units)
Base	1	Hot roll steel	12.7 mm × 127 mm × 203.2 mm
Cleat	2	Hot roll steel	12.7 mm × 101.6 mm × 203.2 mm

www.standardscentre.co.uk.

www.roymech.co.uk.

Line Representation

Different lines are used for various parts of the object being illustrated, as shown in Table 3.2 below. Dimensions are not placed on the surface of the object being illustrated. Instead, they are written to the side, with lines to indicate the part of the object they refer to.

TABLE 3.2   Alphabet of lines

Line Type	Description	Purpose
OBJECT LINE	Solid bold line	To show the intersection of surfaces or the extent of a curved surface.
HIDDEN LINE	Broken medium line	To show the intersection of surfaces or the extent of a curved surface that occurs below the surface and hidden from view.
CENTRE LINE	Fine broken line made up of longer line sections on both sides of a short, dashed line	To show the center of a hole, curve, or symmetrical object.
EXTENSION LINE / DIMENSION LINE	Extension lines (fine line) extending from near the surface of the object	Extension lines extend from an object line or a hidden line to locate dimension points.
	Dimension lines (medium line) extending between extension lines or object lines	Dimension lines touch the extension lines and/or object lines that represent the points being dimensioned.
CUTTING PLANE LINE	Bold broken lines with arrowheads pointing in the direction of the cut surface	These lines extend all the way across the surface that is being imaginarily cut. The arrowhead ends point in the direction in which the cut surface will be shown in the sectional drawing.
SECTION LINES STEEL CAST IRON	Series of fine lines drawn at an angle to the object lines. The line angle usually changes from one part to another. The cast iron section lines are used universally for most sections.	Used to indicate a surface that has been imaginarily cut or broken. The spacing and pattern can be used to indicate the type of material that is being viewed.
LEADER OR ARROW LINE	Medium line with an arrowhead at one end	Leader and arrow lines are used to locate points on the drawing to which a specific note, dimension, or welding symbol refers.
LONG BREAK LINE	Bold straight line with intermittent zigzag	To indicate that a portion of the part has not been included in the drawing either to conserve space or because the omitted portion was not significant to this specific drawing.
SHORT BREAK LINE	Bold freehand irregular line	Used for the same purposes as the long break above except on parts not wide enough to allow the long break lines with their zigzags to be used clearly.

Types of Drawings

The drawings used for most welding projects can be divided into two categories: orthographic projections and pictorial.

● A pictorial view gives a general idea of the shape of the object, but only a limited amount of detailed information.

● Orthographic projection drawings are used to define all the necessary information to enable the object to be manufactured.

These drawings are made as though one were looking through the sides of a glass box at the object and tracing its shape on the glass. If all the sides of the object were traced and the box unfolded and laid out flat, six basic views would be shown.

Pictorial drawings present the object in a more realistic or understandable form and usually appear as one of two types.

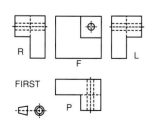

● Isometric drawings are drawn with one axis at 30° to a base line, one axis at 90° vertical to the base line, and the remaining axis at 30° to the base line.

● Oblique drawings have one axis on the base line, one axis vertical to the base line and one axis at 45° to the base line, as shown in Figure 3.11.

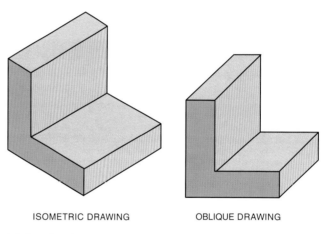

ISOMETRIC DRAWING   OBLIQUE DRAWING

FIGURE 3.11 Pictorial drawing types

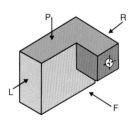

FIGURE 3.12 First angle projection

Orthographic Projection Drawings

Orthographic projection can be produced in either 'first' angle or 'third' angle. You must understand the differences between these two types if you are to avoid costly errors.

First Angle Projection

First angle projection drawings are projected so that the view drawn on the right shows what is seen when looking at the left-hand side (end elevation). The view drawn at the bottom shows what is seen when looking down on the object (plan view). This type of projection has often been referred to as looking through a window; what you see is on the other side of the window.

Third Angle Projection

Drawings done in *third angle* are projected so that the view drawn on the right shows what is seen when looking at the right-hand side and is often compared to looking in the mirror; the view is reflected back and drawn there. The view at the top is again reflected back and drawn above the front elevation, see Figure 3.13.

First angle or third angle projection is indicated on the drawing by a *frustrum* of a cone symbol with concentric circles. If the circles are to the right of the frustrum then the projection is first angle. For most workshop drawings the information required can be shown by drawing front elevation, side elevations and plan views only.

The front view is not necessarily the front of the object but is the view which best shows the object's overall shape.

Drawings come in a variety of sheet sizes depending upon how much detail is provided, from full drawing down to component parts.

When laying out a drawing, start by calculating the size of box that the object will sit in. Draw a 45° line from the bottom or top box, depending on the projection (first or third angle) and with a convenient gap, draw lines parallel to and at right angles to the original box to create additional boxes to house the other views. What you now have is a circuit in which information and detail can be transferred from one view to another without having to measure. This helps the person manufacturing the object to trace a line or dimension in a number of views.

Always check that you have the correct drawing for the job in the most up-to-date issue available. This information is contained in the title block on the drawing. You must check this with your supervisor before starting work as important changes could have been made to the component you are working on. All drawings should be returned to the issuing office to prevent an old issue being used later on.

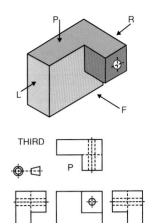

FIGURE 3.13 Third angle projection

Special Views

Special views may be included on a drawing so an object can be made accurately:

- *The sectional view* is drawn as if part of the object were sawn away to reveal internal details. This is useful when the internal details would not be as clear if they were shown as hidden lines. The imaginary cut surface is defined by section lines drawn at an angle on the cut surfaces. The location of this imaginary cut is shown using a cutting plane line.

- *The cut-away view* is used to show detail within a part that would be obscured by the part's surface. Often a free-hand break line is used to outline the area that has been imaginarily removed to reveal the inner workings.

- *The detail view* is usually an external view of a specific area of a part. Detail views show small details and remove the need to draw an enlargement of the entire part.

- *A rotated view* can be used to show a surface of the part that would not normally be drawn square to any of the six normal view planes. If a surface is not square to the viewing angle, then lines may be distorted. For example, when viewed at an angle, a circle looks like an ellipse, see Figure 3.14.

TIP

Remember first right, third left for concentric circles to differentiate between projections.

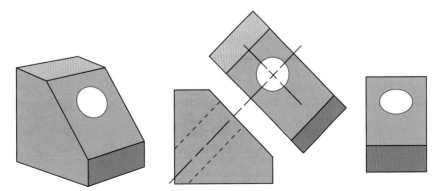

FIGURE 3.14 Distortion of lines

Dimensioning

It is often necessary to look at several views to locate all the dimensions required to build the object. Length dimensions can be found on the front and top views. Height dimensions can be found on the front and right side views. Width dimensions can be found on the top and right side views. This applies to both the first angle and third angle projection layouts.

A properly executed drawing will contain all necessary dimensions. If you cannot find the required dimensions, do not try to obtain them by measuring the drawing itself, contact the person who made the drawing.

Keep the drawing clean and well away from any welding operation. Avoid writing or doing calculations on the drawing. Often a drawing will be filed following the project for use at a later date.

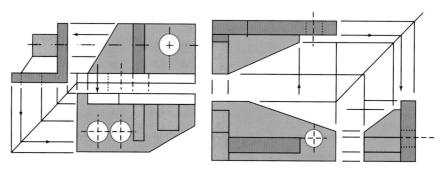

FIGURE 3.15 **Drawing dimension locations**

WELDING SYMBOLS

Welding symbols enable a designer to communicate important, detailed information about the weld, such as:

- length;
- reinforcement or flush;
- weld preparation;
- weld dimensions;
- location;
- process;
- number of welds;
- weld shape.

Welding symbols are a form of language for the welder. They save time and money and serve to ensure understanding and accuracy. Welding symbols have been standardized by BS EN 22553 and the processes by BS EN ISO 4063.

Figure 3.16 shows the basic components of welding symbols. The symbols are based on a reference line with an arrow at one end. Other information is shown by symbols, abbreviations, and figures located around the reference line. A tail is added to the basic symbol as necessary for the placement of specific information.

BS EN 22553 Welded, brazed and soldered joints –
Symbolic representation on drawings

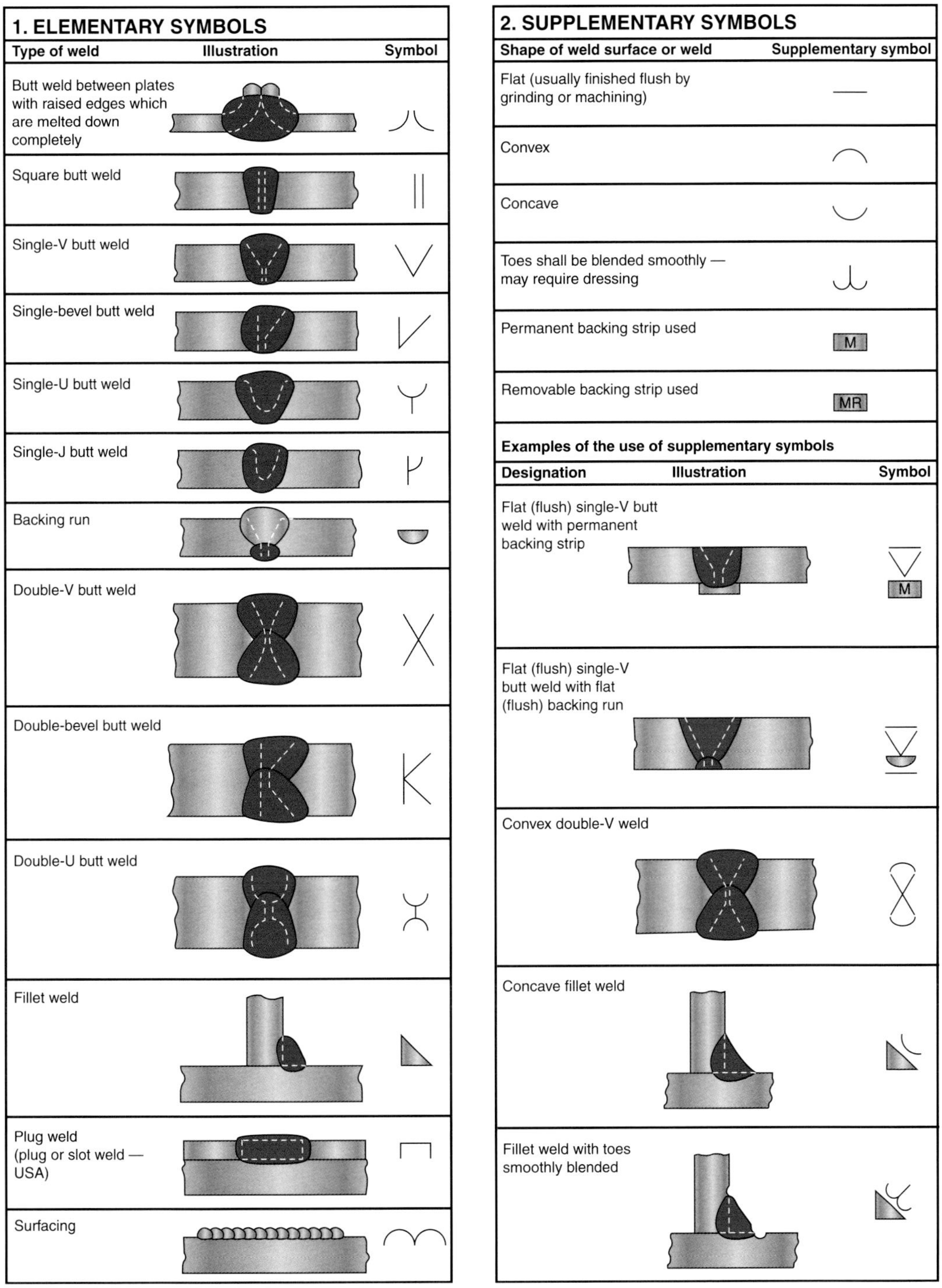

1. ELEMENTARY SYMBOLS

Type of weld	Illustration	Symbol
Butt weld between plates with raised edges which are melted down completely		⅃⅃
Square butt weld		‖
Single-V butt weld		V
Single-bevel butt weld		⌁
Single-U butt weld		Y
Single-J butt weld		Ⱶ
Backing run		⌣
Double-V butt weld		X
Double-bevel butt weld		K
Double-U butt weld		⅄
Fillet weld		◺
Plug weld (plug or slot weld — USA)		⊓
Surfacing		⌒⌒

2. SUPPLEMENTARY SYMBOLS

Shape of weld surface or weld	Supplementary symbol
Flat (usually finished flush by grinding or machining)	—
Convex	⌒
Concave	⌣
Toes shall be blended smoothly — may require dressing	⌣⌣
Permanent backing strip used	[M]
Removable backing strip used	[MR]

Examples of the use of supplementary symbols

Designation	Illustration	Symbol
Flat (flush) single-V butt weld with permanent backing strip		▽ [M]
Flat (flush) single-V butt weld with flat (flush) backing run		▽⌣
Convex double-V weld		⧓
Concave fillet weld		◺⌣
Fillet weld with toes smoothly blended		◺⌣

TIP

All this information would normally be included on the welding assembly drawings.

FIGURE 3.16 **Standard location of elements of a welding symbol**

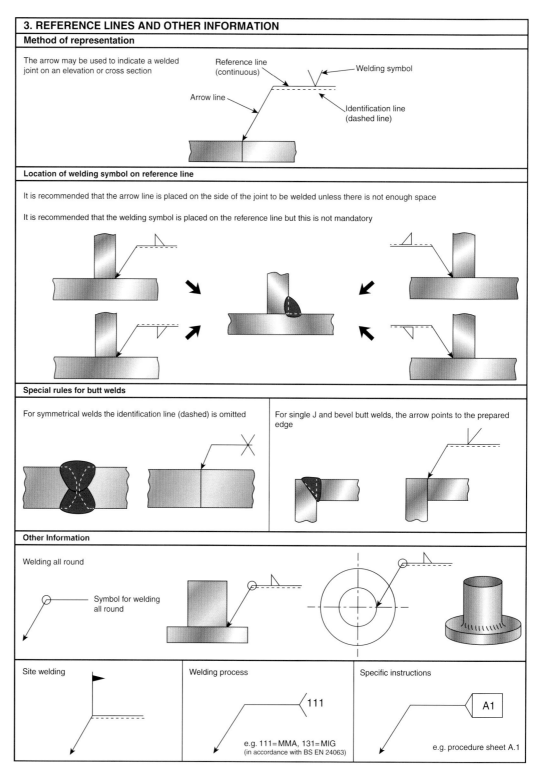

3. REFERENCE LINES AND OTHER INFORMATION

Method of representation

The arrow may be used to indicate a welded joint on an elevation or cross section

Reference line (continuous)

Welding symbol

Arrow line

Identification line (dashed line)

Location of welding symbol on reference line

It is recommended that the arrow line is placed on the side of the joint to be welded unless there is not enough space

It is recommended that the welding symbol is placed on the reference line but this is not mandatory

Special rules for butt welds

For symmetrical welds the identification line (dashed) is omitted

For single J and bevel butt welds, the arrow points to the prepared edge

Other Information

Welding all round

Symbol for welding all round

Site welding

Welding process

111

e.g. 111=MMA, 131=MIG
(in accordance with BS EN 24063)

Specific instructions

A1

e.g. procedure sheet A.1

FIGURE 3.16 **Standard location of elements of a welding symbol (continued)**

4. WELD DIMENSIONS

Butt welds

's' = minimum specified throat (penetration) thickness. If no dimension is shown, the weld is full penetration

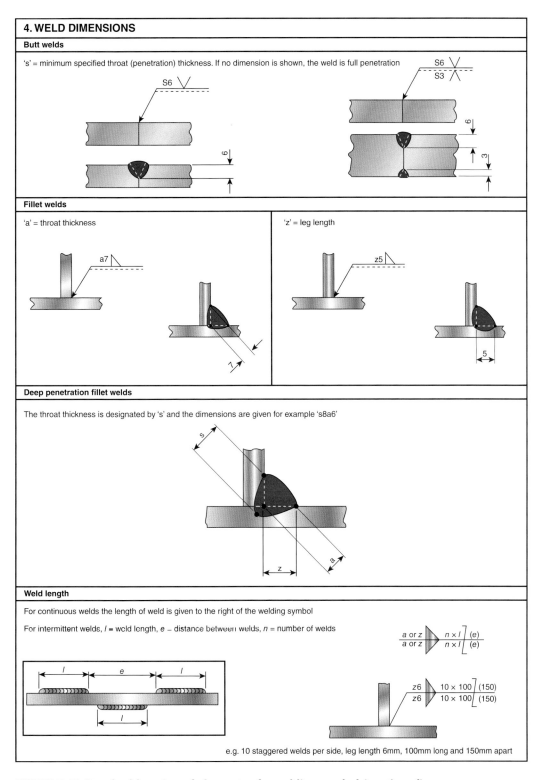

Fillet welds

'a' = throat thickness

'z' = leg length

Deep penetration fillet welds

The throat thickness is designated by 's' and the dimensions are given for example 's8a6'

Weld length

For continuous welds the length of weld is given to the right of the welding symbol

For intermittent welds, l = weld length, e – distance between welds, n = number of welds

e.g. 10 staggered welds per side, leg length 6mm, 100mm long and 150mm apart

FIGURE 3.16 Standard location of elements of a welding symbol (continued)

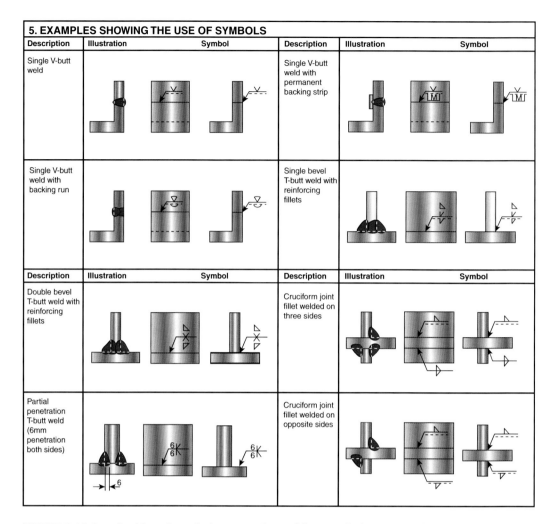

FIGURE 3.16 Standard location of elements of a welding symbol (continued)

Indicating Types of Welds

Weld types are classified as: *fillets, butts, flange, plug* or *slot, spot* or *projection, seam, back* or *backing* and *surfacing.* Each type is indicated on drawings by a specific symbol. A fillet weld, for example, is designated by a right triangle. A plug weld is indicated by a rectangle.

Weld Location

Welding symbols are applied to the joint as the basic reference. All joints have an arrow side (near side) and another side (far side). The terms *arrow side, other side* and *both sides* are used to indicate the *weld location* with respect to the joint. The reference line is always drawn horizontally. An arrow line is drawn from one or both ends of a reference line to the location of the weld. It can point to either side of the joint and extend either upward or downward. If the weld is to be deposited on the arrow side of the joint, the weld symbol is placed below the reference line. If the weld is to be deposited on the other side of the joint, the weld symbol is placed above the reference line. When welds are to be deposited on both sides of the joint, the same weld symbol appears above and below the reference line, along with detailed information.

A tail is added to the basic welding symbol to give welding specifications, procedures or other information required to make the weld. The notation placed in the tail of the symbol may indicate the welding process to be used, the type of filler metal, whether or not peening or root chipping is required, and other information. If notations are not used, the tail of the symbol is omitted.

TIP

For joints that are to have more than one weld, a symbol is shown for each weld.

Significance of Arrow Location

In the case of fillet and butt welding symbols, the arrow connects the welding symbol reference line to one side of the joint. The surface the arrow point touches is considered to be the arrow side of the joint.

For welds designated by the plug, slot, spot, seam, resistance, flash, upset or projection welding symbols, the arrow connects the welding symbol reference line to the outer surface of one of the members of the joint at the center line of the desired weld.

Fillet Welds

The dimensions of fillet welds are shown on the same side of the reference line as the weld symbol and to the left of the symbol. Some key points to remember:

- When both sides of a joint have the same size fillet welds, they are dimensioned. When the two sides of a joint have different size fillet welds, both are dimensioned. When the dimensions of one or both welds differ from the dimensions given in the general notes, both welds are dimensioned.
- The size of a fillet weld with unequal legs is shown in brackets to the left of the weld symbol. The length of a fillet weld, when indicated on the welding symbol, is shown to the right of the weld symbol.
- In intermittent fillet welds, the length and pitch increments are placed to the right of the weld symbol. The first number represents the length of the weld, and the second represents the pitch, or the distance between the centres of two welds.

Plug Welds

Holes in the arrow side member of a joint for plug welding are indicated by placing the weld symbol below the reference line. Holes in the other side member are indicated by placing the weld symbol above the reference line, Figure 3.16 shows the location of the dimensions used on plug welds.

Spot Welds

The dimensions of resistance spot welds are indicated on the same side of the reference line as the weld symbol. The welds are dimensioned either by size or by strength. The centre-to-centre spacing (pitch) is shown to the right of the symbol. When a specific number of spot welds is desired in a certain joint, the quantity is placed above or below the weld symbol in brackets.

Butt Welds

Joint strengths can be improved by weld preparation before the joint is welded. Weld preparation can be made in one or both plates or on one or both sides and can be produced in a number of different ways using oxy-fuel cutting, plasma arc cutting or machining.

The types of butt welds are classified as follows:

- *Single V-butt and symmetrical double V-butt welds that extend completely through the members being joined.* No size is included on the weld symbol.
- *V-butt welds that extend only part way through the parts being joined.* The size as measured from the top of the surface to the bottom (not including reinforcement) is included to the left of the welding symbol.

The size of butt welds with a specified effective throat is indicated by showing the depth of preparation with the effective throat appearing in brackets and placed to the left of the weld symbol. The size of square butt welds is indicated by showing the root penetration. The depth of chamfering and the root penetration are read in that order from left to right along the reference line.

The main purpose of the root face is to minimize the burn-through that can occur with a feather edge (acts as a heat sink). The size of the root face is important to ensure good root fusion, Figure 3.18. The angle of butt welds is considered to extend only to the tangent points of the members. The root opening of butt welds is the user's standard unless otherwise indicated, when it is shown inside the weld symbol.

Backing

A backing (strip) is a piece of metal placed on the back side of a weld joint. It must be thick enough to withstand the heat of the root pass as it is burned in. A backing strip may be used on butt joints and may be left on the finished weld. However, it is often removed because it can be a source of stress concentration and a crevice to promote corrosion. If it is to be removed, the letter *R* is placed in the backing symbol, Figure 3.19.

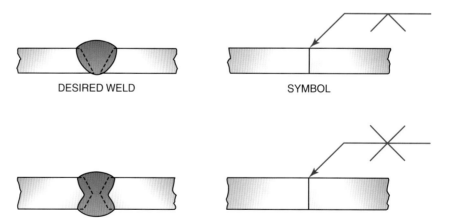

DESIRED WELD   SYMBOL

FIGURE 3.17 Designating single- and double-groove welds with complete penetration

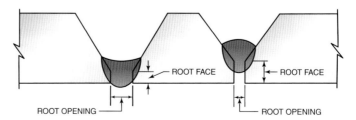

FIGURE 3.18 Effect of root dimensioning on groove penetration

Flange Welds

The following welding symbols are used for light-gauge metal joints where the edges to be joined are bent to form flanges or edge welds:

- Edge flange welds are shown by the edge flange weld symbol.
- Corner flange welds are indicated by the corner flange weld symbol.
- The dimensions of flange welds are shown on the same side of the reference line as the weld symbol and are placed to the left of the symbol. The radius and height above the point of tangency are shown separated by a plus sign.
- The size of the flange weld is shown by a dimension placed outward from the flanged dimensions.

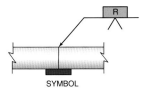
SYMBOL

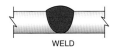
WELD

FIGURE 3.19 Butt weld with backing plate

SUMMARY

As a production welder you will be expected to follow simple or complex drawings in the fabrication of weldments and you must be able to interpret the meaning of welding symbols. A weld that is made excessively large can cause a structural failure as easily as one that is undersized. Welded structures must often flex under load, within limits, so they will not break. Excessive weld deposition can cause a structure to be subjected to high stress concentrations (notch effect) at the toes of the weld and lead ultimately to weld failure. The ideal weld should blend the toes of the weld with the parent metal and so not allow stress concentrations to build up.

ACTIVITIES AND REVIEW

1 List the parameters of a welded joint that can be changed to ensure a sound weld deposit.

2 Identify the following welding positions from the following abbreviations:

   a PA
   b PB
   c PC
   d PD
   e PG
   f PF

3 Sketch the typical layouts for 1st and 3rd angle projection, indicating the projection by symbol representation.

4 Name four pieces of information that should be indicated in a title box.

5 Identify the lines shown in the chart below, giving a brief description and typical application for each.

Type of line	Description	Applications
▬▬▬▬		
——————		
– – – – – – – – ·		
— ·· — · — ·· — ·		
– – – – –→		

6 Draw simple sketches to show the differences between oblique and isometric drawing.

7 Indicate what is represented by the following welding symbols:

Weld Symbol	Definition
◺	
∨	
○	
⚲	
∪	
▭	
⊬	

8 Sketch a V-butt joint, and label all of the joint's features.

9 State how the front elevation is selected in orthographic projection.

10 Sketch a fillet weld in the horizontal-vertical axis.

11 Give a brief description of 'back gouging'.

12 State the advantage of using a sectional view.

13 Name four pieces of information that may appear on the reference line of a weld symbol.

14 Sketch and dimension a double V-butt weld preparation in 25mm plate, with an included angle of 60°. Indicate suggested root face and root gap dimensions.

15 Identify how the removal of the backing strip is noted on a welding symbol.

16 Using first angle projection, prepare simple sketches of a welding table 600mm wide, 1.2m long, 900mm tall, using 6mm mild steel plate for the top and 37mm mild steel angle iron material. Include front elevation, end elevation and plan, weld symbols and materials cutting list.

17 State what type of information would be included in the tail of a weld symbol.

18 Identify why is it usually better to make a weld in the flat position.

19 State what is meant if the weld symbol is placed below the reference line.

20 Identify what is contained in a set of drawings.

Engineering Materials, Measurement and Fitting Skills

LEARNING OBJECTIVES

After completing this chapter, you should be able to:

- list the mechanical properties of materials
- identify common engineering materials
- recognize heat treatments appropriate to materials
- list the common forms of supply
- identify standards and how they apply to manufacturing
- understand how to read and interpret a vernier scale
- recognize and understand the function of marking out tools
- use a range of formulae to calculate area and circumference
- mark out using datums, co-ordinates and dimensioning techniques
- recognize the term 'nesting'
- identify aids to assembly
- use tacking and welding to assemble components.

FUNCTIONAL SKILLS

- Undertaking workshop calculations.
- Interpreting technical data.
- Communicating technical data by use of graphical means.

UNIT REFERENCES

NVQ:

Cutting Materials using Saws and Abrasive Discs.
Assembling Components using Mechanical Fasteners.
Marking Out Components for Fabrication.

VRQ:

Engineering Environment Awareness.
Engineering Techniques.
Producing Components from Metal Plate.

KEY TERMS

back mark a measurement taken to ensure accuracy when marking out holes on structural steel sections.

electrolytic a reaction that takes place between the anode and cathode of an electrical circuit in the presence of some electrically conducting fluid.

fitting the mechanical or hand machining of components or assemblies.

graphite a mineral form of carbon.

strongback a structure used to support the underside of a welded joint and to maintain shape by controlling distortion and act as a 'heat sink'.

witness mark dot punch indentations used to define a line to which a material is to be filed, machined or cut to.

INTRODUCTION

Students of welding need a sound understanding of engineering materials, measurement, marking out and fitting procedures. In this chapter we consider the physical properties of common engineering materials, their composition and appropriateness for particular engineering applications. Fabricators must be able to interpret drawings and welding symbols, follow detailed directions and produce accurate technical drawings or sketches to represent structures or fabrications. Many people who used not to enjoy mathematics find it more interesting when the numbers become part of a plan for something they are going to build with their own hands. Metric measuring techniques are now used by all fabrication shops and manufacturing facilities that need to compete in a global marketplace.

FABRICATION

A weldment may form a completed project or may be part of a larger structure or fabrication. Some weldments are composed of two or three parts; others have hundreds or even thousands of individual parts, Figure 4.1. But even the largest weldments start by placing two parts together. The number and type of steps required to move from a plan to a completed project vary depending on the complexity and size of the finished weldment. The plan can be very simple, perhaps only existing in the mind of the welder, or a complex design that is described in a set of drawings.

FIGURE 4.1 **Large welded oil platform**

FIGURE 4.2a **Space shuttle being prepared for launch**

FIGURE 4.2b **Astronaut making a weld outside a space ship**

Safety

Safety is of primary concern in the fabrication of metallic materials, as it is in any manufacturing operation. Fabrication may present safety concerns not normally encountered during training. For example, larger fabrication work may need to be performed outside an enclosed welding booth. In addition, several welders may be working simultaneously on the same structure, and extra care must be taken to ensure their safety. Chapter 2, Working Safely, Efficiently and Effectively, provides guidance on safety awareness and procedures. It is essential that you are well-prepared for the hazards you may encounter in the workplace.

MATERIAL SELECTION

It is of the utmost importance that the appropriate material is selected for the environment in which it will be used. If this is not done, the structure could fail, causing damage, injury and even loss of life. The choice of material will depend to a large extent on its mechanical properties.

Mechanical Properties

It is essential for the welder to fully understand these characteristics of materials:

- *Malleability* is the ability of a material to be subjected to a compressive force such as a hammer blow and be capable of deforming or shaping without fracturing. This is very important in forging, where the material must flow to shape under the force of the press without fracture.
- *Ductility* is often linked with malleability, and is the ability of a material to be drawn out along its length, bent or twisted under an applied load, as happens in wire, bar or tube manufacture.
- *Hardness* is the ability to resist indentation or abrasion, qualities needed in gear teeth and the contact faces of agricultural machinery.
- *Toughness* is often linked with hardness. It is the ability to withstand a shock load or impact (impact strength), as in the case of a hammer hitting the end of a chisel without fracturing. The softer the material is, the tougher it is. Toughness can be increased by heat treatment.
- *Elasticity* is the ability of a material to return to its original shape and size when an applied load is removed, a property that quite a few engineering materials possess to some degree. A typical application would be in the manufacture of springs which must recover their original shape after compression or extension.
- *Plasticity* is the ability to flow to shape and retain that shape upon the load being removed, a property desirable for forging and presswork.
- *Brittleness* means that a material has little toughness and cannot withstand shock loading or impact. A typical material that displays this property would be cast iron, which is hard but brittle.
- *Strength* is the ability of a material to withstand one or more of the following; tensile, compressive, shear or torsional (twisting) force without fracture. This ability is exploited in the high-tensile bolts, rivets and rotating found in most cars or factory components.
- *Conductivity* is the ability of a material to allow heat or electricity to flow through it.
- *Thermal conductivity* is the ability of a material to conduct heat. In welding this aids the cooling of the weld. Copper and aluminium are good conductors of heat. Materials that do not conduct heat readily are often used to retain heat in a body, such as lagging or insulation on pipe work.
- *Electrical conductivity* is the ability of a material to conduct electricity. Examples of electrical conductors in welding are welding leads, connecting plugs and electrode holders. Materials that do not conduct electricity are known as insulators. They include plastics, ceramics and glass. Rubber and plastic are used to prevent short circuits and electric shock on conductors such as the welding cables used in the welding circuit.
- *Weldability* is not an easy property to define but it can be described in simple terms as the ability of a material to join together under controlled conditions without cracking.

- *Corrosion resistance* is the ability to resist oxidation or chemical attack from substances such as acids, alkalis and solvents. It is of considerable importance if components need to keep operating in a corrosive environment. It can be seen in architecture, cryogenic applications and marine environments.

- *Creep resistance* is the ability of the material's lattice structure to not to slip under load or stress.

COMMON ENGINEERING MATERIALS

The most common engineering material is steel. Today's materials owe a lot to the early pioneers, such as Sir Henry Bessemer who revolutionized the industry by developing the converter process for steel making in 1856.

Production of iron starts with iron ores being melted with coke as a fuel and limestone as a fusible slag in large blast furnaces. At this stage it is often referred to as 'pig iron', from the shape of the ingots in which the hot iron solidifies. It contains approximately 4 per cent carbon derived from the coke, 2 per cent silicon and 1 per cent manganese reduced from the ore, and up to 2 per cent phosphorous and sulphur which are the by-products of the ore.

The pig iron requires further refining before it is useful. This is done by oxidation, by inserting an oxygen lance to reduce the carbon, phosphorous and sulphur content of the molten ingots in a Kaldo converter which is continuously rotated.

Steel Properties

Dead Mild Steel

- The carbon content is left deliberately low 0.1–0.15 per cent so that the steel will have high ductility.

- It is for cold deep pressing such as car panels or for extruded components such as thin wire, rod and drawn tubes.

Mild Steel or Low Carbon Steel (LCS)

- This is relatively soft and ductile, containing 0.15–0.3 per cent carbon.

- It can be forged and drawn in both hot and cold condition.

- It is ideal for constructional sections, bars, sheet or plate.

Medium Carbon Steel

- This contains between 0.3–0.8 per cent carbon.

- It is harder, tougher and less ductile than mild steel.

- It cannot be bent or formed in cold condition to any extent without cracking.

- It is used extensively for forgings, axles, leaf springs and cold chisels.

High Carbon Steel

- This contains between 0.8–1.4 per cent Carbon, and has low ductility and toughness.

- Hot forges well, provided the temperature is closely controlled between 700–900°C.

- Applications range from cold chisels, knives, drills and metal cutting tools.

Other ferrous metals that are commonly used include the following:

Wrought Iron

- This is an iron carbon compound in which the carbon content is so low that steel cannot be formed and in which fibres of slag are trapped.

- It has good corrosion resistance and is easily formed cold.
- It is an ideal material for cold forming garden gates and architectural features.

Alloy Steels

- These are plain carbon steels with other alloying elements added to improve mechanical properties and improve their range of applications.
- Typical alloying additions include nickel (grain refinement and strength), chromium (corrosion resistance), and molybdenum (increase temperature resistance).
- By adding high levels of manganese you increase wear resistance and the inclusion of tungsten and cobalt ('High Speed Steel') improves ability to remain hard at high temperatures.
- They have largely replaced high carbon steels for power cutting tools.

Cast Iron Properties

If the carbon content is increased, cast iron is produced. There are three main categories:

Grey Cast Iron

- This contains 3.2 per cent carbon which is present as 'free' carbon flakes of graphite, giving it its characteristic grey colour when fractured.
- It is self-lubricating which gives it excellent machining characteristics.
- It has a low resistance to tensile loading, but good 'dampening' effect upon vibration which makes it ideal for machine frames.

Malleable Cast Iron

- This has a lower carbon content 2.5–3.0 per cent and because it is more malleable than grey cast iron it is often used to produce threaded pipe fittings.

Spheroidal Graphite (SG) Cast Iron

- This is cast iron in which traces of magnesium or cerium are added to ordinary grey cast iron in order to redistribute the graphite flakes as very fine spheroids or balls of carbon.
- It has improved mechanical properties which allow it to be used for more highly stressed components.

Corrosion Resistant Metallic Materials

These are alloys or natural ores that have a protective self-healing surface oxide. A wide range of corrosion resistant materials are available and more are being developed. Here, we will consider some of the most commonly used examples.

Stainless Steel Properties

'Stainless steel' is a term applied to a group of steels with more than 10–12 per cent chromium. Other elements such as nickel, molybdenum, niobium and titanium are added to provide additional operating characteristics which extend the range of applications that they can be used for. The corrosion resistant properties come from the invisible protective coating which is always present on the metal's surface. If the surface is scratched and the oxide film is damaged it will immediately reform in the presence of oxygen in the atmosphere, maintaining the metal's corrosion resistance. Sheets can be supplied with a protective PVC or polythene coating to protect against damage during storage and handling. Stainless steel is available in four classifications:

Ferritic

- These are stainless steels containing 13–20 per cent chromium and less than 0.1 per cent carbon which may be strengthened by work hardening.
- They are tough and ductile and easily hot or cold worked.
- They are magnetic.
- Their low strength prohibits them from being used for highly stressed components.
- Typical applications include domestic and automotive trim, exhaust systems, food processing and catering equipment and architectural cladding panels.

Martensitic

- These contain between 11–18 per cent chromium with a carbon content between 0.1–1.5 per cent and do not contain nickel in appreciable quantities.
- They possess good corrosion resistance and can be hardened and tempered by heat treatment.
- They are magnetic.
- They are not easily formed and welding is not recommended.
- Typical applications include cutlery, surgical instruments, fastenings, valves, spindles and shafts.

Austenitic

- These contain both chromium and nickel with a carbon content below 0.15 per cent.
- The best known type 18/8 containing 18 per cent chromium and 8 per cent nickel.
- They have corrosion resistance and remain tough and ductile even at sub-zero (cryogenic) temperatures.
- They are capable of being cold worked, though they do work harden, they can be softened by annealing.
- They are non-magnetic in the annealed condition.
- Readily welded especially with the addition of 'stabilizers' such as molybdenum, titanium and niobium.
- Typical applications include food processing equipment, hospital and domestic use, and extensively in the petro-chemical industry.

Duplex

- This is a group of high strength, corrosion resistant steels containing higher percentages of chromium and lower percentages of nickel than their equivalent austenitic steels.
- They have nitrogen and manganese as important alloying elements.
- They are called 'Duplex' because structure consists of 50 per cent ferrite and 50 per cent austenite.
- Typical applications include oil and chemical processing industries to maintain corrosion resistance and toughness under extreme working conditions.

Aluminium Properties

Aluminium is produced from the ore bauxite, a hydrated alumina mineral which has iron oxides and other minerals associated with it. The bauxite has to be purified before electrolytic reduction can take place. Aluminium has a melting point of 660°C and possesses several properties which make it an extremely useful engineering material. Its good corrosion resistance and low density make it particularly useful for transportation purposes by land, sea and air. This resistance is due to the fact that aluminium has a high affinity for oxygen and any fresh metal surface will rapidly oxidise. High purity aluminium (99.5 per cent and above) are too weak to be used for many purposes. The term commercially pure aluminium, is in reality an aluminium-silicon-iron alloy (0.5 per cent) which gives a considerable increase in strength, although there is some reduction in ductility.

Aluminium is commonly alloyed with manganese, magnesium, copper, silicon and zinc to produce useful engineering materials. The alloys of aluminium may be sub-divided into non-heat treatable and heat treatable. 'Non-heat treatable alloys' are made up of three main alloy systems:

Aluminium – Manganese (1.25 per cent)

- This has higher tensile strength than commercially pure aluminium.
- It has good ductility and is easily formed by rolling and pressing operations.
- It can be easily welded by oxy-acetylene, manual metal arc or resistance welding.
- Mechanical properties may be increased by work hardening, and can be softened by annealing at about 380°C.
- It can be used for a range of applications including aeronautical.

Aluminium – Magnesium (2–5 per cent)

- These have higher tensile strength than previous group and extremely good resistance to corrosion by sea water.
- They are to work hardening and may be softened by annealing.
- Typical applications include marine and automobile engineering.

Aluminium – Silicon (10–13 per cent)

- These have considerable strength, ductility and resistance to corrosion, with small contraction on solidification.
- They have good flow characteristics and are ideal for casting.

Heat Treatable Alloy Properties

These contain elements such as copper, magnesium, silicon and nickel. Typical of this group are:

Aluminium – Copper (4 per cent)

- This often goes under trade name of 'Duralumin' and requires heat treatment prior to working the material.
- Copper content needs to be taken into 'solution' to reduce its hardening effect.
- 'Solution treatment' allows the copper to disperse within the structure by heating to 425–540°C and then quenching in water.
- Work on the material is then done within an approximately two-hour time frame after which the constituents start to reform but in a more uniform pattern in the alloy, which will increase strength and hardness. This is known as 'age hardening' or 'precipitation hardening' and can take from a few hours to many months for ageing to take place.

TIP

Typical applications for aluminium are in thin foils for packaging, kitchenware, rod and wire for electrical transmission.

Nickel – Aluminium and Copper – Aluminium

- Alloy with the highest strength is one containing aluminium-zinc (5–7 per cent) with the additions of smaller amounts of magnesium, manganese and copper.
- Subject to stress corrosion (cracking).
- Primarily used in high strength applications such as forgings, piston and cylinder heads, rivets, and extruded sections.
- Have slightly lower corrosion than commercially pure aluminium and work hardened alloys.
- To give increased strength in sheet form, the alloy is often clad with a thin layer of pure aluminium during hot rolling to give an 'Alclad' alloy.

INTERNATIONAL ORGANIZATION FOR STANDARDIZATION (ISO)

The International Organization for Standardization (ISO) has a four digit system series which identifies the alloy group.
e.g. 1050

- The first digit (1) indicates the major alloying elements.
- The second digit (0) shows modifications in impurity limits or the addition of alloying elements. If the second digit is zero, then the alloy has only natural limited impurities.
- The last two digits show the maximum percentage of aluminium.
- Therefore 1050 is an aluminium with a purity (minimum) of 99.5 per cent.

TABLE 4.1   Aluminium Alloy Designations

	Aluminium Alloy Designations
1xxx	Aluminium of 99.00 per cent minimum purity and higher
2xxx	Copper
3xxx	Manganese
4xxx	Silicon
5xxx	Magnesium
6xxx	Magnesium and silicon
7xxx	Zinc
8xxx	Other elements
9xxx	Unused series

Copper Properties

Copper is found in the ore copper pyrites. It is first smelted in a blast furnace and is referred to as 'blister copper'. At this stage it is unsuitable for commercial use since it contains impurities such as sulphur and oxygen. Further refining by the electrolysis process produces a commercial grade of copper. Copper has a melting point of 1083°C and is an excellent conductor of both electricity and heat. It is available as 'tough pitch' or 'de-oxidized' copper. De-oxidized copper is often chosen for welding as the tough pitch copper contains minute particles of cuprous oxide which produce steam under welding conditions, liberating hydrogen and oxygen which lower ductility, cause porosity and increase the liability of cracking. The use of deoxidizers, such as phosphorous, silicon, lithium and magnesium, to the molten metal combine with the oxygen to form slag and thus deoxidize the copper. Copper alloys most frequently encountered in welding are:

Copper – Zinc (Brasses)

- These are available as 'alpha brass' (70 per cent copper and 30 per cent zinc). They have good strength and ductility when cold and are used for sheet, strip, wire and tube.
- 'Beta brass' (60 per cent copper and 40 per cent zinc) is often referred to as 'yellow' or 'Muntz metal'. They have and a higher tensile strength but lower ductility than alpha brass and therefore these materials are usually hot worked. They are used extensively for plumbing fittings.
- Brasses can be cast, rolled, pressed, drawn and machined as they are self-lubricating.
- They have good resistance to atmospheric corrosion and are therefore extensively used in catering, brewing, and marine industries.

Copper – Tin – Bronzes (88 per cent copper, 10 per cent tin, and 2 per cent zinc)

- These have good resistance to corrosion and high strength values.
- When lead is added to the bronze it improves the bearing properties of the material and also increases machineability.
- They are used extensively for bearings, statues and marine environments.

Phosphor Bronzes (8 per cent tin and up to 0.4 per cent phosphorous)

- Phosphorous is a powerful reducing agent and helps to reduce impurities as well as increasing wear resistance.
- These metals have good corrosion resistance combined with reasonably high tensile strength.
- They are used extensively in marine engineering, in rod, strip and wire to form springs and bi-metal strips for kettles and circuit breakers.

Copper – Aluminium – Aluminium Bronzes (9 per cent aluminium)

- These have very good corrosion resistance especially in marine environments.
- Typical applications include marine propellers, pumps, valves, condenser tubes and heat exchangers.

Copper – Nickel – Cupro-nickels

- These are ductile and can be hot or cold worked.
- They have excellent corrosion resistance and are excellent conductors of heat and electricity.
- 90 per cent copper, 10 per cent nickel is used for marine, petro-chemical, power and heat exchangers.
- 80 per cent copper, 20 per cent nickel is used for electrical components, deep drawn pressings and decorative parts.
- 75 per cent copper, 25 per cent nickel is used for coinage.
- 68 per cent nickel alloy with the addition of approximately 2 per cent iron is known as Monel and has good resistance at moderately high temperatures which lends itself to turbine blades, valves and surface coating applications.

MATERIAL TREATMENT

Many materials are supplied after having undergone some form of treatment and these processes can improve their operational efficiency.

- *Hardened* means that they have been heated to a predetermined temperature and quenched in water, oil or brine. Water produces the hardest surface. Not all metallic materials can be hardened by heat treatment but all can be work hardened.

- *Tempered* refers to the material being re-heated after hardening to reduce some of the hardness and increase the toughness. This is done by heating to a predetermined temperature, observing a colour change and then quenching in water, oil or brine.

- *Cold rolled* refers to one method of work hardening. The material is passed back and forth through rollers to increase its tensile strength, temper or hardness.

- *Close annealed* means that the material is heated in a container in which an inert rich gas (argon) is circulated, and then allowed to cool at a controlled rate within this inert environment; this leaves a bright surface finish.

- *Annealed* means that the material is heated to above the upper critical temperature, to bring about re-crystallization, which will also reduce the stresses within the material and return some ductility. It is held at this temperature for one hour per 25mm thickness of the material and then allowed to cool slowly. There is also a 'process anneal' in which the material is heated to re-crystallize the structure to enable further working of the material and prevent cracking occurring. The temperature is lower than that for full annealing and various workshop methods can be used to identify that the temperature has been reached. For example, the use of wood or soap passed on the heated surface of aluminium will leave a brown or black marking, indicating the annealing temperature has been reached. To anneal mild steel, heat to cherry red and then allow to cool slowly either in a bed of sand or out of cooling draughts.

- *Normalized* refers to another heat treatment process where the material is heated to above upper critical temperature to bring about re-crystallization and help relieve stresses. It is then allowed to soak for one hour per 25mm thickness followed by slow cooling in air, to bring the structure of the material back to what would be described as its 'normal' structure and properties (tensile strength, hardness and toughness) without greatly affecting its ductility.

MARKET FORMS OF SUPPLY

Metal stock can be purchased in a wide variety of shapes, sizes and materials. Weldments may be constructed from combinations of 'material shapes' and sizes. Only a single type of metal is used in most weldments, unless a special property such as corrosion resistance is required, when dissimilar metals may be joined into the fabrication where they are needed. The most common metal used is carbon steel, but non-ferrous metals are also available in the following forms:

- *Plate* is usually 3mm or thicker and measured in mm. It is available in widths from 300mm up to 2.4m, lengths from 2.4m – and thickness ranges up to 300mm.

- *Sheets* are usually up to 3mm and are measured in mm or Standard Wire Gauge (SWG).

- *Pipe* is dimensioned by its diameter and schedule, or strength. Pipe that is smaller than 300mm is dimensioned by its inside diameter, and the outside diameter is given for pipe that is 300mm in diameter and larger, Figure 4.3. The strength of pipe is given as a schedule. Schedules 10 through 180 are available; schedule 40 is often considered a standard strength. The wall thickness for pipe is determined by its schedule (pressure range). The larger the diameter of the pipe, the greater its area. Pipe is available in welded ('seamed') and extruded ('seamless') forms.

FIGURE 4.3 Inside diameter (ID) and outside diameter (OD)

- *Hollow section* sizes are always given as the outside diameter. The desired shape of hollow section, such as square, round or rectangular, must be listed with the ordering information. The wall thickness is measured in millimetres, although conversion can be done for imperial measures. Tubing should also be specified as rigid or flexible. The strength of hollow section may also be specified as the ability to withstand compression, bending or twisting loads.

- *Structural sections* are available in a range of sections which include beams, joists, channels, angles, T-stalks and z-sections. All of these are produced by hot rolling and rely on their shape to resist loads or buckling. Available in a wide range of dimensions and thicknesses.

- *Angles* are dimensioned by giving the length of the legs of the angle and their thickness. Stock lengths of angles are 6m, 9.1m, 12.2m, and 18.3m.

- *Extrusions* offer a lightweight alternative to the sectional sections and again the shape provides strength under load. Used extensively where weight is a critical factor especially in automobile and aircraft manufacture.

- *Wire and rod* produced by drawing down the material to produce a range of wire, rods and bars for electrical purposes, filler wires and electrodes, as well as a multitude of engineering applications.

- *Forgings* meet an increasing range of applications from aircraft, automobile and marine manufacturing industries.

MEASUREMENT AND MARKING OUT

At all levels of work, some form of measurement is required in order that a component complies with a standard or specification.

Measuring instruments can be 'indicating', which means that a readable measurement of a component can be made. Examples include a rule, vernier height gauge, protractor, calliper or micrometer.

'Non-indicating' instruments merely indicate acceptance or rejection of a component. Examples include plug gauges (used to check a hole or gap), weld profile gauges and gap gauges (used to check shafts).

Standards have become an essential part of manufacturing and are now part of every contract. They are controlled by a host of regulating bodies, including International Organization for Standardization (ISO), British Standards Institution (BSI), and Conforms to European Standards (CE). For measurement, the Systeme International d'Unites (International System of Units SI units) is recognized internationally.

A system of standards ensures a level of conformity which means parts from different sources can be combined and it is possible to use standard off the shelf bolts, bearings, wire sizes or electrodes. This reduces cost and ensures that parts are readily available.

Tolerances

Most drawings state a dimensioning tolerance, which is the amount by which the part can be larger or smaller than the stated dimensions and still be acceptable. Tolerances are usually expressed as plus (+) and minus (−). If the tolerance is the same for both the plus and the minus, it can be written using the symbol ±. In addition to the tolerance for a part, there may be an overall tolerance for the completed weldment.

Rounding Numbers

When multiplying or dividing numbers, we often get a whole number followed by a long decimal fraction. When we divide 10 by 3, for example, we get 3.3333333. For all practical purposes, we need not lay out weldments to an accuracy greater than the second decimal place. We would therefore round off this number to 3.33, a dimension that would be easier to work within the welding shop.

GOLDEN RULE

When rounding off a number, look at the number to the right of the last significant place to be used. If this number is less than 5, drop it and leave the remaining number unchanged. If this number is 5 or greater, increase the last significant number by 1 and record the new number. For example, 15.6549 rounded off to the second decimal place is 15.65.

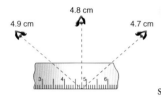

FIGURE 4.4 Parallax error

'*Parallax Error*' is the result of taking a measurement when one's line of sight is off to one side. It is important to ensure that the eye is at 90° to the work surface in order to get a true reading.

MARKING OUT TOOLS

Precision Steel Rules are made from hardened and tempered stainless steel. They are available in 150mm, 300mm, 600mm and 1000mm lengths with graduations on both edges.

When marking out it is essential to establish a datum edge from which all measurements are taken. When measuring out a distance, line the rule up so that the distance to mark is in line with the datum edge. With both graduation lines now aligned, and with a scriber angled to scratch the end of the rule at the designated distance, mark the material and, using an engineer's square, continue the line to the drawing destination.

When you are measuring, always take the reading between graduations on the rule to ensure accuracy.

Where approximate measurements are required over greater distances both steels and fabric tapes are available. The disadvantage of fabric tapes is that they can deteriorate and stretch, which makes accurate measurement more difficult.

A *line of chords rule* has a graduated scale for calculating circumferences. For example to find the circumference of a 150mm circle, place a square on the rule at 150mm and check across on to the circumferences scale to read the correct value. The rule also has two dot punch marks to place dividers on to set a specific angle to use for checking folds or marking out sectors for polar co-ordinates.

Scribers are used to mark lines on metal surfaces. They are made from hardened and tempered steel, ground to a fine point which should be kept sharp to give well defined lines. They should always be used so that the point is in direct contact with the edge of a square or template, whichever is being used. Like all marking out equipment, they should be used and stored with safety procedures in mind.

Engineers squares consist of a stock and blade made from hardened steel and ground on all faces and edges to give a high degree of accuracy in straightness, squareness and parallelism.

Sheet squares are made from a hardened steel and ground on all faces and edges, with a graduated scale, to mark out and read measurements as well as to check squareness. They are used predominately in laying out material and for checking alignment and accuracy of folding/welding on brackets etc.

Combination sets consist of a graduated steel rule on which is mounted three separate heads: 90° square with 45° mitre face and bubble level, centre finder and a protractor. The rule has a slot in which each head can be locked at any position along its length. This makes it possible to mark accurately any dimension lines at 90° to the datum edge, 45° to the edge and any angle between 0–180° accurately. The centre finder relies on the V faces touching the cylindrical surface to identify a common centre. This tool is especially useful when you need to mark square, mitred and angular measurements, and to check folds against designated angles. The blade can also be used to measure depth measurements against the stock of the square. Care needs to be exercised in storage especially with the locking screw mechanisms, which can easily fall on the floor if not being used in conjunction with the rule.

Hermaphrodite callipers (odd-legs) are often used to transfer dimensions parallel to the datum edge. They consist of straight pointed divider leg with a stepped leg which is pushed up against the datum edge to ensure accuracy when marking out. They do have some limitations, the biggest being they are limited by the distance they can be opened out and maintain accuracy. They also rely on a central pivot which can become slack.

Scratch gauges are used by fabricators but are based on a woodworkers tool. Most consist of a stock end through which a diameter bar is passed with a scribing point at one end, locked in place by a setscrew. Used for transferring measurements on sheet or plate and for the marking out of back marks for drilling on constructional members.

Dividers have two pointed steel legs hardened and ground to a fine point. They are available in a spring bow or a central pivot construction and are made in a range of sizes. They can be used to mark circles, arcs or to mark off a series of lengths such as whole centres. When adjusting the dividers, place one point in the graduation of the rule, and adjust to the required measurement, taking in to account where measuring from. For example, if you require a 50mm dimension, put the point in the 10mm graduation on the rule and adjust to 60mm.

Trammels are used where large arcs or circles are required to be marked out. They consist of a beam or bar with two pointed scribers mounted in machined stocks which have a clearance for the bar or beam to pass through and a setscrew to lock the points in position. One point is placed in a dot punch and the second point scribes the arc or line, they are adjusted for measurement as previously described and some also have a fine adjustment screw for increased accuracy.

Vernier tools are a type of fine precision measuring tools. They include height gauges, callipers and protractors. All vernier instruments consist of two scales; one fixed and one moving. The fixed scale is graduated in millimetres, every ten divisions equalling 10mm, and is numbered 0, 1, 2, 3, up to the capacity of the instrument. The moving scale allows you to measure with greater precision. When taking a measurement, note how many millimetres is on the main scale and then check on the vernier scale for a line that coincides with a line on the main scale and add together the two dimensions to get the total dimension.

For example if the main scale registers 40mm and the vernier scale registers 5 we have:

$$5 \times 0.02 = 0.10 + 40 = 40.10\text{mm}.$$

Vernier Tool Types

Vernier Calipers

Vernier calipers are used to measure internal, external and depth measurements. They are available from 150-1000mm. Measurement is taken by slackening both locking screws and moving the sliding jaw until it touches the work surface to be measured. Tighten the locking screw on the main scale and adjust the wheel nut on the vernier scale until the correct 'feel' is obtained and lock the second locking screw before reading the measurement. A variant is available with a dial caliper. Vernier calipers are also available with a digital readout, but take care to zero the gauge before taking readings.

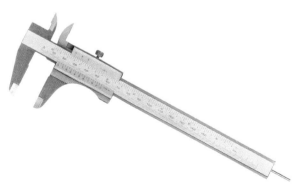

FIGURE 4.5 **Vernier caliper**

FIGURE 4.6 Vernier height gauge

Vernier Height Gauge

A vernier height gauge consists of a machined base with a perpendicular graduated blade on which a slide mechanism with two locking nuts is attached. The scriber point is attached by a coupling mechanism and can be used to measure heights and scribe lines on components. It is also available in digital format.

Vernier Protractors

Vernier protractors have a main scale marked off in degrees and a vernier scale marked off in divisions of 1 degree 55 minutes apart. They can be read to the nearest five minutes but must be read in the same direction as the main scale. Like all vernier tools, scales are satin chromed to prevent corrosion and to reduce glare when working in artificial light.

Vernier Depth Gauge

Supplied with a machined T-bar, in which a graduated blade slideson to which an adjustment mechanism slides up and down to record depth.

Dial Test Indicators

Dial test indicators are used in conjunction with surface tables and plates to measure very small discrepancies in surface texture or parallelism. They magnify small movements by means of a plunger or lever linked to a pointer on a graduated dial. They are also available in digital format.

A *scribing gauge or scribing block* is used in conjunction with a scriber to mark out lines on the work piece parallel with a reference surface table. The height of the scriber is adjustable and is set in conjunction with a steel rule giving an accuracy of 0.3mm.

Punches are used to define points of interest in the marking out schedule and are available in two forms *dot* and *centre* punch. Dot punches have a shallower point which results in less indentation. They are used to witness mark a line or curve so that the components can be worked to these marks in order to achieve an accurate dimension. Centre punches have a broader angle which produces a deeper indentation, and are used to define centres for drilling or centre points in scribing circles or co-ordinates.

In fine limit work, *surface tables and plates* are used to establish a defined datum from which all measurements can be made or taken. They are made from precision ground cast iron to give high levels of accuracy. These are used in conjunction with vee-blocks, angle plates, straight edges and parallels, which are also usually made from cast iron and precision machined. The components to be marked are clamped to the angle plate or vee block and scribed off a known datum to ensure accurate dimensioning.

Slip gauges are hardened and ground steel which are lapped to super finish and a high degree of accuracy. Because these surfaces are so nearly perfect, they can be wrung (slid over one another) together and become one piece. These then become the reference from which a lot of measurements can be taken to set up accuracy of height gauges, vernier callipers or micrometers.

Feeler gauges are made from hardened and tempered steel with blades (or leaves) covering a range from 0.05–1.0mm in steps of 0.05mm. They are used to check gaps, error between two objects or surfaces. Care must be taken to avoid bending or wrinkling the thinner blades. A thin film of oil is used to protect against corrosion.

Spirit levels are used extensively to check components such as flanges on pipework for straightness, flatness and parallelism.

TIP

Most vernier type instruments are now provided with a digital readout to at least three decimal places offering a high degree of accuracy.

MARKING OUT

Terminology Associated with Marking Out

- *Engineers blue* is a dye that can be painted or sprayed on bright surfaces to make scribed lines more visible. It can be easily removed with a solvent cleaner.

- *Datums* are used to establish a reference point from which all dimensions are taken. The datum may be a point, an edge or a centre line, depending upon the shape and complexity of the work piece. Datums are usually at right angles to each other, to ensure that lines that are marked-out are perpendicular to one another.

- *Co-ordinates* are often used in conjunction with datums and fall into two distinct groups.

- *Rectangular co-ordinates* have dimensions which are at right angles to one another.

- *Polar co-ordinates* have a dimension which is measured in conjunction with a radial line from the datum. Marking out polar co-ordinates requires particular care as even small inaccuracies in angles can greatly affect the final dimension.

- *Chain dimensioning* consists of taking each dimension from the previous dimension. It can result in *cumulative error* in which a discrepancy in one dimension is magnified along the plate, giving a substantial error.

- When marking out it is recommended that you use *progressive dimensioning* in which each dimension is taken from a *known* datum. If there is a discrepancy, it will only impact on that particular dimension and therefore produce a greater degree of accuracy.

- *Flatness* is a workshop standard where the flatness of a component measured in two parallel planes is compared with a known datum, which in most cases is a surface table or plate.

- *Parallelism* is a measure of two surfaces with a constant distance apart which can be measured along its length or across its surface. This is usually done by micrometers, vernier callipers, height gauges or by using a surface table with a dial test indicator.

- *Squareness* is a function of angular measurement, in which two surfaces are measured at 90° to one another, which is usually identified by an engineer's square.

- *Roundness* is measured at a number of diametrically opposite points around the circumference using a micrometer or vernier calliper.

- *Concentricity* is the relationship between two diameters. To be concentric both diameters should lie on the same centreline. It is checked by using a vee block and dial test indicator on the surface plate or table.

- *Profile* is the outline of a component and is usually checked with a profile gauge such as a radius or weld profile gauge.

- *Surface finish* is usually defined in the manufacturing drawings. For most fabrication purposes we work to a surface finish free of indentations or marking which would affect performance or aesthetic appearance. For machining performances surface finish is measured against comparison specimens.

Because of the increasing cost of materials and the need to keep projects on target, it is important that you maximise the use of the materials. One commonly used method is *Nesting*, which is the grouping of components in such a way that only a small amount of scrap is produced, while allowing sufficient material to remove the components without distortion occurring.

When marking out large components many workshops use a floor space specifically for this purpose, often referred to as a *template loft*. Here large components are marked out using a combination of French chalk, chalk lines, trammels and sheet squares.

Chalk lines are held in tension between two points and *snapped* to release chalk on to the metal surface being marked out. These lines may then be *witness marked* to define cutting lines. It is good practice to identify which are cutting lines or fold lines; all cutting lines should have the scrap side clearly indicated to reduce the risk of cutting on the wrong line.

Useful Workshop Formulae

Perimeters and Areas of Rectangle

The perimeter is the distance around the outside of the figure: the sum of the sides
2 x 3 + 2 x 4 = 6 + 8 = 14 cm

The area is found by multiplying the lengths of the adjacent sides:
3 x 4 = 12 cm²

The volume of a rectangle is the length x width x height : 2 x 3 x 5 = 30 cm³

Circumference and area of a circle

The circumference of a circle is πD :
3.142 x 30 = 94.26 cm

The area is found by multiplying the radius squared by π.
15 x 15 x 3.142 = 706.95 rounding up to 707 cm²

The volume of a cylinder is the area x height
707 x 10 = 7070 cm³

Triangles

The perimeter is calculated by adding all the lengths if they are known.

In a right-angled triangle, if two of the sides are known, the third may be calculated using

Pythagoras theorem:

The sum of the two adjacent sides is equal to the square root of the hypotenuse.

FIGURE 4.7 Useful welding formulae

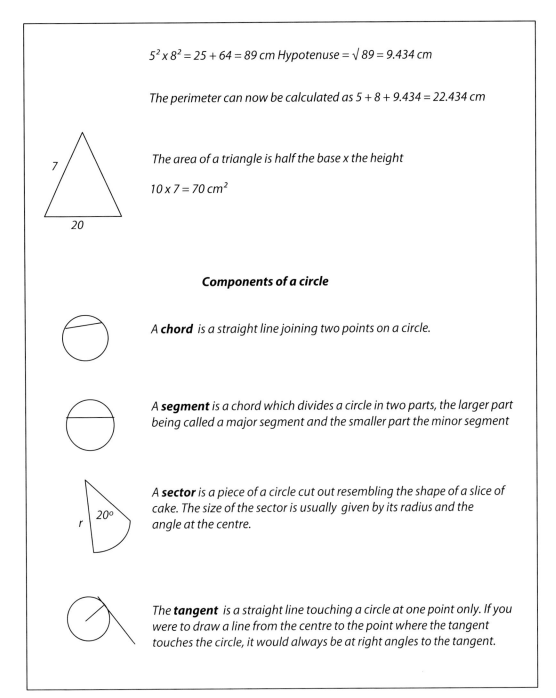

$5^2 \times 8^2 = 25 + 64 = 89$ cm Hypotenuse $= \sqrt{89} = 9.434$ cm

The perimeter can now be calculated as $5 + 8 + 9.434 = 22.434$ cm

The area of a triangle is half the base x the height

$10 \times 7 = 70$ cm^2

Components of a circle

A **chord** is a straight line joining two points on a circle.

A **segment** is a chord which divides a circle in two parts, the larger part being called a major segment and the smaller part the minor segment

A **sector** is a piece of a circle cut out resembling the shape of a slice of cake. The size of the sector is usually given by its radius and the angle at the centre.

The **tangent** is a straight line touching a circle at one point only. If you were to draw a line from the centre to the point where the tangent touches the circle, it would always be at right angles to the tangent.

FIGURE 4.7 Useful welding formulae (continued)

Folding Allowances

When marking out it may be necessary to calculate the length of material that is required. Consideration also has to be given to material allowances. For thin sheet the deducted allowance is generally the thickness of the sheet. For example, if a fold is to finish at 50mm when folded, mark a line at 49mm if the thickness of plate is 1mm. For an outside dimension involving two folds to finish at 50mm the work piece would be marked 48mm and the two folds.

On thicker plate a neutral line formula needs to be applied. When forming on thick plate, the outside of the material stretches and the inside of the bend compresses. The neutral line formula is based on the material being bent, its mechanical properties, the thickness of plate and the inside radius of the bend. The neutral line generally lies between 0.35–0.45 x material thickness from the inside edge. For most applications the bending radius is usually between 2 x and 4 x metal thickness, therefore the general value of 0.4 x material thickness is acceptable, anything less than this and the plate is likely to crack. When calculating cylinder diameters, the neutral line is taken as 0.5 x material thickness (at the centre of the material thickness).

TABLE 4.2   Workshop Calculation Samples

Calculate the bending and bend allowance for the following examples:

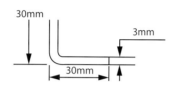

All dimensions are external.

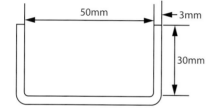

All dimensions are internal.

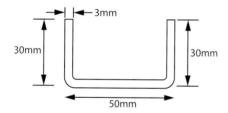

Diagram shown is the same as above but with external dimensions.

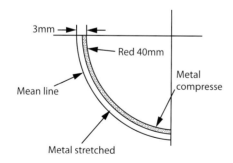

When forming to a radius consideration is given to the 'mean line', for calculation use 0.5t where 't' is the metal thickness.

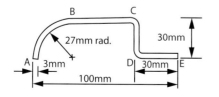

When calculating cutting size of component shown divide the component into parts for easier calculation.

Calculating the areas for the following figures:

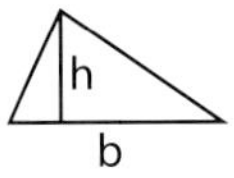

$\text{Area} = \frac{1}{2} \times b \times h$

$= \frac{bh}{2}$

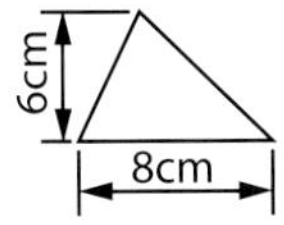

Fig. 1

Fig. 2, the area $= \frac{6 \times 8}{2}$

$= 24\text{cm}^2$

Fig. 2

Problems

Determine the area of each of the following figures.

1

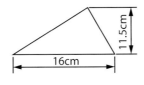

16cm

11.5cm

(82; 87; 92; 97)

2

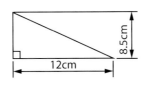

12cm

8.5cm

(102; 51; 90; 45)

3

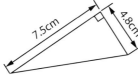

7.5cm

4.8cm

(18; 36; 20; 40)

4

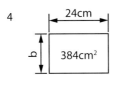

24cm

b   384cm²

The area of the rectangle is 384cm². Find the length of the side `b`.
(16; 15; 14; 13)

5

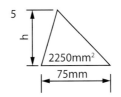

h

2250mm²

75mm

The area of the triangle is 2250 mm². Find its height `h`.
(30; 40; 50; 60)

6

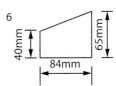

40mm

84mm

65mm

Determine the area of this figure.
(4200; 4410; 4850; 5460)

7

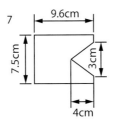

9.6cm

7.5cm

3cm

4cm

Calculate the area of this figure.
(72; 66; 60; 54)

Using 3.142 as a value for π calculate the following:

Dimensions in millimetres, unless-otherwise stated

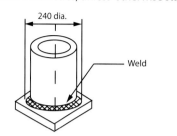

Fig. 1

Calculate the length of weld and the number of electrodes required. The weld consists of a root run using 2.5mm electrodes with 200mm weld metal per electrode and 3.25mm capping run depositing 250mm per electrode.

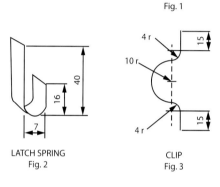

LATCH SPRING
Fig. 2

CLIP
Fig. 3

Calculate the length of strip required to form Figures 2 and 3.

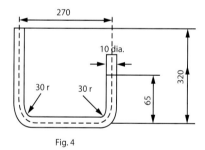

Fig. 4

Calculate the length of rod in metres required to form Fig 4.

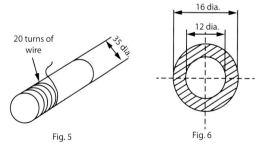

Fig. 5

Fig. 6

Fig 5 Calculate the length of wire required in metres to make 20 complete turns.

Figure 6. Calculate the area of the shaded part.

Transposition of formulae

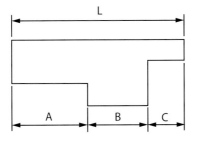

1 From the figure shown, it can be seen that
L = A + B + C

Suppose that L, A, and C were known, to obtain the value of B the formula could be transposed so that to obtain B on one side we subtract A and C from

BOTH sides of the formula. Then: L – A – C = B.

This use of transposition is used to determine values both in length calculations and operating values.

2 A good example of this is in determining the values for current, voltage and resistance in Ohm's Law. In an electrical circuit, given that   V = I R   where *V = Voltage, *I = Current and *R = resistance.

Transpose the formula to get values for I and R and then calculate the resistance if the voltage is 240 v and the current flow is 22 amps.

3 Given that 5cp = 3b find a formula for p.

4 Given that L x B x D = V transpose the formula for a value for D

ASSEMBLY

The assembly process, in which all the parts of the weldment are brought together, requires proficiency in several areas. It is necessary to read and interpret the drawing to properly locate each part. An assembly drawing contains the necessary graphic and dimensional information to allow the various parts to be properly located as part of the weldment. Pictorial or exploded views makes this process much easier for the novice assembler, but, most assembly drawings are given as two, three or more orthographic views, see Figure 4.8.

On very large projects such as buildings or ships, a corner or centre-line is established as a baseline. This is the point where all measurements for part location begin. With smaller weldments, a single part may be selected as a starting point. Often, the selection of the base part is automatic because all other parts are to be joined to this central part. On other weldments, however, the selection is made by the assembler.

Select the largest or most central part to be the base for the first part of the assembly. All other parts will then be aligned to it. Using a base also helps to prevent location and dimension errors. Otherwise, a slight misalignment of one part, even within tolerances, will be compounded by the misalignment of other parts, resulting in an unacceptable weldment.

Identify each part of the assembly and mark them for future reference. If it helps, hold the parts together and compare their orientation to the drawing. Locate points on the parts that can be easily identified on the drawing, such as holes and notches, Figure 4.9. Now mark the location of these parts top, front, or other orientation so you can locate them during assembly.

Using a consistent method of marking helps prevent mistakes. One method is to draw parallel lines on both parts where they meet, Figure 4.10.

After the parts have been identified and marked, they should be clamped in place. Holding the parts in alignment by hand for tack welding is fast, but it often leads to errors. Experienced assemblers recognize that clamping the parts in place before tack welding is a much more accurate

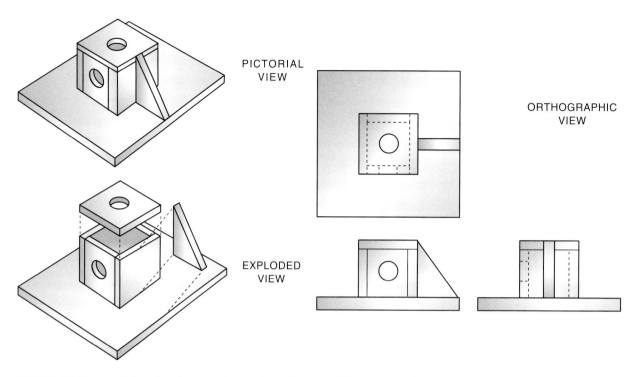

FIGURE 4.8 Types of drawing that can be used to show a weldment assembly

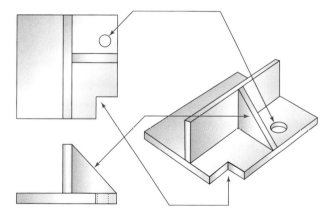

FIGURE 4.9 Identifying unique points to aid in assembly

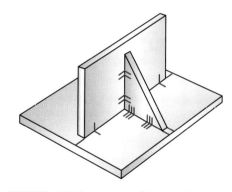

FIGURE 4.10 Layout markings to help locate the parts for tack welding

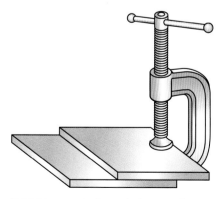

FIGURE 4.11 G-clamp being used to hold plates for tack welding

method, Figure 4.11. If thermal joining is to be used, care must be given to the thermal conductivity of the material and the tendency of the materials to distort. This will be covered in further depth within this book, as it may have an impact upon the dimensional accuracy of the finished assembly.

ASSEMBLY TOOLS

Clamps

A variety of clamps can be used to temporarily hold parts in place so that they can be tack welded:

● *G-clamps,* among the most commonly used types, come in a variety of sizes, Figure 4.12. There are G-clamps specifically designed for welding. Some have a spatter cover over the screw, and others have screws made of spatter-resistant materials such as copper alloys.

- *Bar clamps* are useful for clamping larger parts. They have a sliding lower jaw that can be positioned against the part before tightening the screw clamping end, Figure 4.13. They are available in a variety of lengths.
- *Pipe clamps* are very similar to bar clamps, with the advantage that the ends can be attached to a section of standard 12mm pipe. This allows for greater flexibility in the length of the clamp, and the pipe can easily be changed if it becomes damaged.
- *Locking pliers* are available in a range of sizes with a number of jaw designs, Figure 4.14. Their versatility and gripping strength make locking pliers very useful. Some have a self-adjusting mechanism that allows them to be moved between parts of different thicknesses without the need to readjust.
- *Cam-lock clamps,* also known as *toggle clamps,* are specialty clamps that are often used in conjunction with a jig or a fixture. They can be preset, allowing for faster work, Figure 4.15.

TIP

Specialty clamps such as those for pipe welding, Figure 4.16, are available for many different types of jobs and make it possible to do faster and more accurate assembling.

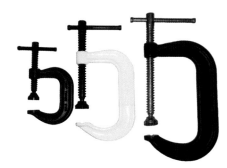

FIGURE 4.12 **G-clamps**

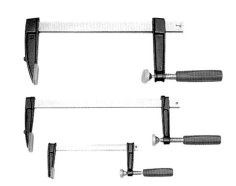

FIGURE 4.13 **Bar clamps**

FIGURE 4.14 **Three common types of locking jaw pliers**

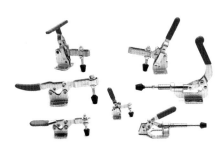

FIGURE 4.15 **Toggle clamps**

FIGURE 4.16a **Pipe alignment clamps and level**

FIGURE 4.16b **Pipe alignment clamp**

FIGURE 4.16c **Pipe gap adjustment tool**

Jigs and Fixtures

Jigs and fixtures are specially made to aid in assemblies and the fabrication of weldments. When a number of similar parts are to be made, fixtures can increase speed and accuracy in the assembly. Fixtures must be strong enough to support the weight of the parts and withstand the rigours of repeated assemblies, while remaining within tolerance. They may have clamping devices permanently attached to speed up their use. Often, locating pins or other devices are used to ensure proper part location. A well-designed fixture allows adequate room for the welder to make the necessary tack welds. Some parts are left in the fixture throughout the welding process to reduce distortion.

FITTING

HEALTH & SAFETY ⊘

Hand grinders should never be used without appropriate PPE.

Not all parts fit exactly as they were designed to. There may be slight imperfections in cutting or distortion as a result of welding, heating, or mechanical damage. Some fitting problems can be solved by grinding away the problem area. Hand grinders are most effective for this type of defect, Figure 4.17. Other situations may require that the parts be manipulated into alignment.

A simple way of correcting slight alignment problems is to make a small tack weld in the joint and then use a hammer and possibly an anvil to dress the part into place, Figure 4.18. Small tacks applied in this way become part of the finished weld. Be sure not to strike the part in a location that will damage the surface and render it unsightly or unusable.

Greater force can be applied using cleats or strongbacks with wedges or jacks. These are pieces of metal that are temporarily attached to the weldment's parts to enable them to be forced into place. Jacks are more effective if the parts must be moved more than about 12mm, Figure 4.19. Anytime cleats or strongbacks are used, they must be removed and the area ground smooth.

Some codes and standards do not allow temporary attachments to be welded to the fabrication or structure. In these cases more expensive and time-consuming fixtures must be constructed to help align the parts if needed.

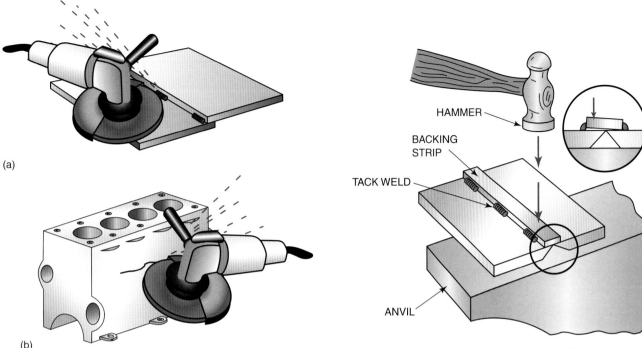

(a)

(b)

FIGURE 4.17 **Abrasive grinding disk**

FIGURE 4.18 **Using a hammer to align the backing strip and weld plates**

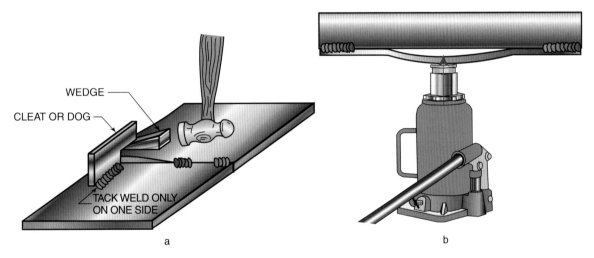

a

b

FIGURE 4.19 **Alignment**

TACK WELDING

This is a temporary method of holding parts in place until they can be completely welded. Usually, all of the parts of a weldment should be assembled before any finishing welding is started. This will help reduce distortion. Tack welds must be strong enough to withstand any stresses during assembly and any forces caused by weld distortion during final welding. They must also be small enough to be incorporated into the final weld without causing an imperfection in its size or shape, see Figure 4.20.

Tack welds must be made with an appropriate filler metal, in accordance with the welding procedure. They must be located well within the joint so that they can be fused again during the finish welding. Post tack welding cleanup is required to remove any slag or impurities that may cause flaws in the finished weld. Sometimes the ends of a tack weld must be ground down to a taper to improve the pick-up on the finished weld metal.

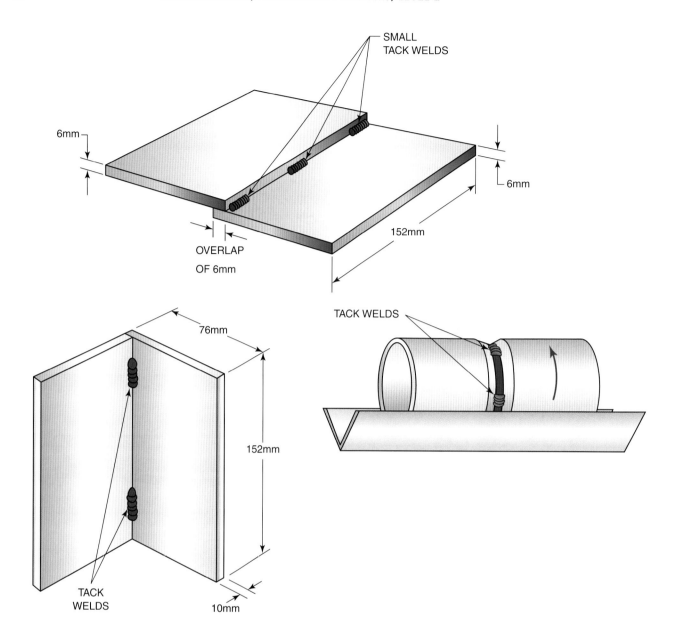

FIGURE 4.20 Tack welds

A good tack weld is one that does its job by holding parts in place yet is undetectable in the finished weld.

WELDING

Good welding requires more than just filling up the joints with metal. The order and direction in which welds are made can significantly affect distortion and residual stress in the weldment. Generally, welding on an assembly should be staggered from one part to another. This allows the welding heat to dissipate and welding stresses to be controlled.

Keep the arc strikes in the welding joint so that they will be re-melted as the weld is made. This will make the finished weldment look neater and reduce post-weld cleanup. Some codes and standards do not allow stray arcing or arcing outside of the welding joint.

Striking the arc in the correct location on an assembly is more difficult than working on a welding table, because welding will often be in an awkward position. Several techniques will help improve accuracy. Use a free hand to guide the electrode or weld gun to the correct spot. Resting an arm, shoulder, or hip against the weldment to assist balance can also help. It is sometimes helpful to practise starting the weld with the power off.

Be sure that that there is enough freedom of movement to complete the weld joint. Check that welding leads will not snag on anything that would prevent the making of a smooth weld. Make sure that sparks will not fall on those working on the structure. If the weldment is too large to fit into a welding booth, portable welding screens should be used to protect other workers in the area from sparks and welding light, and local extraction should be used at all times.

Follow all safety and setup procedures for the welding process. Practise the weld to be sure that the machine is set up properly before starting on the weldment.

FINISHING

Depending on the size of the workshop, the welder may be responsible for some or all of the finishing work, from chipping, cleaning, or grinding the welds to applying paint or other protective surfaces.

If possible, avoid grinding of welds by properly sizing the weld as it is made. Grinding can be an expensive process, adding significant cost to the finished weldment. If it is necessary to grind for fitting purposes or appearance, it should be kept to a minimum.

Guidance on the safe use of grinders is covered in Chapter 7. Make sure that you have been instructed in the safe use of this equipment and are fully aware of the hazards before using it.

FIGURE 4.21 Wire brushes and grinding stones used to clean up welds

SUMMARY

Being able to identify and understand the mechanical properties of common engineering materials and recognize their common forms of supply helps the engineer to select the most appropriate material for the job in hand. Welders need to be able to interpret drawings and joint design in order to follow instructions about a weldment and work independently in the workshop or on site. Plans, drawings and sketches that will be used on the fabrication or structure may be produced anywhere in the world. A strong grasp of workshop mathematics will allow the welder to quickly and accurately produce weldments from drawings received from different sources. Being able to recognize tools and use them effectively all contribute to accurate work.

ACTIVITIES AND REVIEW

1 Identify the property from the following descriptions:

Property	Description
	The ability of a material to be subjected to a compressive force such as a hammer or press and be capable of deforming or shaping without fracture occurring. A property very much sought after in forging where the material must flow to shape under the force of the press without fracture.
	The ability to resist indentation or abrasion. The qualities that would be expected of gear teeth and the contact faces of agricultural machinery.
	The ability to flow to shape and retain that shape upon the load being removed, a property desirable for forging and presswork.
	The ability to be drawn out along its length, bent or twisted under an applied load as in the case of wire, bar or tube manufacture.
	The ability of a material to return to its original shape and size when an applied load is removed, a property that quite a few engineering materials possess to some degree. A typical application for this property would be in the manufacture of springs which require this property after compression or extension.
	The ability to withstand a shock load or impact (impact strength) as in the case of a hammer hitting the end of a chisel. Both surfaces need to be hard but one must have a slightly lower value otherwise it would shatter. Toughness can be controlled by heat treatment or by work hardening.

2 Select a suitable steel for the following applications and state your reason for selection.

Application	Material Selection	Reason for Selection
Structural section		
Garden gate		
Hacksaw blade		
Machine bed		
Metal cutting tools		

3 List the four groups of stainless steel and identify a typical application for each.

4 State from which ore aluminium is produced.

5 Indicate what defines a plain carbon steel.

6 Define what pig iron is in conjunction with steel production.

7 Identify what element added to carbon steel allows it to be called a stainless steel.

8 Aluminium alloys are defined by a four-digit system. What does the first digit identify?

9 Identify what the code 1050 aluminium refers to.

10 State what the following terms relate to:
  a Solution treatment
  b Artificial ageing
  c Precipitation hardening.

11 Define what the term 'blister' refers to with copper.

12 Name three copper alloys and give a typical application for each.

13 From the following definitions identify the material treatment processes:

Material treatment process	Definition
	Has been heated to a predetermined temperature and quenched in a medium such as water, oil or brine.
	Method of work hardening where the material is passed back and forth through rollers to increase its tensile strength, temper or just to increase the hardness.
	Where the material is heated in a container, in which an inert rich gas is circulated, and then allowed to cool at a controlled rate within this inert environment.
	Refers to the material being re-heated after hardening to reduce some of the hardness and increase the toughness. This is done by heating to a predetermined temperature and observing a colour change and then quenching in some medium.
	A heat treatment process where the material is heated to a predetermined temperature to bring about re-crystallization and relieve stresses. It is then allowed to soak followed by rapidly cooling in air.
	Where the material is heated to a predetermined temperature, to bring about re-crystallization, which will reduce the stresses within the material and return some ductility. It is held at this temperature to ensure full soaking of the material and then allowed to cool in 'still air' (free of draughts).

14 Sketch four forms of material supply.

15 State the difference between indicating and non-indicating methods of measurement.

16 Give a typical application for cupro-nickels.

17 List three precision instruments that can be used to measure accurately.

18 State what the term 'witness mark' refers to?

19 Indicate what the following terms relate to:
  a Chain dimensioning.
  b Progressive dimensioning.
  c Cumulative error.
  d Parallax error.

20 State the formulae for:
  a Circumference of a circle.
  b Area of a triangle.
  c Volume of a cylinder.

21 State what the term 'nesting' refers to?

22 State what co-ordinates are and what are they used for?

23 When manufacturing a cylinder, 100mm high by 100mm diameter, calculate the size of material required if the joint is to be butt welded.

24 List four tools used to assemble components.

Oxy-fuel Gas Welding, Brazing and Soldering

LEARNING OBJECTIVES

After completing this chapter, you should be able to:

- identify all of the components and equipment found in a typical oxy-fuel welding station
- demonstrate the proper assembly, testing, lighting, adjusting and dismantling of an oxy-fuel system
- list the proper safety procedures for setting up and operating an oxy-fuel system
- discuss the uses of oxy-acetylene welding and its advantages and disadvantages
- list factors that affect the weld
- discuss some commonly occurring problems associated with oxy-acetylene welding
- explain what factors are affected by adjusting the flame on mild steel
- state how changes in the torch angle and torch height affect the molten weld pool
- demonstrate tack welds and weld beads
- make welds on outside corner joints, T-joints and lap joints in the horizontal-vertical position and butt joints in the flat position
- compare the differences between soldering and brazing
- list the advantages of soldering and/or brazing
- explain why flux is used in soldering and brazing
- describe what factors must be considered when selecting a filler metal
- discuss the applications for common soldering and brazing alloys
- describe the preparation needed for a part before it is soldered or brazed.

UNIT REFERENCES

NVQ units:

Complying with Statutory Regulations and Organizational Safety Requirements.
Joining Materials by the Manual Gas Welding Process.
Producing Fillet Welded Joints using a Manual Welding Process.
Joining Materials by Manual Torch Brazing and Soldering.

VRQ:

Engineering Environment Awareness.
Manual Welding Techniques.
Non-fusion Thermal Joining Methods.

KEY TERMS

blowback occurs when the nozzle end is partially blocked, which results in a build-up of pressure to clear the obstruction and a large bang. This condition can be remedied by cleaning the nozzle orifice.

flashback a serious occurrence in which the flame travels back up the torch and possibly the hoses by back pressure towards the gas bottles. This occurrence results in a loud squealing noise with sparks emitting from the end of the nozzle.

fusion the flowing together or deposition into one body of the materials being welded.

fusion welding any welding process or method that uses fusion (joining metals to form one) to complete the Metal Inert Gas Shielded (MIG) and Metal Active Gas Shielded (MAGS). Arc welding processes that use an arc between a continuous filler metal electrode and the weld pool. The process is used with shielding from an externally supplied gas and without the application of pressure. In the case of MAGS the shielding gas is mixed with another gas to make it 'active' which gives improved mechanical properties such as increased surface deposition.

gouging removal of metal by oxy-fuel or the plasma arc processes.

nozzle the end piece that directs the flow of gas of a, welding torch or gun.

orifice a term used to describe the hole at the end of a tube or nozzle.

snifting the rapid opening and closing of a gas bottle to remove any residue that may have accumulated during storage of the gas cylinders. This procedure should *never* be carried out with hydrogen as it is possible to produce an explosion by the sudden release of hydrogen in the atmosphere.

tufting a surface deposition of flux on the end of the filler material when brazing or silver soldering. The 'tuft' is produced by heating the end of the filler wire and dipping into the flux which causes adhesion of the flux.

INTRODUCTION

Oxy-fuel gas welding uses a selection of fuel gases, although the preferred choice is often acetylene as this gives the hottest flame temperature of 3200ºC with a 'neutral' flame. The oxy-acetylene process has been extensively used since its origins in Europe around the 1900s. It was one of the major joining processes used in World War Two and is used for a wide range of applications today. This is because a diverse range of materials can be joined, using fusion welding, brazing, and hard soldering within one process, with relatively inexpensive equipment.

It is a versatile process and the equipment can be used for cutting, brazing, hard soldering, flame cleaning, gouging, pre-heating and post-heating. It is also a mobile process, which means that it is popular with maintenance engineers for onsite applications, such as using heat to free off seized components.

Many professionals consider it be one of the best processes to start your welding career on, as it teaches discipline in waiting for the molten pool to form and observing its characteristics. It also teaches you how to manipulate the filler wire and speed of travel to ensure weld integrity, an essential of TIG welding. For many welders, it is the premier welding process because of the dexterity it allows and the fact that it can be used with virtually all metals. To many it is an art form, and it does in fact attract a number of artists who use it to bring their creations to life. This is because a diverse range of materials can be joined, using fusion welding, brazing and hard soldering within one process, with relatively inexpensive equipment.

EQUIPMENT

Gases

Acetylene is a colourless, flammable gas, which is slightly lighter than air and has an unpleasant odour (garlic) due to its impurities. Acetylene is usually made commercially by the reaction of calcium carbide and water. Explosions can occur when acetylene gas is present in air in any proportions between 2 per cent and 82 per cent. It is also liable to explode when under unduly high pressure, even in the absence of air, therefore the working pressure of acetylene should *never* exceed 0.62 bars.

Acetylene is usually stored in maroon high pressure cylinders, under pressure and dissolved in liquid acetone to make it more stable. The cylinder is filled with a porous mass of kapok or prepared charcoal, which are both highly absorbent, to retain the acetone. Acetylene cylinders are charged to a pressure of 1552 kN /m (15.5 bar). As the acetylene is drawn off for use it evaporates out of the acetone. If this happens too quickly, some of the acetone can flow out with the gas, which will cause contamination of the weld deposit and attack the rubber seals on regulators and hoses. Because the stability of the gas is affected by discharging acetone in the gas stream, it could result in the cylinder exploding, For this specific reason it is recommended that the discharge rate for any acetylene cylinder should not exceed 20 per cent. If a high discharge rate is required then cylinders should be manifolded to achieve the desired rate within the recommended limits. Dissolved acetylene (DA) for distribution systems uses steel pipes. *Do not* use copper pipe tubing or fittings because it forms a copper acetylide, an extremely explosive compound. You should also not use any copper or silver alloys containing more than 70 per cent copper or 43 per cent silver respectively which can also produce explosive compounds.

To prevent the interchange of fittings between cylinders containing combustible and non-combustible gases, the cylinder valve outlets are threaded in opposite hands. Non-combustible gases like oxygen, nitrogen, argon and air all have conventional *right-hand* threads. Combustible gases such as acetylene, hydrogen, propane and mixtures containing fuel gases all have *left-hand* threads. These precautions mean that oxygen and fuel gas pressure regulators are not interchangeable. Spindle keys are interchangeable and are used to open and close all cylinders. Turn anti-clockwise to open and clockwise to close.

Most acetylene cylinders are protected by one or more safety devices:

- The bulk of cylinders with kapok or charcoal masses have a *bursting disc* on the back of the cylinder valve opposite the spindle, which discharges the gas should pressure rise to unacceptable rates.
- The industrial monolithic mass welded cylinders have a bursting disc and older cylinders filled with prepared charcoal mass have an inverted plug in the base of the cylinder designed to release pressure should an incidence occur.

Acetylene cylinders should be transported in an upright and secure position in open back trucks. If this is not possible and they are transported lying down in an enclosed vehicle, they must be secured and have air circulating throughout the vehicle. They must stand for a minimum of one hour before use, to ensure that any acetone that may have come out of solution in the cylinder is absorbed back, thereby stabilizing the gas again.

Oxygen is a colourless, odourless and tasteless gas. It is produced commercially by the electrolysis of water and the distillation of liquid air and is supplied in black high pressure cylinders with a white shoulder. Oxygen vigorously supports combustion of many materials which do not normally burn in air, including fireproofing materials. It is highly dangerous with oil, greases, tarry substances, common solvents and many plastics. The initiation, speed, vigour and extent of these reactions depends in particular upon:

- the concentration, temperature and pressure of the reactants;
- ignition energy and type of ignition.

If oxygen is withdrawn from a cylinder at too great a rate of consumption, a rapid drop in pressure will occur and the cylinder valve may freeze. When flame cutting heavy sections, which involves high rates of consumption, it is advisable to manifold cylinders.

When using gas cylinders in the workshop they *must* be secured to an appropriate cylinder trolley, wall, bench, or manifold from an external storage facility.

REGULATORS

All regulators work on the same principle regardless of whether they are high or low pressure, single or two stage, cylinder, manifold or line and the type of gas they regulate. The regulator reduces a high cylinder pressure down to a safe working pressure which must be held over a range of flow rates and volumes.

HEALTH & SAFETY ❯

Only open the acetylene cylinder a maximum half a turn so that in an emergency you can close the cylinder in one movement.

TIP ❯

Oxygen is present in the atmosphere at 21 per cent by volume.

Leave the pressure adjustment knob/screw fully out when the regulator is not in use; this ensures a minimum of tension on the springs and diaphragms. The pressure adjustment must remain captive on the regulator to comply with the British Compressed Gas Association (BCGA) Code of Practice 7 and the relevant British and European standard. Air or nitrogen regulators must not be used with oxygen and vice versa, because if air is supplied directly from compressors, it could contain traces of oil which can contaminate the regulator and cause an explosion if used with oxygen. All regulators for use with combustible gases are left-handed thread and this is denoted by 'nicks or grooves' on the flats of the gland nut (connection to cylinder).

The golden rule for regulators is only use the regulator for the gas it was designed to be used with.

Single-Stage Regulators

When the cylinder valve is opened, gas flows from the cylinder into the high pressure chamber of the regulator (Figure 5.1). While the pressure adjustment screw is fully backed off (in the closed position), no gas will flow.

When the pressure adjusting screw is turned, the spring is compressed. This moves the diaphragm across and pushes the valve open (Figure 5.2) allowing gas into the low pressure chamber and out to the torch.

The degree of compression on the spring and diaphragm will determine the level of pressure on the outlet from the regulator. This will remain fairly constant until the pressure in the cylinder starts to fall. When this happens the existing pressure in the low pressure chamber starts to exert a greater force on the valve, opening it even further and causing the outlet pressure to rise. This is known as rising outlet pressure characteristic which will have an impact upon the welding/cutting operation, with the operator having to make periodic adjustments to maintain flame characteristics.

While the accuracy of the outlet or working pressure is subject to variation, single-stage regulators are suitable for general purpose cutting, heating and for small gas sets as well as for certain welding applications such as MIG/TIG. When the torch valve is closed, the pressure in the outlet for the regulator rises and the increased force causes the valve to close.

HEALTH & SAFETY ❯

The BCGA Code of Practice states that the regulator MUST NOT be left under pressure when the system is not in use.

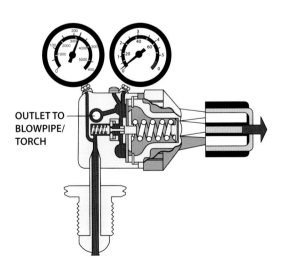

FIGURE 5.1 Single-stage regulator with cylinder valve open

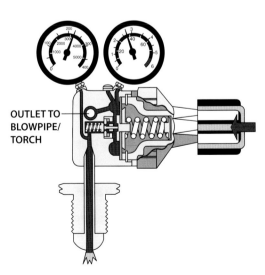

FIGURE 5.2 Single-stage regulator after the pressure adjusting screw has been turned

Multi-stage Regulators (two-stage)

Any application or process which relies on a constant accurate measurement being carried out should use multi-stage regulator in which the inlet pressure is reduced in two stages. This regulator has a manufacturer's spring loading preset which reduces the cylinder pressure to a lower or medium pressure which then passes on to a second stage (Figure 5.3). The pressure adjusting screw is adjusted to give the desired working pressure.

Following an inspection, if a regulator shows any signs of any of the following faults it should be replaced immediately:

- pressure relief valves removed and connections blanked off;
- damaged or inaccurate gauges;
- pressure adjusting knob/screw not permanently attached to the body;
- cylinder connection or regulator outlet showing signs of physical damage (denting or pitting);
- soot in the regulator outlet;
- worn inlet or outlet connections;
- evidence of sealing compounds.

HEALTH & SAFETY ⊙

Check the equipment every time it is used to ensure you are working to safe working practices.

Regulators are precision instruments containing machined components which need to be handled with care. Rough treatment can damage sensitive springs, diaphragms, valve seats, safety valves, etc. Regulators are now manufactured to the new European Standards. Equipment should display the relevant European/British Standard number and the pressure up to which it can operate. You should also refer to the manufacturer's operating instructions and the age of the equipment. Legislation lays down a maximum five-year life, as after this period the rubber seals will start to deteriorate.

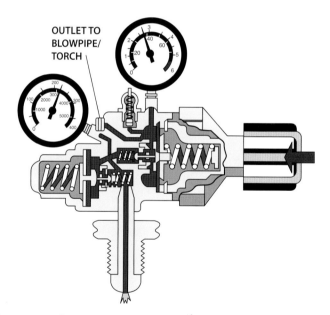

OUTLET TO
BLOWPIPE/
TORCH

FIGURE 5.3 Multi-stage regulator

Safety Release Valve

Regulators may be equipped with either a safety release valve or a safety disc to prevent excessively high pressures from damaging the regulator. The safety release valve consists of a small ball held tightly against a seat or spring. The valve will re-seat itself after the excessive pressure has been released. The safety disc valve contains a thin piece of metal held between two seals, which is designed to blow to release excessive pressure. The disc must be replaced before the regulator can be used again.

Pressure Gauges

The two gauges on a regulator have different functions: one indicates the cylinder pressure and the other shows the line pressure at the regulator, not at the torch. The torch pressure will always be lower than the working pressure due to 'line drop'. This is caused by the resistance of the gas as it flows through the line and will be greater with a smaller diameter or longer line.

Within the gauge lies a Bourdon tube, which is a hemispherical elliptical tube sealed at one end to the gauge body and linked to an indicating needle and gear mechanism. As the gas enters, the tube tries to straighten and in doing so carries the gear and indicating needle around a graduated panel, showing the gas pressure in the tube. This is usually shown in bars and pounds per square inch. The back of the gauges have a thin back plate which is designed to blow open and release the pressure if the Bourdon tube ever ruptures; this ensures that the front glass does not explode and injure the operator.

The second element, the pressure wave, lifts a pressure plate which cuts off the gas flow valve.

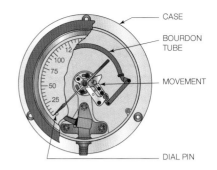

FIGURE 5.4 Pressure gauge

Flashback Arrestors

A flashback is the result of a mixture of fuel gas and oxygen burning within the hose. The flame travels towards the gas source at great speed. Flashbacks can result in fire or explosion in either or both oxygen and fuel gas cylinders. Flashback arrestors are required by law to be fitted on both cylinders to comply with the BCGA Code of Practice 7 and they must be designed to comply with the European Standard EN 730.

If the gas/oxygen mixture leaves the nozzle at a velocity that is slower than the combustion velocity or flame speed of the fuel then the flame will tend to burn back along the mixture and will 'backfire'. If the mixture within the hoses ignites then we have a flashback. Which hose the flashback occurs in depends on the velocity the gases are travelling at. The flame will travel either back towards the fuel, pulling behind it the oxygen needed to sustain combustion, or vice versa.

The factors which can lead to a flashback are:

* incorrect purging of the hoses and torch prior to use;
* incorrect gas pressures;
* incorrect nozzle size;
* damaged torch valves allowing cross flow feeding within the torch;
* blocked gas passages within the torch;
* kinked or trapped hoses.

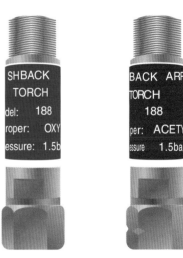

FIGURE 5.5 Pressure wave

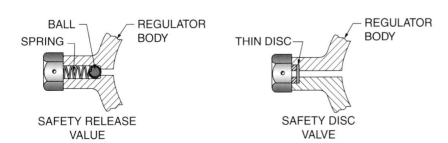

FIGURE 5.6 Pressure release valves

A flashback has two components, the flame front and a pressure wave. When the flame enters a flashback arrestor it passes through a fine sintered metal flame filter. This absorbs the heat of the flame, thus extinguishing it. The second element, the pressure wave, lifts a pressure plate which cuts off the gas flow valve and renders the cylinder safe. On some models this would be indicated by a 'pop up' button, which can be re-set when the hoses and flashback arrestor have been cleaned of any soot deposit.

Flashback arrestors are a major safety feature of any oxy-acetylene setup and must be regularly checked for any signs of damage. They should be replaced if they show any of the following conditions:

- external damage, such as a badly dented body, indicating that at some time the arrestor has been dropped;
- reset button or lever broken or bent;
- evidence of PTFE tape or some sealing compound.

An arrestor should also be checked for compliance with BCGA Code of Practice 7 and should have the following:

- the name of the gas for which it can be used;
- maximum inlet pressure;
- direction of flow of gas;
- British/European Standard Number;
- manufacturer's name or logo.

If you suspect that a flashback has occurred close both valves on the torch, *oxygen first* to reduce the velocity and intensity of the flashback. Close both cylinder valves on the gas bottles. Then check the temperature of the acetylene cylinder with the back of your hand for any indication of warming of the cylinder wall. If the temperature of the cylinder has not risen, check the torch has not overheated. If necessary you can cool the torch by plunging it in water with the oxygen valve fully open to prevent water entering the nozzle and equipment. Check the nozzle is not damaged, and then open both valves on the torch, oxygen first to vent the system. Finally, unwind the pressure adjusting screws to relieve the pressure on both regulators. Dismantle and inspect equipment for any signs of damage, and replace any faulty equipment.

Other common problems associated with oxy-acetylene include:

- *Intermittent backfire.* This happens when the flame at the nozzle is inadvertently extinguished, usually by touching the nozzle on the work. It usually gives a pop sound, but in itself is not dangerous. It may occur as a single backfire or a succession of backfires with the flame re-igniting at the nozzle.
- *Sustained backfire.* This is potentially very dangerous as the flame retreats within the body of the torch and continues to burn. There is a loud bang followed by a high pitched whistling sound, not unlike a squealing pig, with sparks coming from the nozzle. Follow the procedure previously described for a flashback before re-commencing work.

Hoses

The hose conveys the gas to the welding or cutting torch and may be the weak link in the oxy-fuel process, so it is imperative to select the correct hose for the job. Hoses are made of two layers of synthetic rubber (neoprene lining and cover) to prevent ballooning of the hose under high gas pressures. Between these layers is a canvas reinforcement layer which helps to bind the two layers of rubber and also acts as a safety indicator. If you flex or bend the hose during routine inspection and you can see the canvas, the hose *must* be replaced.

The correct bore size, pressure rating, length, composition and colour coding are essential for safety. EN 559 sets the requirements for the manufacture of hoses, including the colour for different gases. Hoses are available in lengths of 5–20m and bore diameters of 4.5–10mm (1/4 in BSP – 3/8 in BSP connections). The maximum working pressure must be marked on the hose and must *never* be exceeded, this pressure is typically 20 bar.

Colour coding identifies the gas or range of gases with a particular hose:

- Blue   Oxygen
- Red   Acetylene and other fuel gases (except LPG)
- Black   Inert and non-combustible gases
- Orange   Liquid petroleum gases

Hoses are designed to be lightweight, oil resistant and to be able to cope with normal working conditions. They are resistant to burns, but are not burnproof and should be kept out of direct flames and sparks. A few simple guidelines will ensure a long life and safe operation:

1 Ensure hoses are free from burns, cuts and cracks.
2 Avoid dragging hoses over sharp edges and objects.
3 Do not wrap hoses around cylinders when in use or stored.
4 Do not use hoses longer than necessary, the longer the hose, the more potential for damage.
5 Avoid contact with oil or grease.
6 Protect the hoses by 'ramping' or 'routing' them away from traffic.
7 Keep hoses clear of sparks especially when oxy-fuel cutting.
8 Always use hoses fully extended, as a fire in a coiled hose will be very intense. The heat generated makes it virtually impossible to extinguish with a water extinguisher.
9 Hose assemblies must incorporate a non-return valve (hose check valve) at the torch end. There should be only one non-return valve to each hose in order for gas to flow.
10 The further the cylinders are from the point of work, the less control you have if an incident occurs. It is better to move the cylinders closer to the work, using shorter length hoses and protecting the cylinders with appropriate shielding.

HEALTH & SAFETY ⊙

You should NEVER attempt a running repair by wrapping insulation tape around a damaged hose, or use copper piping to repair acetylene hoses as the explosive compound copper acetylide will form.

Hose Connectors

For connecting the hoses to the torch at one end and the flashback arrestor at the other, two types of connections are available:

- The nipple and nut type (Figure 5.8) is a straight through connector used to attach the hose to the flashback arrestor.
- The non-return or hose check valve is fitted to the welding or cutting torch.

HEALTH & SAFETY ⊙

Do not join hoses along their length with tape, especially when cutting, since debris or hot slag can get into the space between hoses and if a fire starts, the hoses may be difficult to separate.

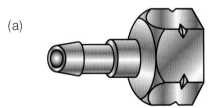

(a)

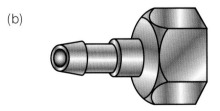

(b)

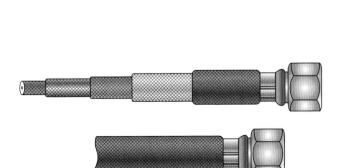

FIGURE 5.7 **Reinforced hose construction**

FIGURE 5.8 **Nipple and nut type hose connectors (a) grooved hexagon nut for acetylene (b) plain hexagon nut for oxygen**

The connections are available in sizes to suit the bore diameters of the hose and have a grooved or nicked nut to indicate a left-hand thread for the fuel gas, which in this case is acetylene. Oxygen has no nick and is right-hand threaded.

Non-Return Valves (Hose Check Valves)

The hose check valve or 'hose protector' as it is commonly called, is used to attach the hoses to the welding or cutting torch. It incorporates a non-return valve, and automatically prevents the backfeeding of gases (premature mixing of the gases in the hoses) as shown in Figure 5.9. Backfeeding is one of the main causes of a backfire; fitting a hose protector will therefore reduce the risk of fire, damage to the equipment and possible injury to the operator. The hose check valve is also designed to reduce the risk of a flashback from a partially blocked nozzle or leaking torch. Although it will prevent a backfire and reduce the risk of a flashback it will not actually stop a flashback. For full protection against the dangers of a flashback, arrestors must be fitted.

Welding Torch

HEALTH & SAFETY ⊙

The majority of welding applications are done with high-pressure torches. They should *never* be used with a low-pressure system.

The welding torch consists of a shank which incorporates the oxygen and acetylene contol valves, a mixer/injector and a range of copper nozzles. Two types of torch are available:

- *High-pressure torches* are designed for use with high-pressure gases supplied from cylinders. The torch acts as a mixing device, supplying approximately equal pressures of oxygen and acetylene which are mixed prior to being burnt at the nozzle. One mixer can accommodate different nozzle sizes, to enable the welding of a wide range of metal thicknesses from 1–25mm. The mixer system ensures that there is the least amount of explosive mixed gas in the system, prior to being burnt.

- *Low-pressure torches* use an 'injector' system where the higher pressure oxygen draws the lower pressure acetylene into the mixing chamber. See Figure 5.10. The injector is usually included as part of the nozzle, so in this respect each nozzle has its own injector. This makes these nozzles far more expensive than those used with the high-pressure system.

Welding torches are subject to EN 5172. Leaks from any part of the system are a serious hazard, given that the torch is the closest object to the flame and operator. Poor purging and bad lighting up techniques are the most frequent cause of flashbacks. Always check the torch and nozzle for source of leaks. Any signs of heat damage around the torch valves may indicate that the equipment has suffered internal damage and could be leaking, and must be replaced.

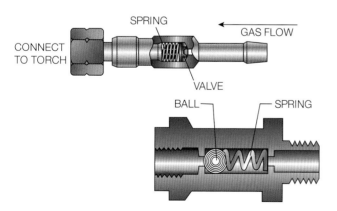

FIGURE 5.9 Hose check valve

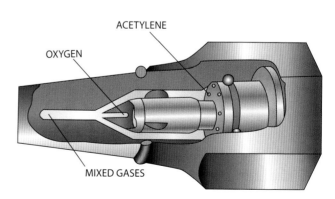

FIGURE 5.10 Detailed view of an injector

Nozzles

Nozzles provide a safe and convenient method of varying the amount of heat supplied to the weld and of directing the flame and heat to the exact place the operator chooses. The heat value is governed by the size of the orifice (hole) in the nozzle. The heat value (calorific value) of the flame must be regulated to match the thermal conductivity of the material, thickness, joint design and melting point

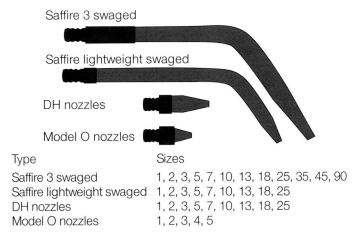

Type	Sizes
Saffire 3 swaged	1, 2, 3, 5, 7, 10, 13, 18, 25, 35, 45, 90
Saffire lightweight swaged	1, 2, 3, 5, 7, 10, 13, 18, 25
DH nozzles	1, 2, 3, 5, 7, 10, 13, 18, 25
Model O nozzles	1, 2, 3, 4, 5

FIGURE 5.11 Nozzles

of the parent metal. This should be achieved by changing the nozzle size and not by increasing the gas pressure. For example, material 3mm thick can be welded by changing the nozzle size without excessive alteration to pressure. A flame with too low a heat value will result in a lack of penetration and fusion, and backfiring may also occur. If you increase the gas pressure, there comes a point where the flame leaves the end of the nozzle. This causes a very noisy flame which indicates that the pressure is too high to support the flame. If the flame heat is too high, overheating, lack of control of the molten pool and excessive penetration will be experienced. If you try to weld with a large sized nozzle and reduced gas pressure, small explosions will occur at the nozzle, because the gas tends to build up around the nozzle. Effective flame adjustment and heat values can only be maintained if the nozzle orifice is clean and square with the end of the nozzle.

Nozzles are stamped with a number that corresponds to the size of the orifice; a number 2 nozzle will use 0.14 bar (2psi) of fuel gas in one hour with a steady neutral flame. When selecting your nozzle always check the seating to ensure that the machine part is parallel and not 'flared' in any way as this will prevent a good gas seal. Nozzles should *never* be modified in any way to accommodate a job. Bending the nozzle is likely to result in cracking in the threaded portion of the nozzle with subsequent gas leaks or possible explosion. Using a pair of pliers to change the angle is likely to result in a 'restriction' within the internal bore and the potential for gas pressure build up with explosive potential.

 TIP

Check the thread pattern for signs of cross threading or damage which can harm the mixer.

Welding Goggles and Visors

Welding goggles/visors are designed to protect the eyes from sparks, heat and light radiated from the work. They are available in various styles such as goggles, ski-type mask which may have a lift up lens with a clear lens for grinding or a full-face visor. The goggles have a filter lens corresponding to the amount of light being produced, and a protective cover glass. The cover glasses should be replaced when they obscure your vision. When fluxes are used to assist in the removal of some surface oxides, notably aluminium, cast iron, brass or stainless steel, more glare is produced, resulting in additional shielding being required. Filter lenses for use with fluxes are identified by a suffix 'F' for example 4 GWF is a shade four to be used with gas welding using a flux.

TABLE 5.1 GWF

3 GWF	For aluminium and its alloys
4 GWF	For brazing and braze welding
5 GWF	For copper and its alloys
6 GWF	For thick plate and pipe

Gas Economizer

The gas economizer is used with manifold distribution systems. It contains two valves, which are normally held open by a spring, but are closed by the weight of the torch being hung on the control lever arm, as shown in Figure 5.12. The oxygen and acetylene supplies are coupled to the inlet side of the fitting, and the gas hoses to the torch are coupled to the outlet side. When the torch is lifted off the arm, the two valves open and gas flows to the torch. The economizer includes a pilot light which stays alight even when the torch hangs on the lever arm. To correctly light the torch, first purge the hoses away from the pilot light, and then close the valves on the torch. Open the acetylene valve a small amount and pass the nozzle back through the pilot light, this will ensure a smooth lighting up technique. If you push the nozzle into the pilot light you may have excessive gas pressure which will manifest itself as a 'roar' and make it difficult to ignite the flame. Once ignited, the flame can be adjusted and you are now ready to weld. The gas economizer allows you to stop welding briefly, and safely place the torch on the lever arm to extinguish the flame. Without the use of gas economizers, a lot of gas can be wasted while the flame is being adjusted or while the torch is laid aside with the flame burning. Installing gas economizers can reduce gas consumption by as much as 20 per cent. Gas economizers are designed for short intervals of time between welding operations. If a torch is not going to be used for some time it should be shut down completely.

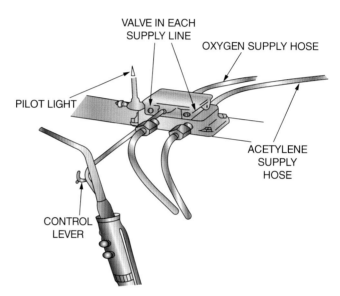

FIGURE 5.12 Gas economizer

TIP

Do not use wire brushes or odd bits of wire to clear any blockages, as this could lead to packing in the nozzle, with possible explosive results and damage to the rest of the equipment.

Reamers or Cleaners

Due to poor manipulation of the torch, the wrong angle or possibly getting too close to the weld pool, the nozzle may get partially blocked and need cleaning. This can be done with a set of reamers or cleaners. Using the file provided in the set, file the end of the nozzle smooth and square. Select the size of reamer that fits easily into the orifice. Move it in and out of the orifice a few times to clear any particles that may have adhered to the inside. Keep the reamer straight and hold it in a steady position to prevent it from bending or breaking off. Excessive use of reamers enlarge the orifice.

Manifold Systems

Where the total required discharge rate would be above the safe maximum (20 per cent) for a single cylinder, use is made of manifolded cylinders. This involves interconnecting two or more cylinders which will reduce by 50 per cent the discharge rate from a single cylinder. This may be done by using the correct type of fittings: a three-way valve adaptor and connecting arm. For permanent installation, where a number of fixed welding stations are served from a central bank of cylinders, flashback arrestors, non-return valves and main-line control valves and gauges are essential features. The system should be designed and installed by one of the specialist firms who are approved and registered with the BCGA. The principle advantages of a central system are:

- Increased standards of safety are provided as the cylinders can be located outside of the welding shop where fire and explosions are much less likely to occur, with the possibility of mechanical damage considerably reduced.
- A constant uninterrupted supply of gas is maintained at each welding point.
- Gas pressure regulators are not exposed to damage in the same way as when used on individual cylinders.
- Handling of cylinders is kept to a minimum.
- The absence of cylinders in the workshop also gives an increased floor space as well as an improved safety factor.

Consumables

Filler rods for gas welding are covered in European Standards and have a known chemical composition and tensile strength. For example the A1 type is for 'general purpose low carbon steel filler rods and wires'. These filler rods are intended for applications where a minimum butt weld strength of 340 N/mm is required. Manganese is added to the filler metal to impart strength to the weldment and also as a deoxidant, while nickel acts as a grain refiner. Additional deoxidizing agents can be added to improve mechanical properties, typical of these are aluminum, silicon and niobium. Other alloying elements can be added to improve performance in corrosive environments or at elevated temperatures, typical of these are chromium and molybdenum. Many of the low carbon steel filler rods are copper coated to protect them from atmospheric contamination during storage and use.

Selection of filler wire is critical to the successful completion of a weld deposit. If it is not compatible with the parent material, the weldment may undergo electrolytic corrosion. It is also important to ensure that there are sufficient alloying elements to counter the effects of heat input on the parent metal. Filler wires should be free from rust, scale, oil, grease or moisture, all sources of low melting impurities, which can cause weld failure.

Filler wires are available in the drawn condition or as cast for welding cast iron and other materials.

Handling and Storage

Filler wires should be kept at room temperature and handled with care. All packaging should be clearly marked with composition and diameter. After completion of the weld any surplus filler wires should be returned to their appropriate container for future use. Where it is not possible to store filler wires under heated conditions, a moisture absorbant such as silica gel should be used to keep them dry.

To manipulate the filler wire, open up the first two fingers of your hand and lay the filler wire across them. Using your thumb, press down on the wire to give it some tension and rigidity, then slide it between your fingers to supply the molten pool.

- Always bend the end of the filler rod to ensure you do not poke yourself in the eye and so that you can identify the hot end.
- Avoid potential fire hazards by keeping hot filler wires away from combustible materials.
- Always be aware of where the end of the filler wire is and never push the rod against yourself to move your fingers down the filler wire.

SETUP AND ASSEMBLY OF EQUIPMENT

In order to set up and assemble equipment it is important that the following procedure is carried out:

1 Secure both cylinders vertically, either in a cylinder trolley or against the bench/wall with a length of chain.

2 Ensure that the cylinder outlets are free from debris by opening and closing quickly snifting, to eject any debris, standing to one side while doing this with safety glasses on. Do not do this with hydrogen as it could cause an explosive reaction.

3 Inspect the outlets for any signs of damage, such as cross threading or denting of the gland nut. Never use your finger to do this, you could pick up a brass splinter which can cause infection.

4 Check that the regulators are the correct ones for use with the gases, and that no visible damage is evident in the thread patterns or seating, such as dents or pitting.

5 Fit to the appropriate gas, remember acetylene is maroon with a left-hand thread and oxygen is black with right-hand thread.

6 Fit flashback arrestors as appropriate.

7 Uncoil hoses and check connections, ensuring that the non-return valve or hose check valve is at the torch end.

8 If the hoses are new, quickly blow through them with a short burst of gas to dislodge any chalk that may still be present from manufacture as this will affect the other components in the welding setup.

9 Connect the torch to the non-return valve.

10 Select a suitable nozzle and check the condition of the seating, thread pattern and orifice.

11 Ensure all fittings are tight.

12 Open the spindle valve on the oxygen bottle first and then the acetylene so that the spindle key remains on the acetylene cylinder in case of emergency. Check that the cylinder capacity on the gauge is sufficient for the job in hand.

13 Screw the pressure adjusting screws in to obtain appropriate pressure for the nozzle selected on both regulators, as shown in Figure 5.13.

14 Open the torch valve to purge line and to adjust to working pressure for the nozzle selected. Repeat with the other torch valve.

15 Pressure is now charged to the torch. Now carry out a leak detection inspection and tighten any couplings if necessary and then re-test for leaks. If using an aerosol application ensure that you wait till the gas propellant has subsided to test for leaks. (The fluid should be clear.) The equipment is now ready for use.

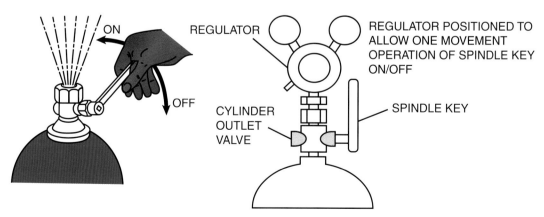

FIGURE 5.13 Regulator positioned to allow one movement operation of spindle key ON/OFF

LIGHTING UP, FLAME VARIATIONS AND SHUTTING DOWN TECHNIQUE

With the system fully charged and leak tested, the equipment is now ready for lighting up. Begin by opening the acetylene torch valve slowly and ignite using a spark lighter. Adjust the acetylene as quickly as possible to reduce the amount of soot being put into the atmosphere. When the black smoke 'just disappears', slowly introduce the oxygen until a 'feathered inner cone' is produced. When this is approximately 25–37 mm long a 'carburising' flame is produced which contains an excess of acetylene and is used for hard surfacing components.

A further increase in oxygen until a 'rounded inner cone' is produced will give a 'neutral' flame in which there are equal volumes of both gases. This is the flame used for the fusion welding of carbon steels and some non-ferrous metals. With a further increase in oxygen a 'pointed inner cone' is produced. This is an oxidizing flame, typified by a characteristic 'hiss', in which there is an excess of oxygen, used for brazing and braze welding. To close down the system, turn off both valves on the torch; acetylene first then oxygen. Close the valves on the bottles, and open both valves on the torch to discharge gas from the regulators and hoses. Only close the valves on the torch when both gauges, on both bottles are reading zero. Remove the nozzle, taking care that it may still be hot, curl up the hoses, hang up the torch to prevent any damage, and reinstate the work area.

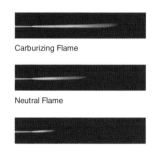

Carburizing Flame

Neutral Flame

Oxidising Flame

FIGURE 5.14 Flame variations

Structure of the Oxy-Acetylene Flame

The welding flame is produced by burning approximately equal volumes of the oxygen and acetylene which are supplied to the torch. Figure 5.15 illustrates the basic structure of the oxy-acetylene flame.

The oxy-acetylene flame, like most oxy-fuel gas flames used for welding, is characterized by the following zones:

(a) The innermost cone of mixed unburnt gases leaving the nozzle. It appears intensely white and clearly defined with the correct neutral flame setting.

(b) A very narrow stationary zone wherein the chemical reaction of the first stage of combustion takes place producing a sudden rise in temperature.

(c) The reducing zone appearing dark to light blue, in which the primary combustion products are concentrated. The nature of these products determines the chemical nature of the flame – neutral, oxidizing or carburizing.

(d) Within the region of primary combustion and approximately 3mm from the tip of a nicely defined inner cone of unburnt gases there is a region of maximum temperature and this is the zone used for welding.

(e) The yellow to pinkish outer zone or 'plume' around and beyond the other zones represents the chemical reaction of the 'second stage of combustion'. Here the two combustible gases carbon monoxide and hydrogen, which are the products of primary combustion, combine with oxygen from the surrounding atmosphere. This part of the flame is always oxidizing and contains large amounts of nitrogen, due to the fact that the atmosphere contains approximately 80 per cent nitrogen and 20 per cent oxygen by volume.

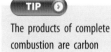

TIP

The products of complete combustion are carbon dioxide and water vapour.

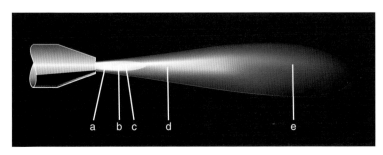

FIGURE 5.15 Good fusion weld

WELDING TECHNIQUES

The key to successful oxy-acetylene welding is manual dexterity and observation of the characteristics of the molten pool. When you apply the welding heat to the parent material three distinct features will occur:

● a change in colour on the surface of the material indicating the effects of the application of heat;

● a form of 'sweating' which is the early stage of the formation of a molten pool;

● the formation of the molten pool in which fusion conditions exist.

We are now ready to put a deposit of weld down. For oxy-acetylene welding, two techniques are well established.

Leftward Welding

This is used for welding sheet and steel plate up to 4.8mm thick and for the welding of non-ferrous metals. 'Leftward' refers to the relationship of the filler rod, torch and solidifying weld metal. In leftward welding, the rod is in front at a slope angle of 20–30º, followed by the torch at a slope angle of 60–70º and this is followed by the solidifying weld metal. This applies regardless of whether you are right-handed or left-handed. The torch is moved in a side-to-side or circular movement to ensure even fusion on the sides of the joint, and the filler rod is dipped in and out at regular intervals into the centre of the molten pool. The tip of the inner cone should be approximately 3–5mm above the metal, and the flame of the cone should not leave the weld pool. Ideally the rod and the torch should make a 90º angle, and when coming to the edge of the run, the torch should be lowered and played to allow the end of the weld (crater) to be built up to the same profile as the rest of the weld and also to prevent burnthrough.

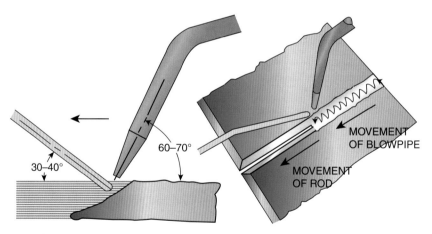

FIGURE 5.16 Leftward technique

TABLE 5.2 Leftward Welding: High Pressure Torch Data

Thickness of plate – mm	0.8	1.6	2.4	3.2	4.0	4.8
Regulator Pressure – Oxygen (bar/psi)	0.14bar/2psi	0.14bar/2psi	0.21bar/3psi	0.21bar/3psi	0.21bar/3psi	0.21bar/3psi
Regulator Pressure – Acetylene (bar/psi)	0.14bar/2psi	0.14bar/2psi	0.21bar/3psi	0.21bar/3psi	0.21bar/3psi	0.28bar/4psi
Nozzle Size	1	2–3	5	7	10	13
Root Gap – mm	—	1–1.6	1.6	2	2	2.5
Filler Wire Diameter – mm	1.6	1.6	2	2.4	3.2	3.2
Welding Rate/cm per hr	610	750	610	550	450	350

Rightward Welding

Rightward welding is recommended for steel plate over 4.5mm thick and enables better fusion at the root of the joint for thicker materials or for welding heavy sections and cast iron. With this technique, the torch is in front at a slope angle of 40–50°, this is followed by the filler rod at a slope angle of 30–40° with the solidifying weld metal behind the filler rod. See Figure 5.17. The torch moves in a straight line and it is the filler rod which is rotated as it is fed into the molten pool. The advantage of this technique is that there is no likelihood of the molten metal being pushed forward over any unheated surface due to the fact that the torch is pointing backwards towards the part that has been welded.

A larger nozzle is required, because the torch has no side motion and the molten metal is controlled by the torch and rod. The larger flame gives greater welding speed and thick plate can be welded in one pass. Because the torch does not move, the molten metal is agitated and very little oxidation is produced. The flame playing on the metal just deposited helps to anneal it, while the smaller volume of filler metal in a V-butt reduces the amount of expansion. In addition, a better view of the *keyhole* and molten pool is obtained, resulting in improved penetration. The rod diameter is approximately half the thickness of plate. If the rod diameter is too large, it melts too slowly resulting in poor penetration and poor fusion. However if the rod is too small it melts too quickly, and reinforcement of the weld is difficult to control, resulting in underfill and possible crater cracking.

The advantages of the rightward technique are:

- less cost per metre run due to less filler rod being used and increased speed;
- less expansion and contraction;
- better view of the molten pool, giving better control of the weld;
- annealing action of the flame on the weld metal.

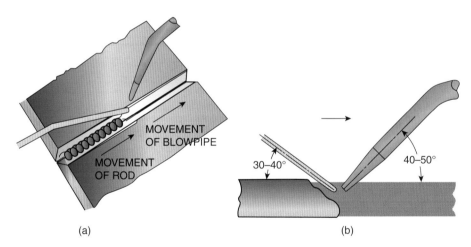

(a)   (b)

FIGURE 5.17 Rightward technique

TABLE 5.3 Rightward Welding: High Pressure Torch Data

Thickness of plate – mm	4.8	6.0	8.0	10.0	12.0
Regulator Pressure – Oxygen (bar/psi)	0.21bar/3psi	0.21bar/3psi	0.28bar/4psi	0.28bar/4psi	0.42bar/6psi
Regulator Pressure – Acetylene (bar/psi)	0.28bar/4psi	0.49bar/7psi	0.63bar/9psi		
Nozzle Size	13	18	25	35	45
Root Gap – mm	2.5	3	4	3	3
Filler Wire Diameter – mm	2.4	3.2	4	4.8	6
Filler Wire Used/cm per hr	3.5	3.5	3.5	4	4.75
Welding Rate/cm per hr	35	28	20	15	12
Edge Preparation/V			60	60	60

TIP ❯

Information on the correct nozzle size and filler rod selection is readily available from welding suppliers and manufacturers.

Before attempting any welding operations, the beginner should acquire a sense of fusion (speed of travel), a knowledge of torch and filler rod slope and tilt angles, and manipulation techniques. The best way to do this is by depositing fusion runs with and without the filler rod on thin-gauge sheet. It is also important to know how to pick-up from a previous deposit, which is done by going back approximately 12mm, reheating until the molten pool is established and then moving forward, adding filler wire as required. Successful welding depends on the correct selection of nozzle size and filler rod to match the thickness of plate, joint design and welding position.

It is also important to identify the number of heat paths in the joint design: two paths for butt, lap and outside corner and three paths for a T-fillet. To compensate for this additional heat loss, it is common practice to move up a nozzle size, to that for the same thickness/plate. The end of the filler rod should always be kept within the protective envelope of the flame, as the rod oxidizes each time it is removed. This oxide is deposited in the weld pool, causing porosity and a weakened weld. When the rod is added to the pool, try to keep it at a shallow angle to prevent touching the nozzle causing a possible blowback. It is also good practice to move the flame back, to allow the end of the rod to be dipped into the weld pool. If the torch is not moved back, the rod may melt and drip into the weld pool, resulting in the following problems:

- The drop of metal tends to overheat, resulting in important alloys being burned out.
- The metal cannot always be added where it is needed.
- The method only works in the flat (PA) position.

The main features of a good fusion weld are:

- good fusion over the whole side surface of the V or square edge preparation;
- penetration of the weld metal to the underside of the parent plate;
- slight reinforcement of the weld above the parent plate;
- no entrapped oxide or blowholes.

Indications of poor weld deposition are:

- poor flame manipulation and too large a flame causing molten metal to flow forward on to un-fused plate, giving adhesion instead of fusion;
- poor positioning of work, incorrect temperature of molten metal, and poor manipulation, causing oxide to become entrapped and grooves to be formed on each line of fusion resulting in undercutting;
- too small a flame, or too rapid a speed of travel, combined with a lack of skill in manipulation, resulting in a lack of penetration.

TIP ❯

Reinforcement on the face of a weld will NOT make up for a lack of penetration on the root of a weld.

Welding Practice

We will now introduce one of the most important weld parameters (something you can change) in all welding processes: slope and tilt angles. We can define these angles in very simple terms: 'Slope' is the angle that the torch, welding gun, or electrode makes with the work surface, whereas 'tilt' is the angle that bisects the joint. These angles vary according to the welding position and joint design but in the practices that follow the slope and tilt angles quoted will be for the flat (PA) position and horizontal/vertical (PB) only. A closed outside corner joint should have an even weld ripple, running parallel along the joint with little evidence of undercut, and full penetration which does not exceed a maximum of 3mm.

A lap joint should have an even leg length and there should be a smooth transition between the two plates as shown in Figure 5.21. The weld ripple should be even with little evidence of undercut and an even height to the weld profile. The ideal profile for this joint in this position should be slightly convex to distribute the stresses applied to the members of the joint.

WORKSHOP TASK 5.1

Fusion Runs With and Without Filler Wire

Using a properly set-up oxy-acetylene plant, appropriate PPE, pieces of 150 x 50 x 1.6mm low carbon steel you will deposit single runs with and without the filler rod.

- The flame should be adjusted to a neutral flame and the steel sheet placed on firebricks on the welding bench, to allow air to circulate on the underside.
- Using the leftward technique, hold the torch at a slope angle of 60–70° to the sheet surface, with the inner cone 3–5mm from the metal surface.
- Heat the material and observe the surface to look for sweating, which is an indicator that the melting point is very close. This will be followed by the molten pool becoming fully established.
- At this point move forward and try to maintain the same diameter molten pool, using a circular motion of the welding torch to encourage the molten pool to flow.
- Maintain a steady rate of travel to ensure fusion into the plate (indicated by a regular continuous bead on the underside and a gouging or groove effect on the top side of the plate).
- The sparks that occur as the weld progresses are due to components of the metal that are being expelled out of the weldment. Silicon oxides make up most of the sparks, and extra silicon can be added by the filler metal so that the weldment retains its desired qualities.
- A change in sparks given off by the weld as it progresses can indicate changes in weld temperature. An increase in sparks on clean metal means an increase in temperature, often a sign that burnthrough is about to take place. If this happens, pull the torch back to allow the metal to cool.
- When you have mastered the fusion runs, practise using the filler rod to fill the groove made earlier. Dip the rod at a uniform rate into the molten pool to give a deposit within the limits for good weld deposition – no more than 3mm reinforcement (excess weld metal).
- When you have completed the task, close down and drain the equipment, and reinstate the work area.

Complete a copy of the 'Student Welding Report' listed in Appendix 1 or provided by your instructor.

WORKSHOP TASK 5.2

Closed Outside Corner Weld

With properly set-up equipment, and appropriate PPE, and pieces of 150 x 50 x 1.6mm low carbon steel (LCS), filler wire and any jig you wish to use, you will produce a closed outside corner weld in the flat (PA) position. See Figure 5.18 for guidance on a good weld.

- Tack up the two pieces of low carbon steel at a 90° angle, either by using a jig or some arrangement ensuring that the edges just touch and do not overlap each other. Tacking should be done at both ends and at regular intervals according to thickness.
- Dress the edges using a light hammer to ensure close fit-up prior to welding the joint.
- Using the leftward technique, with the torch at a slope angle of 60–70° and a tilt angle that bisects the joint, establish an even molten pool over both edges.
- When this has been achieved move along the line of the joint adding filler wire as required.
- Observe a consistent speed of travel and monitor fusion of both edges.
- When coming to the end of the weld lower the torch angle to play the flame and add additional filler wire to terminate the end of the weld correctly and prevent crater cracking.
- Continue with further pieces of metal until you are satisfied with your test piece.
- Allow joint to cool naturally, turn off plant and go through the shut-down procedure, place any scrap in the appropriate bins and reinstate the work area.

Complete a copy of the 'Student Welding Report'.

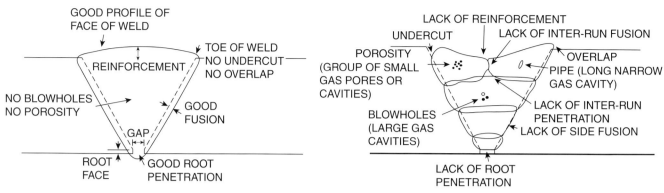

FIGURE 5.18 Good fusion weld

FIGURE 5.19 Poor fusion weld

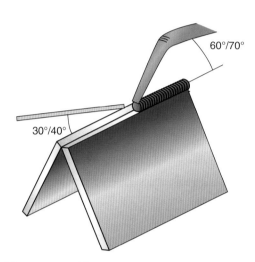

FIGURE 5.20 Closed outside corner weld

WORKSHOP TASK 5.3

Lap Fillet Weld

With properly set-up equipment, and appropriate PPE, and pieces of 150 x 50 x 1.6mm low carbon steel (LCS), filler wire and any jig you may wish to use, you will produce a lap fillet weld in the flat (PA) position.

- Tack up two pieces of low carbon steel overlapping by 10mm, at the ends and at regular intervals according to thickness of plate.
- 'Dress' the edges using a light hammer to ensure close fit-up prior to welding the joint.
- Using the leftward technique, with the torch at a slope angle of 60–70° and a tilt angle of 45° concentrating the bulk of the heat into the bottom plate (greater mass of metal), establish a molten pool.
- When this has been achieved move along the line of the joint adding filler wire to the top edge of the upper plate, and allow it to flow into the molten pool. By using this technique you can counteract the natural tendency for the top edge to melt away before the bottom plate is molten.
- It is important to ensure that both surfaces melt into the corner of the joint and only at this point is filler wire added. This ensures total fusion rather than adhesion of the joint.
- Observe a consistent speed of travel and monitor fusion of both edges.
- When coming to the end of the weld lower the torch angle to play the flame and add additional filler wire to terminate the end of the weld correctly and prevent crater cracking.
- Continue with further pieces of metal until you are satisfied with your test piece.
- Allow joint to cool naturally, turn off plant and go through the shut-down procedure, place any scrap in the appropriate bins and reinstate the work area.

Complete a copy of the 'Student Welding Report'.

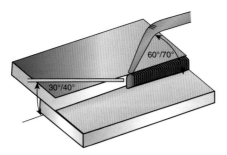

FIGURE 5.21 Lap fillet weld

T-fillet welded joints should again have an even leg length with a consistent weld ripple and profile. There should be little evidence of undercutting, and the ideal profile should be of a flat or slightly convex face in appearance. If the profile is concave in this position, it may well suggest poor technique, possibly too high a flame temperature and insufficient filler wire diameter or insufficient filler wire addition.

WORKSHOP TASK 5.4

T-fillet Weld

With properly set-up equipment, and appropriate PPE, and pieces of 150 x 50 x 1.6mm low carbon steel (LCS), filler wire and any jig you may wish to use, you will produce a T-fillet weld in the horizontal/vertical (PB) position as shown in Figure 5.22.

- Tack up two pieces of low carbon steel, with one plate at 90º to the horizontal plate, at the ends and at regular intervals according to thickness of plate.
- 'Dress' the joint using a light hammer to ensure close fit-up prior to welding the joint.
- Using the leftward technique, with the torch at a slope angle of 60–70º and a tilt angle of 45º concentrating the bulk of the heat into the bottom plate (greater mass of metal), establish a molten pool.
- When this has been achieved move along the line of the joint adding filler wire to the top edge of the molten pool, and allow it to flow down on to the bottom plate. By using this technique you can counteract the natural tendency for the top edge to undercut before the bottom plate is molten.
- It is important to ensure that both surfaces melt into the corner of the joint and only at this point is filler wire added. This ensures total fusion rather than adhesion of the joint.
- Observe a consistent speed of travel and monitor fusion of both edges of the joint.
- When coming to the end of the weld lower the torch angle to play the flame and add additional filler wire to terminate the end of the weld correctly and prevent crater cracking.
- Continue with further pieces of metal until you are satisfied with your test piece.
- Allow joint to cool naturally, turn off plant and go through the shut-down procedure, place any scrap in the appropriate bins and reinstate the work area.

Complete a copy of the 'Student Welding Report'.

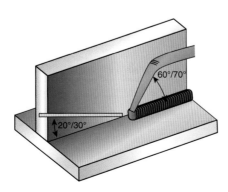

FIGURE 5.22 T-fillet weld

The butt joint profile should have a slightly raised weld bead profile with even ripple and evidence of penetration which does not exceed 3mm. By using a 'tapered joint' setup, allowances can be made for expansion and contraction of the joint and reduce distortion.

WORKSHOP TASK 5.5

Square Edge Butt Joint

With properly set-up equipment, and appropriate PPE, and pieces of 150 x 50 x 1.6mm low carbon steel (LCS), filler wire and any jig you may wish to use, you will produce a square edge butt joint in the flat (PA) position, as shown in Figure 5.23.

- Set up two pieces of low carbon steel with a root gap of approximately 1.6mm at the beginning and tapering to 2mm at the end, commence tacks 10mm in from the 1.6mm root gap end and at regular intervals according to thickness of plate.
- Lightly 'dress' the joint using a light hammer to ensure alignment of plates prior to welding the joint.
- Using the leftward technique, with the torch at a slope angle of 60–70° and a tilt angle of 90°, establish a molten pool and observe the formation of a keyhole at the leading edge of the joint, this ensures penetration.
- When this has been achieved move along the line of the joint adding filler wire to the back edge of the molten pool and no filler wire should be added until the keyhole reforms.
- If the weld pool becomes too hot, momentarily pull up the torch to allow it to cool, thus maintaining the keyhole and weld deposition.
- It is important to ensure that both surfaces melt and fusion takes place over both edges of the weld joint with penetration on the underside of the joint.
- Observe a consistent speed of travel and monitor fusion of both edges of the joint.
- When coming to the end of the weld lower the torch angle to play the flame and add additional filler wire to terminate the end of the weld correctly and prevent crater cracking.
- Continue with further pieces of metal until you are satisfied with your test piece.
- Allow joint to cool naturally, turn off plant and go through the shut-down procedure, place any scrap in the appropriate bins and reinstate the work area.

Complete a copy of the 'Student Welding Report'.

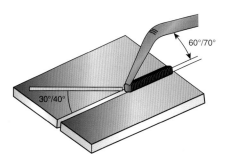

FIGURE 5.23 Square edge butt joint

Brazing and Braze Welding

Flux	Flux type	Material
Hydrochloric acid	Corrosive	Zinc galvanished mild steel
Zinc chloride	Corrosive	Plain carbon steel Brass Copper Tin-plate Terne plate
Phosporic acid	Corrosive	Stainless steel
Tallow	Non-Corrosive	Lead sheet or pipe
Resin-cored solder Resin paste	Non-Corrosive	Electrical connections

Type	Temperature range °C
Borax and fluoroborate	Above 750
Fluoride	Below 750
Alkali halide	Below 580

FIGURE 5.24a Brazing table

Brazing and braze welding are both done using an oxidizing flame with a flux and a common filler material often referred to as a spelter. However the method of deposition is entirely different. In brazing the filler material is drawn through closely fitting joints by capilliary attraction, using heat to draw the filler material into the joint. Braze welding, on the other hand, is the surface deposition of material, in much the same way as that of fusion welding. In both cases, the strength of the process comes from the forming of an inter-metallic bonding of the surface and no fusion takes place. There are several advantages to brazing and braze welding:

- *Low temperature.* Since the base metal does not have to melt, a lower heat source or fuel gas (propane) can be used.
- *May be permanently or temporarily joined.* Since the base metal is not damaged, parts may be dismantled and reused at a later date by simply reapplying heat. However the joint is solid enough to be permanent.
- *Joints have the ability to flex.* This makes them an ideal choice for conditions which are subject to shock loading such as in bicycle frames.
- *Dissimilar materials can be joined.* Examples include copper to steel, aluminium to brass and cast iron to stainless steel – or tungsten carbide tips to carbon steel drills for masonry drills.
- *Speed of joining.* Parts can be preassembled and dipped or furnace soldered or brazed in large quantities. Also, a lower temperature reduces heating times.
- *Slow rate of heating and cooling.* Because it is not necessary to heat a small area to its melting temperature and then allow it to cool quickly to solidify, reducing internal stresses caused by rapid temperature change.
- *Parts of varying thicknesses can be joined.* This can be achieved without overheating or burning.
- *Easy realignment.* Parts can be easily realigned by reheating the joint and repositioning.

TIP

Oxy-acetylene is popular with engineers and welders because it can be used with many different processes.

TIP

See Figure 1 and Figure 2 in the Appendix for information on brazing for different metals.

Filler Materials

The type of filler material used for any specific joint depends upon the base materials, service requirements and mechanical properties needed. The filler metals are alloys of two or more metals, in varying percentages. Some alloys give greater strength, and some melt at a lower temperature range. Each has specific properties and almost all of the alloys have a paste range in which the alloy is partly solid and partly liquid as it is heated or cooled. As the joined parts cool through the paste range, it is important that they are not moved.

Copper–zinc alloys are the most popular brazing alloys, and are available as regular and low-fuming alloys. The zinc has a tendency to burn-out if it is overheated, typified by the presence of a white smoke and spitting coming from the deposit, which is in fact zinc oxide. This oxide is dangerous if breathed in as it can cause metal fume fever and zinc poisoning. Metal fume fever is somewhat similar to a heavy dose of influenza with headache, aching limbs and respiratory problems. The use of an extraction system or respirator with dual canisters is to be recommended when using these materials.

Nickel and nickel alloys are increasingly used as a substitute for silver-based alloys. Nickel is generally more difficult to use than silver because it has lower wetting (surface coating) and flow characteristics. However, nickel has a much higher strength than silver and higher heat resistance, hence its use in jet engine parts and other similar applications. Nickel alloys have a higher surface tension which allows larger fillets and poor fit-ups to be joined. They also have a high corrosion resistance, ideal for petro-chemical and marine applications.

Aluminium-silicon filler metals can be used to join sheet and cast alloys. It is important not to overheat the parent material and the torch should be played on the metal with an elliptical movement to prevent heat build-up. The filler wire is then added in the fluxed condition in a wiping movement and the parent metal provides the heat to melt the filler wire. They are used in conditions where resistance to shock, fatigue and sea-water corrosion is required.

Hard Soldering

Hard soldering refers to any soldering process in excess of 300ºC. Strictly speaking it includes brazing and braze welding, but in this book we shall only include silver soldering and copper-phosphorous alloys. You may have come across some form of silver soldering, perhaps in the making and repair of jewellery, where it is the preferred choice because of the ease with which it flows and the fact that the colour can be matched to gold or silver. If you have ever had a ring size increased or decreased it will have been joined by silver soldering; if you hold the ring up to the light, you may be able to detect a very subtle change in colour.

Silver-copper alloys can be used to join almost any metal, ferrous or non-ferrous, except aluminium, magnesium, zinc and a few other low-melting point metals. This alloy is often referred to as silver soldering and is one of the most versatile materials available. It is also one of the most expensive alloys being second to gold.

Copper-phosphorous alloys, sometimes referred to as copper flow, have good fluidity and wetting ability on copper and copper alloys. Joints must have close fitting tolerances to benefit from the solder; heavy build ups may cause brittleness. Because copper-phosphorous can form brittle iron phosphide at brazing temperatures on steel no copper clad fittings with ferrous substrates should be joined. Copper can be easily burned off leaving the substrate exposed to phosphorous embrittlement. Copper-phosphorous rods are used in refrigeration, air conditioning and plumbing applications (all plumbing in Australia is hard soldered). The phosphorous in the rods makes them self-fluxing on copper, meaning there is no expensive clean-up operation to worry about. This coupled with the speed of deposition makes them extremely desirable. However care needs to be taken in the application of heat and the torch needs to be constantly on the move so as not to oxidize the copper. If this occurs the solder will not flow and you will need to clean the area before recommencing the joint. The movement of the torch also prevents collapse of the copper or burn-through due to its high conductivity. The addition of 2 per cent silver helps with wetting and flow characteristics.

Fluxes

Fluxes used in brazing and soldering have three major functions:

● They must *remove* any oxides that form as a result of heating the parts.
● They must *promote* wetting.
● They should be of a *low* viscosity in order to assist in capilliary attraction.

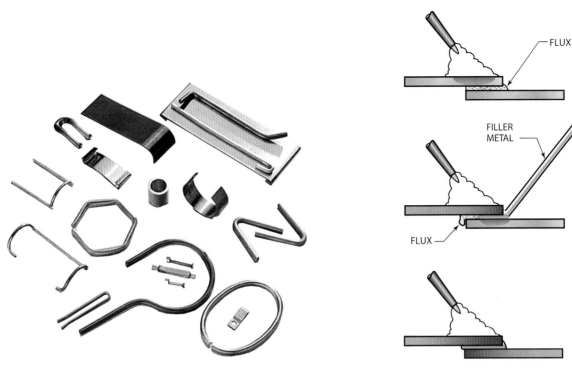

FIGURE 5.24 Braze/solder forms that can be pre-placed in a braze/solder joint

FIGURE 5.25 Flux flowing into a joint reduces oxides to clean the surfaces and gives rise to a capillary action that causes the filler metal to flow behind it

When the flux is heated to its reactive temperature it must have a low enough viscosity to flow through close fitting surfaces of a joint. As it flows through, the flux must absorb and dissolve oxides, allowing the molten filler metal to be pulled in behind it, Figure 5.25. After the joint is complete, the flux residue should be easily removable.

Fluxes are available in many forms such as solids, powders, pastes, liquids, sheets, rings and pre-forms. They are also available mixed with the filler metal, inside the filler rod or on the outside in notches on the filler rod or surface coating (pre-fluxed). Sheets, rings and pre-forms may be placed within the joints of an assembly before heating so that a good bond inside the joints can be ensured, Figure 5.24. Pastes and liquids can be injected into the joint from tubes using a special gun, Figure 5.26. Paste, powders and liquids may be brushed on

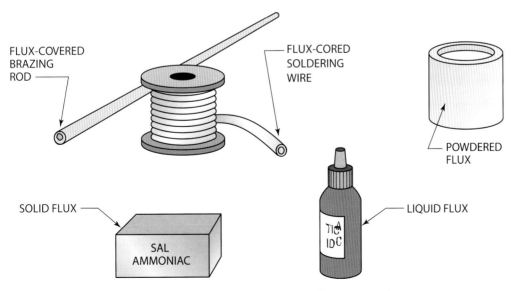

FIGURE 5.26 Fluxes can be purchased with the filler metal or separately

the joint before or after the material is heated. Paste and powders may also be applied to the end of the rod by tufting, heating the rod and dipping it in the flux. Most powders can be made into a paste, or the paste can be thinned by adding distilled water (see manufacturers' specifications for details). Some powdered or liquid fluxes may be added to the gas 'gas fluxing', when using an oxy-fuel gas torch for brazing and soldering. The flux is picked up by the fuel gas as it is bubbled through the flux container and is carried to the torch where it becomes part of the flame.

Flux and filler metal combinations are the most convenient and easy to use. As previously discussed, care should be taken when storing filler wires. Where the flux covers the outside of the filler metal, it may be damaged by humidity or chipped off during storage.

Using excessive flux may result in flux inclusion within a joint, weakening it or causing it to fail. Before disposing of any brazing and soldering fluxes, read the manufacturer's safety data sheet (MSDS) carefully and follow the recommended procedures. Keeping our environment clean and safe is everyone's responsibility.

Fluxing Action

The use of fluxes does not eliminate the need for good joint cleaning. Fluxes will not remove oil, dirt, paint, glues, heavy oxide layers or other surface contaminants.

Soldering fluxes are chemical compounds such as muriatic acid (dilute hydrochloric acid), sal ammoniac (ammonium chloride) or rosin. Brazing fluxes are chemical compounds such as fluorides, chlorides, boric acids and alkalis. These compounds react to dissolve, absorb or mechanically break up thin surface oxides that are formed as the parts are being heated. Fluxes must be stable and remain active through the entire temperature range of the solder or braze filler metal. The chemicals in the flux react with the oxides as either acids or alkalis (bases). Some dip fluxes are salts.

The reactivity of a flux is greatly affected by temperature. As the parts are heated to the soldering or brazing temperature, the flux becomes more active. Some fluxes are completely inactive at room temperature and most have a temperature range within which they are most effective. Care should be taken to avoid overheating fluxes as this will stop them working as fluxes, and they become a contamination in the joint. If overheating has occurred, stop and clean off the damaged flux before continuing.

Fluxes that are active at room temperature must be neutralized (made inactive) or washed off after the job is complete. If these fluxes are left on the joint, corrosion and premature failure may result. Fluxes that are inactive at room temperature may not have to be cleaned off the part. However, if the part is to be painted or auto body filler is to be applied, fluxes must be removed.

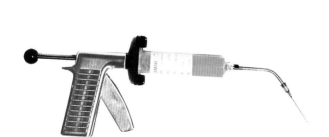

FIGURE 5.27 Gun for injecting flux into a joint

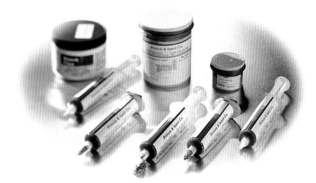

FIGURE 5.28 Tubes that contain flux filler metal mixtures

Flux Residue Removal

Safeguard against corrosive attack and poor joint weld appearance by removing ALL flux residue immediately once the welded component is cooled.

- *Aluminium and its alloys.* Place the welded parts in hot water and scrub vigorously to remove flux residues. Dip the parts into a 5 per cent nitric acid solution. Wash off the acid solution in hot water and dry. Test for complete removal of the flux residue by the use of an indicator (acidified 5 per cent silver nitrate solution) which will react visibly if residue is still present.
- *Magnesium alloys.* Allow to cool and after washing in water, chromate the part in a hot 5 per cent potassium dichromate solution in water.
- *Copper and brass.* Brush the parts after immersing in boiling water. The glass-like deposits may be removed by dipping the part into a suitable acid solution – pickling, followed by rinsing in water.
- *Stainless steel.* Immerse the article in boiling 5 per cent caustic soda solution. Wash in hot water to remove the effects of caustic soda. Alternatively, descale in a solution of equal parts hydrochloric acid and water, with the addition of 5 per cent of the total volume of nitric acid. A restrainer (0.25 per cent of the total volume) should be added.
- *Cast iron.* Flux residues are easily removed by scraper or wire brush, immediately after welding or brazing, while the residue is still hot.

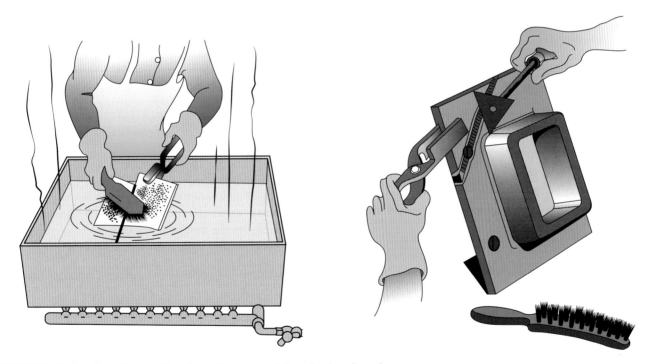

FIGURE 5.29 Cast iron flux residue is easily removed by wire brush and scraper

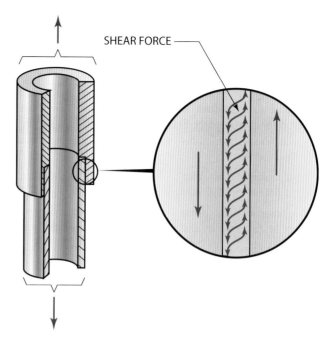

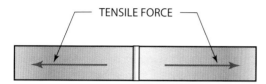

FIGURE 5.30 **Tensile strength of joint**

FIGURE 5.31 **Effect of shear force on a joint**

Physical Properties of the Joint

The tensile strength of a joint is its ability to withstand being pulled apart, Figure 5.30. A brazed joint can have a tensile strength four to five times higher than the filler metal itself.

As the joint spacing decreases, the surface tension increases the tensile strength of the joint. The shear strength of a brazed joint is its ability to withstand a force parallel to the joint, Figure 5.31. For a solder or braze joint, the shear strength depends upon the amount of overlapping area of the base parts. The greater the area that is overlapped, the greater is the strength.

The ductility of a joint is its ability to bend without failing. Most soldering and brazing alloys are ductile metals, so the joint made with these alloys is also ductile.

The fatigue resistance of a metal is its ability to be bent repeatedly without exceeding its elastic limit and without failure, Figure 5.32.

The corrosion resistance of a joint is its ability to resist chemical attack. It is determined by the compatibility of the base materials to the filler metal, as discussed in this chapter. Using filler metals on base materials that are not recommended may result in a joint that looks good but will eventually corrode. For example, a brass brazing rod that contains copper (Cu) and zinc (Zn) will make a nice looking joint on stainless steel. But the zinc in the brass will combine with the nickel in the stainless steel if the part is kept hot for too long, forming a corrosive, brittle structure and reducing strength in the joint.

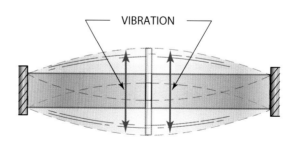

FIGURE 5.32 **Fatigue resistance**

Joint Design

The spacing between the parts being joined greatly affects the tensile strength of the finished part. As the parts are heated, the initial space may increase or decrease, depending upon the joint design and fixturing. The strongest results are obtained when the parts use lap or scarf joints, where the joining area is equal to three times the thickness of the thinnest joint member, Figure 5.33. The strength of a butt joint can be improved if the area being joined is increased.

Some joints can be designed so that the flux and filler metal may be preplaced. When this is possible, visual checking for filler metal around the outside of the joint is easy and indicates an acceptable joint.

Joint cleaning is very important to a successful soldered or brazed part. The surface must be cleaned of all oil, dirt, paint, oxides, or any other contaminants.

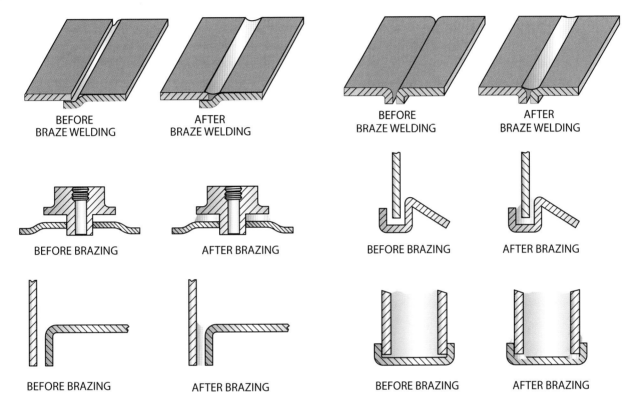

FIGURE 5.33 **Examples of brazing and braze welded joints**

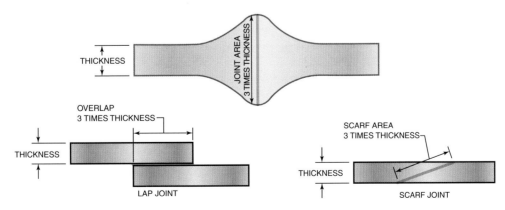

FIGURE 5.34 **Joint design**

SUMMARY

Oxy-acetylene welding is one of the most versatile methods of joining a wide range of materials. It is mobile and to many represents the culmination of observation, manipulation and hand and eye co-ordination. It can be both frustrating and rewarding, and is the ideal method to introduce new learners to welding. Welding is very much about developing a feel, and these skills are transferrable to other disciplines, where they are rarely forgotten.

ACTIVITY AND REVIEW

1   Identify four advantages that oxy-acetylene has over other welding processes.

2   List the equipment that you would need to set up an oxy-acetylene plant.

3   Identify how a flashback arrestor works.

4   State why it is good practice to purge the lines prior to use.

5   Name three types of regulator.

6   State why the pressure at the nozzle is always lower than that shown on the gauge.

7   Indicate why the regulator pressure adjusting screw should be backed off each time the oxy-fuel system is being shut down.

8   State why you should never use oil on regulators.

9   List four possible causes of backfiring.

10   Name the type of safety pressure release that reseals once the excessive pressure has been released.

11   State what colour cylinder and thread pattern is associated with acetylene gas.

12   Identify the procedure to be carried out if a nozzle overheats.

13   State the ideal distance between the inner cone and the metal.

14   Indicate where the flame should be directed when heating a lap joint.

15   Draw a simple diagram to show Leftward Technique indicating the relationship of the filler wire, torch and solidifying weld metal and include slope and tilt angles.

16   List five advantages of brazing.

17   Identify the difference between brazing and braze welding.

18   Name the three main functions of a flux.

19   Identify the factors that affect the selection of a suitable filler metal.

20   State what the suffix 'F' mean with regards to filter lenses.

21   Name the potential weld problems associated with a T-fillet weld.

22   Describe the procedure adopted in terminating a weld.

23   State the flame temperature for a neutral flame.

24   Name a typical application for an oxidizing flame.

25   State what features you would expect to find on a good butt joint.

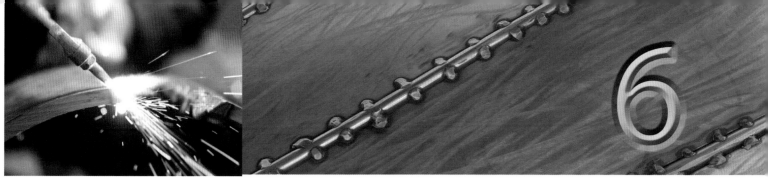

Thermal Cutting Techniques

LEARNING OBJECTIVES

After completing this chapter, you should be able to:

- describe the oxy-fuel gas cutting process and list the most commonly used fuel gases
- determine the correct size and type of cutting nozzle for a specific job
- set up, light and clean the nozzle of a cutting torch
- explain the exothermic process that takes place during the burning away of the metal when an oxy-fuel gas cutting torch is used
- explain what the kerf surface can reveal about what was correct or incorrect with the pre-heat flame, cutting speed and oxygen pressure
- make a machine cut and then evaluate the results
- make a manual flat, straight cut in thin plate, thick plate and sheet metal
- make a flame-cut hole
- explain how a plasma cutting torch works
- set up and use a plasma cutting torch.

UNIT REFERENCES

NVQ:

Complying with Statutory Regulations and Organizational Safety Requirements.

Cutting Materials using Hand Operated Thermal Cutting Equipment.

Cutting and Shaping Materials using Gas Cutting Machines.

VRQ:

Engineering Environment Awareness.

Producing Components from Metal Plate.

Thermal Cutting Techniques.

KEY TERMS

ampere a unit of electrical current.

coupling distance the distance to be maintained between the inner cones of the cutting flame or plasma cutter and the surface of the metal being cut.

cutting tip the part of an oxygen cutting torch from which the gases issue.

delamination an imperfection such as oxide formation that develops inside the material, without being obvious on the surface which will open up in the presence of heat.

drag (thermal cutting) the off-set distance between the actual and straight line exit points of the gas stream or cutting beam measured on the exit surface of the base metal.

drag lines high-pressure oxygen flow during cutting forms lines on the cut faces. A correctly made cut has up and down drag lines (zero drag); any deviation from the pattern indicates a change in one of the variables affecting the cutting process; with experience the welder can interpret the drag lines to determine how to correct the cut by adjusting one or more variables.

dross a mass of solid impurities (iron oxide) attached on the underside of a cut.

electrode setback the distance the electrode is recessed behind the constricting orifice of the plasma arc torch, measured from the outer face of the nozzle.

electrode tip the end of a welding electrode that is closest to the work.

exothermic reaction a reaction in which heat is given off as in oxy-fuel cutting.

groove an opening or a channel in the surface of a part or between two components, that provides space to contain a weld.

hard dross a form of dross caused by oxy-fuel cutting that is difficult to remove.

heat-affected zone the area of base material which has had its microstructure and properties altered by welding or heat intensive cutting operations.

high-speed cutting tip a special cutting tip usually constructed in two pieces that is designed for mechanized oxy-fuel cutting operations.

ionized gas see Plasma.

joules the SI unit of energy.

machine mounted cutting torch cutting with equipment that requires manual adjustment of the equipment controls in response to visual observation of the cutting operation, with the torch held by and controlled by a mechanical device such as a profile mechanism.

nozzle the end piece of a welding torch or gun that directs the flow of gas.

nozzle insulator a non-conductive piece that separates the nozzle from the contact tube.

nozzle tip the end of a nozzle, sometimes replicable on heavy duty MIG, MAG or FCAW equipment.

oxy-fuel gas cutting a group of oxygen cutting processes that use heat from an oxy-fuel gas flame.

plasma a gas that has been heated to an at least partially ionized condition, enabling it to conduct an electric current.

plasma arc gouging a thermal gouging process that removes metal by melting with the heat of a plasma arc torch.

pre-heat flame brings the temperature of the metal to be cut above its ignition point, after which the high-pressure oxygen stream causes rapid oxidation of the metal to perform the cutting.

pre-heat holes the cutting tip has a central hole through which the oxygen flows. Surrounding this central hole are a number of other holes called pre-heat holes. The differences in the type or number of pre-heat holes determine the type of fuel gas to be used in the tip.

soft dross the most porous form of dross which is most easily removed.

stack cutting thermal cutting of stacked metal plates arranged so that all the plates are severed by a single cut.

standard an established norm or requirement. A technical standard is usually a formal document that establishes uniform engineering or technical criteria, methods, processes and practices.

stand-off distance the distance between a nozzle and the workpiece.

tip cleaners tools used to clean the orifices of oxy-fuel torch tips made of abraded steel wires of specific sizes.

volt a unit of electrical pressure.

water table a special table designed for plasma arc cutting operations, where the torch head is submerged under water in order to reduce smoke and noise.

INTRODUCTION

Thermal cutting covers a wide range of processes for cutting materials. It includes modern lasers and water jet cutting and also more traditional processes such as oxy-fuel and plasma cutting. In this chapter we will focus on the latter two processes as these are readily available in fabrication and welding workshops worldwide and are the most predominant methods of material removal and weld preparation.

OXY-FUEL GAS CUTTING

Oxy-fuel gas cutting describes a group of processes that use an oxy-fuel gas flame to heat metal to its ignition temperature and bring about oxidation. A high-pressure stream of oxygen is then directed onto the metal. The pre-heated area rapidly becomes oxidized and the oxide is blown away by the pressure of the jet of oxygen, severing the metal. The ignition temperature of steel is 900–950°C.

The processes are identified by the type of fuel gas used with oxygen to produce the pre-heat flame. Oxy-fuel gas cutting is most commonly performed with oxy-acetylene or propane. Both gases produce good quality cuts. For general purpose use acetylene is the most economical and versatile but requires more safety precautions due to the unstableness of the gas. In situations where storage and use could be hazardous, propane is often the selected choice. Most welders use the oxy-fuel cutting torch more than any other cutting process, and it is extensively used in manufacturing, maintenance, automotive repair, railway work and agriculture. Unfortunately, it is also one of the most misused processes. Most workers know how to light the torch and make a cut, but their cuts are of very poor quality and they use unsafe torch techniques. A good oxy-fuel cut should be straight and square and should require little or no post-cutting clean up.

Manual, mechanized and computer numerical controlled (CNC) processes are used in industry. Hand-controlled, manual cutting is used for short-run production, one-off fabrication, demolition and scrapping operations on site for steel construction. Mechanized or CNC cutting is widely used in production work where a large number of identical cuts must be made or where great precision is required. More than one cutting head may be mounted so that several cuts can be made at the same time.

Various oxy-fuel cutting specialities exist, including flame cutting, gouging, beveling, washing and scarfing.

Oxy-fuel gas cutting is used to cut iron based alloys. Low-carbon steels (up to 0.3 per cent carbon) are easy to cut. Any metal that requires pre-heating for welding, such as high strength and high alloy carbon steels, should also be pre-heated before cutting. Failure to pre-heat some high-strength alloys before cutting can result in a very thin hard zone on the cut surface which can cause cracks to start in the finished part. If high-strength steel flame-cut parts are bent, formed or welded, the hardened edge may cause cracks to form which can cause the part to fracture and fail. High-nickel steels, cast iron, and stainless steel are more difficult to cut. Most non-ferrous metals – such as brass, copper and aluminum – cannot be cut using oxy-fuel gas cutting and plasma cutting is used instead.

Personal Protective Equipment

As well as flame retardant overalls, boots and gauntlets, the operator should consider wearing additional protective clothing. A spark can travel up to thirty feet, and additional leather protection in the form of full welding jacket, apron, spats and leggings and a skull cap is recommended. Metal tongs should be available for handling small cut components with cooling facilities nearby.

Goggles or other suitable eye protection must be worn for flame cutting. Goggles should have vents near the lenses to prevent 'fogging' or steaming up. Cover lenses should be provided to protect the filter lens, which must be marked so that the shade number can be readily identified, Table 6.1. Recent developments in head shields now incorporate filter lens ranges for oxy-fuel gas cutting applications.

TABLE 6.1 A general guide for the selection of eye and face protection equipment

Type of Cutting Operation	Hazard	Suggested Shade Number
Light cutting, up to 2.54cm	Sparks, harmful	3 or 4
Medium cutting, 2.54-15.24cm	rays, molten metal,	4 or 5
Heavy cutting, over 15.24cm	flying particles	5 or 6

Cutting Torches

The cutting torch may be part of a combination welding and cutting torch set, or a cutting torch only, Figure 6.1. The combination welding / cutting torch offers more flexibility because a cutting head, welding nozzle, flame cleaning or heating nozzle can be attached quickly to the same torch body, Figure 6.2. Combination torch sets are often used in colleges, automotive repair workshops, welding workshops and for any job where multi-purpose equipment is needed. A cut made with either type of torch has the same quality, although the dedicated cutting torches are usually longer and have larger gas flow passages. This added length helps keep the operator further away from the heat and sparks and may allow thicker material to be cut.

Oxygen is mixed with the fuel gas to produce a high-temperature pre-heating flame. The two gases must be completely mixed before they leave the nozzle and burn. Two methods are used

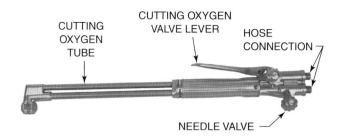

FIGURE 6.1 Dedicated oxygen cutting torch

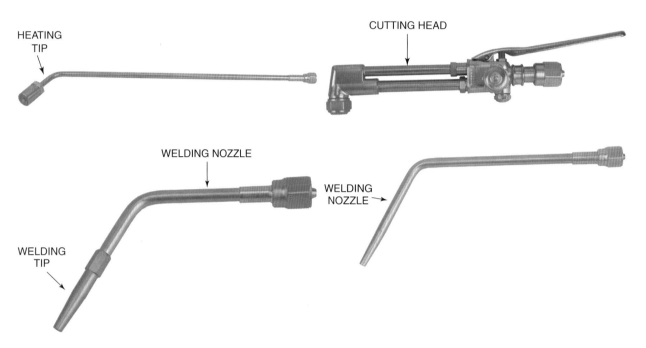

FIGURE 6.2 Torch attachments

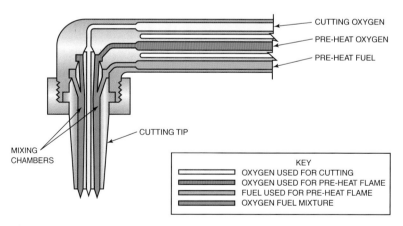

FIGURE 6.3 Mixing chamber located in tip

to mix the gases. One method uses a mixing chamber, and this is the most common in the UK. The mixing chamber may be located in the torch body or in the nozzle, Figure 6.3. Torches that use a mixing chamber are described as equal-pressure torches, because the gases must enter the mixing chamber under the same pressure. The mixing chamber is larger than both the gas inlet and the gas outlet, which causes turbulence in the gases, resulting in them mixing thoroughly.

The second method uses injector torches, which work with both equal gas pressures and low-fuel gas pressures, Figure 6.4. The injector allows the oxygen to draw the fuel gas into the chamber even if the fuel gas pressure is low. The injector works by using a 'venturi' or slipstream effect, and uses the high pressure oxygen to pull the fuel gases in and mix them.

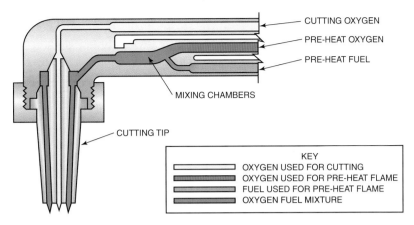

FIGURE 6.4 **Injector mixing torch**

The cutting head normally holds the cutting nozzle at a right angle to the torch body. Torches with the nozzle slightly angled make it easier for the welder to cut flat plate. Torches with a right-angle nozzle make it easier to cut pipe, angle iron, I-beams and other uneven material shapes. Both types of torches can be used to cut, but practice is needed to keep the cut square and accurate. The location of the cutting lever varies from one torch to another, Figure 6.5. Most cutting levers pivot from the front or back end of the torch body. Personal preference will determine which one a welder uses.

A machine mounted cutting torch, sometimes referred to as a straight line or track cutter, operates in a similar manner to a hand cutting torch. It may require two oxygen regulators, one for the pre-heat oxygen and the other for the cutting oxygen stream. The addition of a separate cutting oxygen supply allows the flame to be more accurately adjusted. It also allows the pressures to be adjusted during a cut without disturbing the other parts of the flame. Various machine cutting torches are shown in Figures 6.6–6.8.

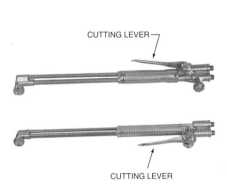

FIGURE 6.5 **Cutting lever**

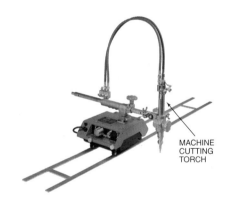

FIGURE 6.6 **Portable oxy-fuel cutting machine**

FIGURE 6.7 **Multiple-head cutting machine**

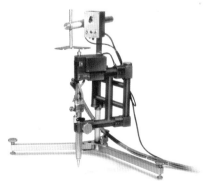

FIGURE 6.8 **Portable cutting machine for highly complex shapes**

Cutting Nozzles

Most cutting nozzles are made of copper alloy, but some have a chrome finish. Chrome plating helps prevent spatter from sticking to the nozzle, thus prolonging its useful life. Nozzle designs change for the different types of uses and gases, and from one torch manufacturer to another.

Nozzles used with acetylene (ANM) are a one piece nozzle with six or more outer holes for pre-heating and a larger central hole for the oxygen stream. Propane nozzles (ANP) are of a two-part construction consisting of a fluted or grooved brass inner part with a central hole for the oxygen stream. This fits into an outer copper surround, providing the pre-heating ring.

A quality cut can only be produced with well-maintained equipment. This process expels small particles of molten metal that could easily block the ports of the nozzle, which should be inspected regularly throughout the cutting process. The acetylene nozzle face can be cleaned by using the file part of a set of nozzle reamers, Figure 6.9, or by using light abrasion with emery cloth on a flat surface, ensuring that the nozzle maintains a 90° angle to the abrasive surface. This should be followed by cleaning the nozzle ports with nozzle reamers/nozzle cleaners, choosing the appropriate size to clean the bore in a twisting motion, Figure 6.10, it should fit easily in to the bore without undue pressure.

When this has been completed, open the oxygen valve momentarily to release anything that may have been loosened during cleaning. See Figure 6.11. When the torch has been lit and adjusted, observe the shape and overall impression of the pre-heating ports and central oxygen stream orifice. Poorly maintained nozzles produce uneven and insufficient pre-heating of the plate, 'kerf width' (width of cut) reduction, and overheating of the nozzle.

When changing nozzles, refrain from tapping the nozzle to free it from its seating. The correct procedure is to use a plastic mallet to the back of the torch housing, not the side of the nozzle, to free it, as shown in Figure 6.12.

HEALTH & SAFETY ⊙

The cutting torch must NEVER be used as a hammer to release severed metal as this can damage the nozzle seating and lead to gas leakage and a potentially dangerous situation.

FIGURE 6.9 **Filing the end of the tip flat**

FIGURE 6.10 Tip cleaner

FIGURE 6.11 **Opening the oxygen valve**

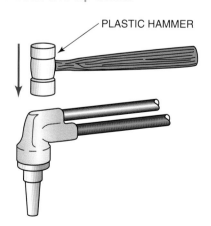

PLASTIC HAMMER

FIGURE 6.12 **Removing a tip that is stuck**

For light cutting of sheet or thin plate there is a sheet nozzle (ASNM) which has one pre-heating orifice and the cutting orifice. The nozzle is shaped with a standard stand-off distance for the cutting oxygen. This gives a very fine cut, ideal for bodywork, thin sheet or box section up to 3mm, see Figure 6.13. Always inspect the seating of the nozzle for damage, if there are indentations or any signs of corrosion do not use. Any deformation at the seating could result in a flashback or potential fire hazard.

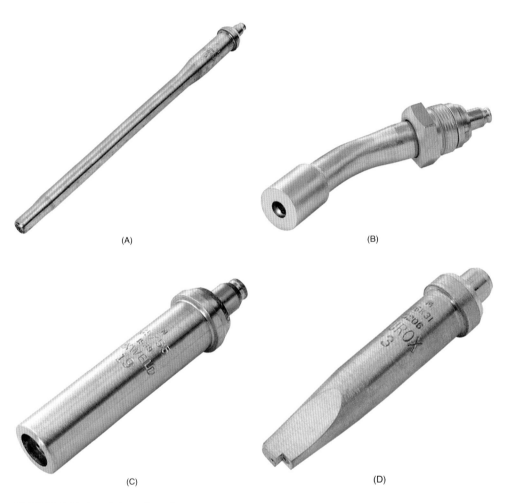

(A)

(B)

(C)

(D)

FIGURE 6.13 Special cutting tips

All equipment manufacturers produce data sheets recommending nozzle size and gas pressures required for each thickness of material as a guide to cutting. As the plate thickness increases, so the size of nozzle will increase.

A number of factors determine gas pressures, including the equipment manufacturer, the condition of the equipment, and hose length and diameter. In all cases, start out with the pressure recommended by the manufacturer of the equipment being used. Adjust the pressure to fit the particular application.

A wide variety of nozzle shapes are available for specialized cutting jobs. Each nozzle, also comes in several sizes, see Figures 6.14 and 6.15.

- *Gouging nozzles.* These are designed to produce a neat controlled back gouging of welds and defects as well as removing cleats, lugs and weld defects. They incorporate a hard wearing tungsten shoe that prolongs the life of the nozzle when running along a gouge and makes the operation easier.

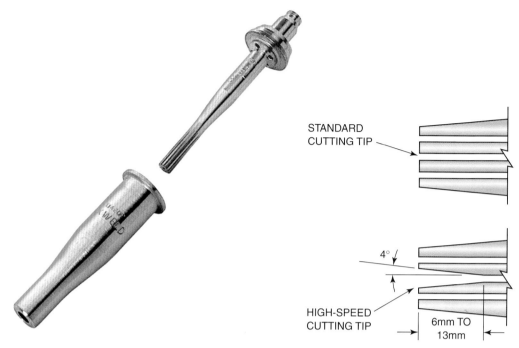

STANDARD
CUTTING TIP

4°

HIGH-SPEED
CUTTING TIP

6mm TO
13mm

FIGURE 6.15 Comparison of standard and high-speed cutting tips

FIGURE 6.14 Parts of a two-piece cutting tip

- *Rivet cutting nozzles.* These are designed with a curved flattened portion of the nozzle in which there are three holes, it is used for removing the heads of rivets and cleats.
- *Rivet washing nozzles.* These are an alternative to the rivet cutting nozzle. They look somewhat similar to the normal cutting nozzle but with a very large central hole designed to blow away the molten oxide in the centre of the rivet.
- *Washing and scarfing nozzles.* Specially shaped cutting nozzles are available that are designed to remove metal from a flat surface or from fillets or contoured joints, although it is also common to use a regular cutting nozzle for some washing or scarfing operations, Figure 6.16 and Figure 6.17.

Different means are used to attach the cutting nozzle to the torch head, the majority are a push fitting with locking collar nut. There are also different designs used for manual, mechanized, and CNC cutting machine nozzles. Mechanized and CNC cutting tips are designed for high-speed cutting with high-speed oxygen flow. Always choose the correct type and size of nozzle for each

FIGURE 6.16 Scarfing tips

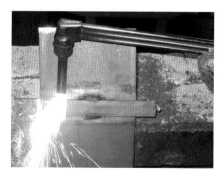

FIGURE 6.17 Washing operations

cutting application. Check the manufacturer's literature for recommendations. Make sure the nozzle is designed for the type of fuel gas being used, and always inspect the nozzle before using it.

The setting of the pre-heat flame required to make a perfect cut is determined by the type of fuel gas used and the material thickness. Materials that are thick, round or have surfaces covered with rust, paint, oil, etc. require more pre-heat flame.

Health and Safety

Propane and natural gas should be used in two-piece nozzles that are typically deeply recessed. The flame burns at such a slow rate that it may not stay lit on any other nozzle.

To check the assembled torch nozzle for a good seal, open the oxygen valve and spray the nozzle with an oil free leak-detecting solution. If the cutting nozzle seat or the torch head seat is damaged, it can be repaired by a recognized supplier. In some cases a repair may not be economical.

Cutting Table

Because of the nature of the manual cutting process, special consideration is given to the flame cutting support. Any piece being cut should be supported so the flame will not cut through it into the table. Special cutting tables are used that expose only a small metal area to the flame. Some tables use parallel steel bars of metal and others use cast iron pyramids. All cutting should be set up so the flame and oxygen stream runs between the support bars or over the edge of the table.

If an ordinary welding table or another steel table is used, special care must be taken to avoid cutting through the tabletop. You may be able to support the piece being cut above the table with a firebrick. Another method is to cut the metal over the edge of the table.

Torch Guides

In manual torch cutting, a guide or support is frequently used to allow for better control and more even cutting. It takes a very skilled welder to make a straight, clean cut even when following a marked line and it is even more difficult to make an accurate radius cut. Guides and supports allow the height and angle of the torch head to remain constant. The speed of the cut, which is important to making a clean, even kerf, must be controlled by the welder.

Since the torch must be held in an exact position to make an accurate cut, the welder normally supports the torch weight with the hand. This allows for more accurate work and cuts down on fatigue. A rest, such as a firebrick, can also be used to support the torch.

Various types of guides can be used to keep the torch in a straight line. Figure 6.18 shows an angle iron guide. The edge of the angle is followed to give the straight cut. Bevel cuts can be made freehand with the torch, but it is very difficult to keep them uniform. More accurate bevel cuts are made by resting the torch against the angle side of an angle section.

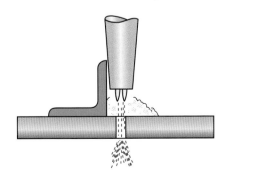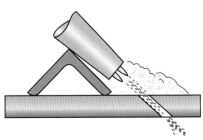

FIGURE 6.18 Using angle irons to aid in making cuts

FIGURE 6.19 Devices used to improve hand cutting

Special roller guides can also be attached to the torch head, which hold the cutting nozzle at an exact height. A circle-cutting attachment is used to cut circles. Figure 6.19 shows how the attachment fits on the torch head. The radius can be preset to any required size. The cutter revolves around the centre point when the cut is made. The roller controls the height of the torch tip above the plate surface.

OXY-FUEL CUTTING, SETUP AND OPERATION

Setting up a cutting torch system is exactly like setting up oxy-fuel welding equipment described in Chapter 5, except for the adjustment of gas pressures.

TIP ●

Acetylene and propane fittings have a left-hand screw thread – oxygen and non-flammable gases have a right-hand screw thread.

Setting Up a Cutting Torch

In order to set up a cutting torch, the following stages should be carried out:

1 The oxygen and acetylene cylinders must be securely chained to a trolley or wall before the safety collars are removed.

2 After removing the safety collar, stand to one side and crack (open and quickly close) the cylinder valves, making sure there are no sources of possible ignition nearby. This blows out any dirt in the valves.

3 Visually inspect all of the parts for any damage and repair or cleaning requirements. Do not use fingers for this purpose as you could pick up a brass splinter which could cause blood poisoning.

4 Inspect the regulators to make sure they are appropriate for the gases to be used. Also check for signs of damage or pitting especially on the seating. Attach the regulators to the cylinder valves and tighten them securely with an appropriate spanner.

5 Check the flashback arrestors for function prior to fitting, by blowing through to make sure they work correctly. Check the condition of the high pressure gas hoses. Attach to the regulator and connect hoses with the hose check valve connected to the torch to ensure correct gas flow, tighten up using the correct spanner. Acetylene to the left-hand thread (notched nut) of the torch. If the hose check valve is connected to the flashback arrestor end, instead of the torch end, no gas will pass through the hoses.

6 If the torch you will be using is a combination torch, attach the cutting head.

7 Fit a cutting nozzle to the torch.

8 Before you open the cylinder valves, back off the pressure-regulating screws so that when the valves are opened the gauges will show zero working pressure.

9 Stand to one side of the regulators as you open the cylinder valves slowly.

10 The oxygen and acetylene valves are opened with the cylinder key remaining on the acetylene cylinder. This is so that in an emergency it can be closed quickly.

11 Open the fuel gas torch valve and then turn the regulating screw in slowly and set the recommended gas pressure (approximate working pressure) showing on the working pressure gauge. Open the valve and adjust to get true working pressure and then close the valve. This also allows the line to be completely purged.

12 When using a combination welding and cutting torch, the oxygen valve nearest the hose connection must be opened before the flame-adjusting valve or cutting lever will work.

13 Open the torch valve and set the recommended pressure for oxygen – showing on the working pressure gauge. Allow the gas to escape so that the line is completely purged.

14 Open the oxygen valve and adjust using the procedure previously described.

15 Be sure there are no sources of possible ignition nearby.

16 With both torch valves closed, spray an oil free leak-detecting solution on all connections, especially on cutting nozzle seatings and the cylinder valves. Tighten any connection that shows bubbles, but do not overtighten. Remember to allow the leak detection fluid to go clear, as the gas used to expel the fluid from the can could give the misleading impression of a gas leak. See Figure 6.20.

Lighting the Torch

Wearing welding goggles, gloves and any other required personal protective clothing, and with a cutting torch set that is safely assembled, you will light the torch.

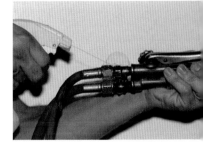

FIGURE 6.20 Leak-checking gas fittings

1 Set the regulator working pressure for the nozzle size. If you do not know the correct pressure for the nozzle, start with the fuel set at 5psi (0.35 bar pressure) and the oxygen set at 25psi (1.8 bar pressure).

2 Point the torch nozzle upward and away from any equipment or other people.

3 Open just the acetylene valve and use only a spark lighter or striker to ignite the fuel gas. The torch may not stay lit. If it goes out, close the valve slightly and try to relight the torch.

4 If the flame is small, it will produce heavy black soot and smoke. In this case, increase the gas flow till the smoke disappears.

5 With the fuel gas flame burning smoke-free, slowly open the oxygen valve and adjust the flame to a neutral setting, Figure 6.21.

6 When the cutting oxygen lever is depressed, the flame may become slightly carbonizing. This can occur because the high flow of oxygen through the cutting orifice causes a drop in line pressure.

7 With the cutting lever depressed, re-adjust the pre-heat flame to a neutral setting. The flame will become slightly oxidizing when the cutting lever is released. Since an oxidizing flame is hotter than a neutral flame, the metal being cut will be pre-heated faster. When the cut is started by depressing the lever, the flame automatically returns to the neutral setting and does not oxidize the top of the plate. Extinguish the flame by first turning off the fuel gas and then the oxygen.

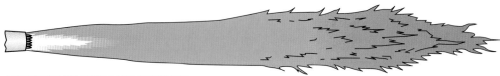

ACETYLENE BURNING IN ATMOSPHERE–
OPEN FUEL GAS VALVE UNTIL SMOKE CLEARS FROM FLAME.

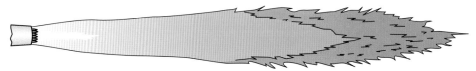

CARBURIZING FLAME–
(EXCESS ACETYLENE WITH OXYGEN). PRE-HEAT FLAMES REQUIRE MORE OXYGEN.

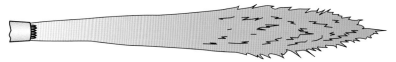

NEUTRAL FLAME–
(ACETYLENE WITH OXYGEN). TEMPERATURE 6300° F. PROPER PRE-HEAT ADJUSTMENT FOR ALL CUTTING.

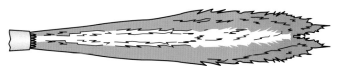

NEUTRAL FLAME WITH CUTTING JET OPEN–
CUTTING JET MUST BE STRAIGHT AND CLEAR.

OXIDIZING FLAME–
(ACETYLENE WITH EXCESS OXYGEN). NOT RECOMMENDED FOR AVERAGE CUTTING.

FIGURE 6.21 Oxy-acetylene flame adjustments for the cutting torch

Layout

A line to be cut can be laid out with a piece of French chalk or a chalk line. To obtain an accurate line, a scribe or a centre punch can be used as this makes it easier to see when cutting. See Figure 6.22 on the next page. If a piece of French chalk is used, it should be sharpened properly to increase accuracy, Figure 6.23.

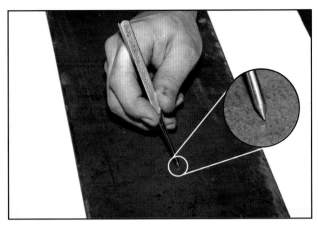

FIGURE 6.22 Punching

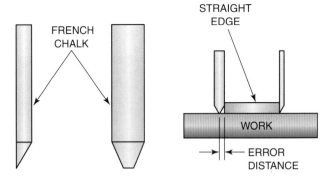

FIGURE 6.23 Proper method of sharpening a french chalk

The Chemistry of a Cut

The oxy-fuel gas cutting process relies upon the principle that when a body is heated, oxygen from the atmosphere is attracted to it and a chemical reaction takes place called 'oxidation'. This reaction takes place when material is heated to ignition temperature, which for steel is 900–950°C. If we then direct a stream of high pressure oxygen on to the hot steel, a magnetic oxide of iron is produced and blown away by the cutting oxygen stream. This process is often referred to as a chemical severance process – an exothermic reaction in which heat is given off. The oxy-fuel gas cutting process will work well on any metal that will rapidly oxidize, such as iron or carbon steel.

On thick sections, once a small spot starts burning (being cut), the heat generated helps the cut continue quickly through the metal. With some cuts, the heat produced may overheat small strips of material being cut from a larger piece. For example, the centre piece of a hole being cut will quickly become red hot and will start to oxidize with the surrounding air, Figure 6.24. This heat produced by the cut makes it difficult to cut out small or internal parts.

Hand Cutting Plate

When a cut is made with a hand torch, the welder must be in a steady position to make the cut as smooth as possible. This involves feeling comfortable and free to move the torch along the line to be cut. It is a good idea to get into position and practice the cutting movement a few times before lighting the torch. Even when the welder and the torch are braced properly, very small physical movements will cause a slight ripple in the cut. Attempting a cut without leaning on the work is tiring and can cause inaccuracies. The torch should be braced with the welder's other hand. It may be moved by sliding it over a supporting hand see Figure 6.25 and Figure 6.26. The torch can also be pivoted on the supporting hand, but care must be taken to prevent the cut from becoming a series of arcs.

A slight forward torch angle helps the flame pre-heat the material, keeps some of the reflected flame heat off the nozzle, aids in blowing dirt and oxides away from the cut, and helps prevent dross from being blown back into the nozzle, Figure 6.27. The forward angle can be used only for a straight-line square cut. If shapes are cut using a slight angle, the part will have beveled sides.

When a cut is made, the inner cones of the flame should be kept 3mm to 5mm for material up to 50mm thick and 6mm for metal 50 to 150mm thick, Figure 6.28. This is sometimes known as the coupling distance. To start a cut on the edge of a plate, hold the torch at a right angle to the surface or pointed slightly away from the edge, Figure 6.29. The torch must be pointed so that the cut is started at the very edge. The edge of the plate heats up more quickly, allowing the cut to be started sooner. Also, fewer sparks will be blown around the workshop. Once the cut is started, the torch should be rotated back to a right angle to the surface or to a slight leading angle.

HEALTH & SAFETY

Some metals and alloys release harmful oxides when they are cut. Extreme caution must be taken when cutting used, oily, dirty or painted metallic material, which can produce dangerous fumes. There may be a need for extra ventilation and a respirator to be safe. Check with the welding workshop supervisor or workshop safety officer or consult the MSDS before cutting any unfamiliar metal or alloy.

TIP

It is advisable to grind back or remove paints or coatings past the heat affected area to reduce the potential for fumes, and repaint or surface coat when the cut has been completed to guard against corrosion.

FIGURE 6.25 Moving the torch – for short cuts, the torch can be drawn over the glove hand

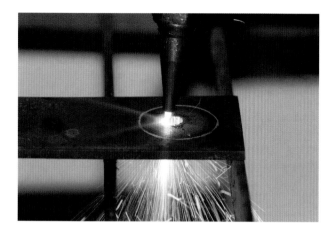

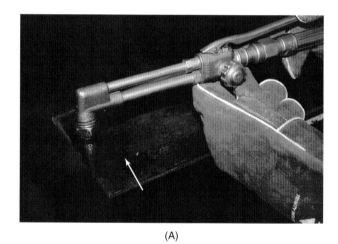

(A)

FIGURE 6.24 Overheating during cutting

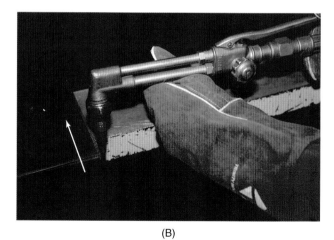

(B)

FIGURE 6.26 Moving the torch – for longer cuts, the torch can be moved by sliding your gloved hand along the plate parallel to the cut (A) start and (B) finish

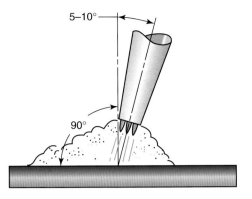

5–10°

90°

FIGURE 6.27 Forward torch angle

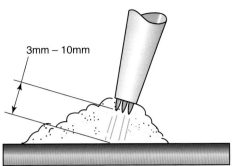

3mm – 10mm

FIGURE 6.28 Inner cone to work distance (coupling distance)

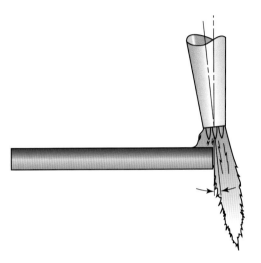

FIGURE 6.29 Starting a cut on the edge of a plate

If a cut is to be started other than at the edge of the plate, the inner cones should be held as close as possible to the metal. Touching the metal with the inner cones will speed up the pre-heat time. When the plate is hot enough to start cutting, the torch should be raised as the cutting lever is slowly depressed. When the metal is pierced, the torch should be lowered again, Figure 6.30. Raising the torch nozzle away from the plate reduces the number of sparks and keeps the nozzle cleaner. If the plate being cut is thick, it may be necessary to move the torch nozzle in a small circle as the hole goes

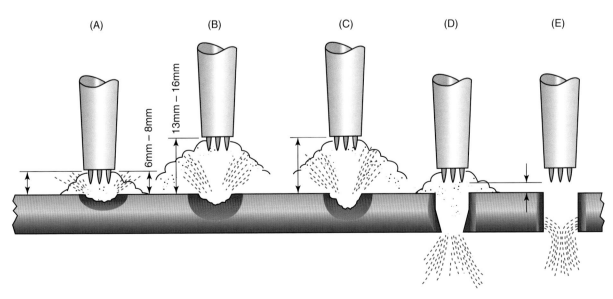

(A)   (B)   (C)   (D)   (E)

6mm – 8mm

13mm – 16mm

FIGURE 6.30 Sequence of piercing plate

through the plate. If the plate is to be cut in both directions from the spot where it was pierced, the torch should be moved backward a short distance and then forward, Figure 6.31. This prevents dross from refilling the kerf (gap) at the starting point and making it difficult to cut in the other direction.

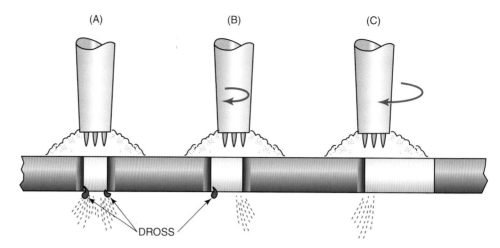

FIGURE 6.31 Cutting in both directions

Starts and stops can be made more easily if one side of the plate being cut is scrap. When it is necessary to stop and reposition before continuing the cut, the cut should be turned out a short distance into the scrap side of the plate, Figure 6.32. This will allow a smoother and more even start with less chance that dross will block the cut. If neither side of the cut is to be scrap, the forward movement should be stopped for a moment before releasing the cutting lever. This action will allow the drag (the distance that the bottom of the cut is behind the top) to be reduced before stopping, Figure 6.33. To restart, use the procedure for starting a cut at the edge of the plate.

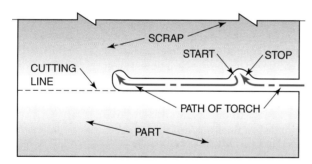

FIGURE 6.32 Turning out into scrap to make stopping and starting points smoother

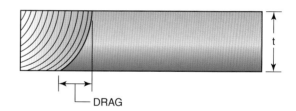

FIGURE 6.33 Drag

Proper alignment of the pre-heat holes will speed up and improve the cut. The holes should be aligned so that one is directly on the line ahead of the cut and another is aimed down into the cut when making a straight-line square cut, Figure 6.34.

When cutting a bevel the flame is directed towards the smaller piece and the sharpest edge. For this reason, the tip should be changed so that at least two of the flames are on the larger plate and none of the flames directed onto the sharp edge, Figure 6.35.

The thicker the plate, the more difficult the cut is to make. Thin plate, 6mm or less, can be cut and the pieces separated even if poor techniques and incorrect pressure settings are used. Thick plate, 13mm or thicker, often cannot be separated if the cut is not correct. Plate that is properly cut can be assembled and welded with little or no post cut clean up. Poor-quality cuts require more time to clean up than it would take to make the adjustments necessary to make a good weld.

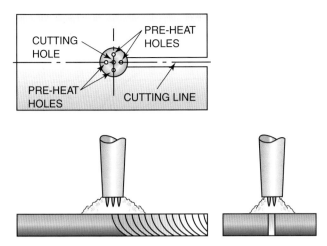

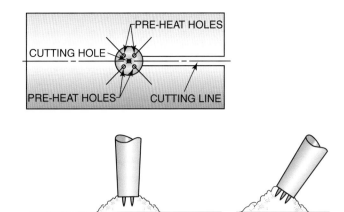

FIGURE 6.34 **Tip alignment for a square cut**   FIGURE 6.35 **Tip alignment for a bevel cut**

Flame Cutting Holes

Using the technique described for piercing a hole, start in the centre and make an outward spiral until the desired hole is achieved, Figure 6.36. The hole should be within + or – 1.5mm of being round and +or- 5degrees of being square. The hole may have dross on the bottom. Repeat this process until small and larger holes can be achieved within tolerance.

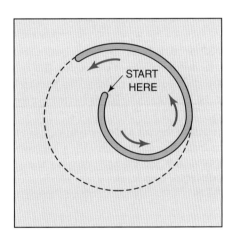

FIGURE 6.36 **Starting a cut for a hole near the middle**

Hand Cutting of Pipe

Freehand pipe cutting may be done in one of two ways. On small-diameter pipe, usually under 75mm, the torch nozzle is held straight up and down and moved from the centre to each side, Figure 6.37. Start the cut at the top of the pipe using the proper piercing technique. Move the torch backward along the line and then forward; this will keep dross out of the cut. If the end of the cut closes in with dross, the oxygen will gouge the edge of the pipe when the cut is continued. Keep the nozzle pointed straight down. When the cut has been progressed as far as it can comfortably go, quickly flip the flame away from the pipe. Restart the cut at the top of the pipe and cut as far as possible in the other direction. Stop and turn the pipe so that the end of the cut is on top and the cut can be continued around the pipe. This technique can also be used successfully on larger pipe.

For large-diameter pipe, 75mm and more, the torch nozzle is always pointed toward the centre of the pipe, Figure 6.38. This technique is also used on all sizes of heavy-walled pipe and can be used on some smaller pipes. The torch body should be held so that it is parallel to the centreline of the pipe, which helps to keep the cut square.

 TIP

An alternative technique that is often used, is to drill a hole of sufficient size on the inner surface of the hole you wish to remove. The flame is then allowed to pre-heat the edge of the hole and follow the internal diameter line.

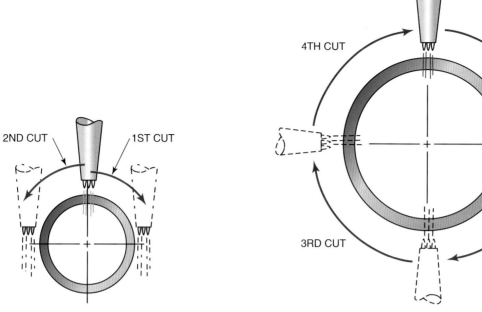

FIGURE 6.37 Cutting small-diameter pipe

FIGURE 6.38 Cutting large-diameter pipe

HEALTH & SAFETY

When pipe is cut, hot sparks can come out of the nearest end of the pipe, causing severe burns. Protect yourself by plugging up the nearest open end of the pipe, put up a barrier to the sparks, or stand to one side of the material being cut.

Distortion

Distortion occurs when the sheet or plate bends or twists out of shape as a result of being heated during the cutting process. If the distortion is not controlled, the end product might be worthless. One method of controlling distortion involves making two parallel cuts on the same plate at the same speed and time, Figure 6.39. Because the plate is heated evenly, distortion is kept to a minimum, Figure 6.40.

A second method involves starting the cut a short distance from the edge of the plate, skipping over short tabs to keep the cut from separating. Once the plate cools, the remaining tabs are cut, Figure 6.41. Another method is to use the zig-zag entry on the scrap metal so that a form of wedge is created which holds the plate in position.

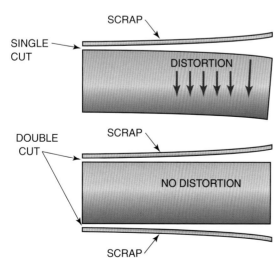

FIGURE 6.39 Making two parallel cuts at the same time to control distortion

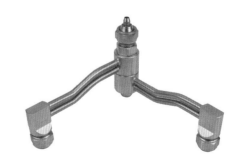

FIGURE 6.40 Slitting adaptor for cutting machine

The adaptor can be used for parallel cuts from 38mm to 500mm. Ideal for cutting test coupons

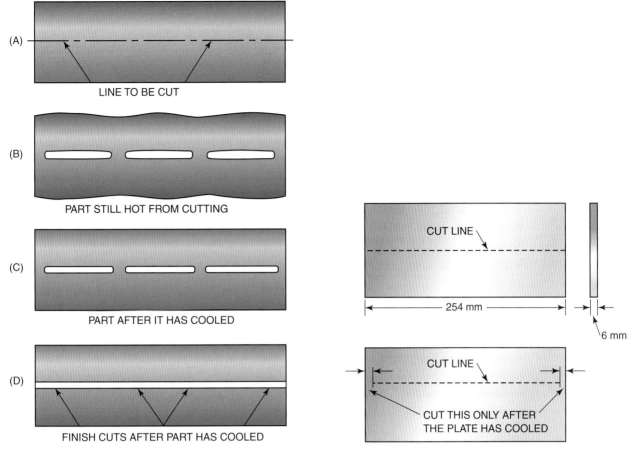

FIGURE 6.41 Steps used during cutting to minimize distortion

FIGURE 6.42 Making two cuts with minimum distortion

Use is often made of *nesting* the components to be cut by using the mass of the material to act in a *chill* effect. This allows the minimal strip between the components to distort, leaving the blanks relatively flat.

Cutting Conditions

Making practice cuts on a piece of metal that will become scrap is a good way to learn the proper torch techniques. Your ability to make a quality cut is affected by:

- *Changing positions*. On large parts, the welder may have to move to complete a cut. Stopping and restarting a cut can result in a small flaw in the cut surface. If this flaw exceeds the acceptable limits, the cut surface must be repaired before the part can be used. To avoid this problem, always try to stop at corners if the cut cannot be completed without moving.
- *Sparks*. Even an ideal cut in a large plate can create sparks that bounce around the plate surface. These sparks often find their way into a glove, or under the arm. With experience you will learn how to angle the torch, direct the cut, and position your body to minimize this problem.
- *Hot surfaces*. As you continue to work on a part, it will begin to heat up and this heat can become uncomfortable. It may be necessary to hold the torch further back from the nozzle, which may affect the quality of the cut, Figure 6.43. Try resting a hand on a block to keep it

TIP ⊙

When making cuts on small pieces of plate, place the metal on top of a large diameter pipe section – this will contain the sparks and molten dross.

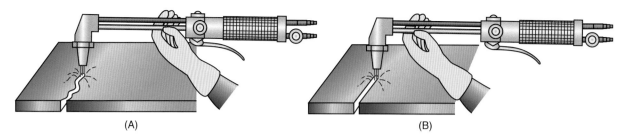

FIGURE 6.43 Bracing the torch

TIP

It is easier to make straight, smooth cuts if you can brace the torch closer to the tip, as in cut B.

off of the plate. Heat buildup may become high enough to affect the cut quality. Heat becomes a problem when it causes the top edge of the plate to melt during a cut, as if the torch nozzle were too large. This can happen when several cuts are being made in close proximity, but the problem can be controlled by planning the cutting sequence and allowing cooling time.

- *Dirty nozzles.* In any cutting, the nozzle will catch small sparks and become dirty or clogged. Stop and clean it. Although time spent cleaning the nozzle reduces productivity, a dirty nozzle will reduce the quality of the cut.

- *Blowback.* As a cut progresses across the surface of a large plate, it may cross supports underneath the plate. During practice cuts this seldom if ever happens, but, depending on the design of the cutting table, it will occur even under the best of conditions in real work. If the support is large the blowback may shower the area with sparks, block the cutting nozzle or cause a major flaw in the cut surface. If the blowback is not clearing quickly, it may be necessary to stop the cut to halt the shower of sparks.

- *Delaminations in plate.* When plate is produced it is passed through pressure rollers which reduce the hot steel ingot down to the required thickness of plate. During this process some surface oxides or impurities can be rolled into the plate and remained trapped within the cooling plate. These imperfections will be revealed if the cutting line lies within them. The oxidation process will stop and the impurities will be expelled as a large volume of sparks or hot metal appearing on the surface of the plate. If this happens, since it is difficult to estimate the size and volume of the defect without non-destructive testing, it would be wise to consider getting alternative material.

Interpretation of a Cut

As a cut progresses along a plate, a record of what happened is preserved along both sides of the kerf. This record indicates to the welder what was correct or incorrect with the pre-heat flame, cutting speed, and oxygen pressure.

The size and number of pre-heat holes in a nozzle has an effect on both the top and bottom edges of the metal. An excessive pre-heat flame results in the top edge of the plate being melted or rounded off and an excessive amount of hard-to-remove dross being deposited along the bottom edge. If the flame is too small, the travel speed must be slower, but reduction in speed may result in the cutting stream wandering from side to side. The torch nozzle can be raised slightly to eliminate some of the damage caused by too much pre-heat, but this causes the cutting stream of oxygen to be less forceful and less accurate.

Speed – The cutting speed should be fast enough that the drag lines have a slight slant backward if the nozzle is held at a 90° angle to the plate, Figure 6.44. If the cutting speed is too fast, the oxygen stream may not have time to go completely through the metal, resulting in an incomplete

FIGURE 6.44 **Correct cut**

FIGURE 6.45 **Poor cut**

FIGURE 6.46 **Poor cut**

FIGURE 6.47 **Poor cut**

cut, Figure 6.45. Too slow a cutting speed results in the cutting stream wandering, causing gouges in the side of the cut, Figures 6.46 and 6.47.

Pressure – A pressure setting that is too high causes the cutting stream to expand as it leaves the nozzle, resulting in the sides of the kerf being slightly dished, Figure 6.48. When the pressure setting is too low, the cut may not go completely through the metal.

Using a variety of nozzles, speeds and oxygen pressures, make a series of cuts on the plate. As each cut is made, listen to the sound it makes. Also look at the stream of sparks coming off the bottom. A good cut should have a smooth, even sound and the sparks should come off the bottom of the plate more like a stream than a spray, Figure 6.49. When the cut is complete, look at the drag lines to determine what was correct or incorrect with the cut, Figure 6.50.

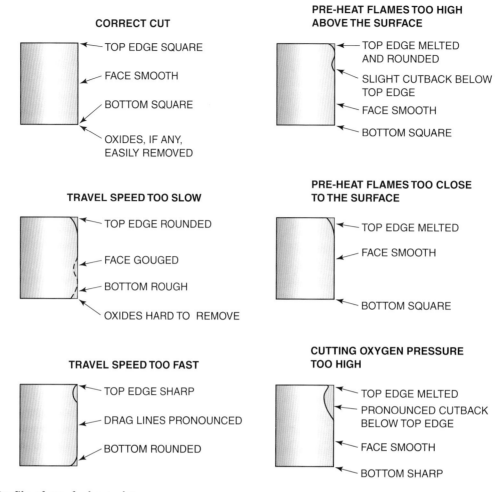

CORRECT CUT

- TOP EDGE SQUARE
- FACE SMOOTH
- BOTTOM SQUARE
- OXIDES, IF ANY, EASILY REMOVED

PRE-HEAT FLAMES TOO HIGH ABOVE THE SURFACE

- TOP EDGE MELTED AND ROUNDED
- SLIGHT CUTBACK BELOW TOP EDGE
- FACE SMOOTH
- BOTTOM SQUARE

TRAVEL SPEED TOO SLOW

- TOP EDGE ROUNDED
- FACE GOUGED
- BOTTOM ROUGH
- OXIDES HARD TO REMOVE

PRE-HEAT FLAMES TOO CLOSE TO THE SURFACE

- TOP EDGE MELTED
- FACE SMOOTH
- BOTTOM SQUARE

TRAVEL SPEED TOO FAST

- TOP EDGE SHARP
- DRAG LINES PRONOUNCED
- BOTTOM ROUNDED

CUTTING OXYGEN PRESSURE TOO HIGH

- TOP EDGE MELTED
- PRONOUNCED CUTBACK BELOW TOP EDGE
- FACE SMOOTH
- BOTTOM SHARP

FIGURE 6.48 Profile of oxy-fuel cut plates

FIGURE 6.49 Good cut showing a steady stream of sparks flying out from the bottom of the cut

FIGURE 6.50 Poor cut

Effect of Flame, Speed and Pressure on a Hand Cut

Two types of dross are produced during a cut. Soft dross is very porous, brittle and easily removed because it contains little or no un-oxidized iron. It may be found on some good cuts. Hard dross may be mixed with soft dross. Hard dross is attached solidly to the bottom edge of a cut, and its removal requires a lot of chipping and grinding. There is 30 per cent to 40 per cent or more un-oxidized iron in hard dross. The higher the un-oxidized iron content, the more difficult the dross is to remove. Dross is found on bad cuts as a result of dirty nozzles, too much pre-heat, too slow a travel speed, too short a coupling distance or incorrect oxygen pressure.

The dross from a cut may be kept off one side of the plate by slightly angling the cut toward the scrap side. The angle needed to force the dross away from the good side of the plate may be as small as 2° or 3°. This technique works best on thin sections; on thicker sections the bevel may show.

WORKSHOP TASK 6.1

Straight Cuts in Plate

Using a properly lit and adjusted cutting torch, with appropriate PPE, cut 12mm strips from low carbon steel in plate from 6 – 12mm thickness, 150mm long and 150mm wide.

- Using a straight edge and French chalk mark several straight lines and pop-mark for greater visibility, allowing sufficient distance for the kerf.
- Starting at one end, make a cut along the entire length of the plate.
- The strip must fall free, be dross free and within + or – 2 mm of a straight line and + or – 5° of being square.
- Repeat this procedure until consistent cuts are made that are straight and dross free.
- Turn off equipment and re-instate the work area cool off any hot components and place in appropriate containers.

  Complete a copy of the 'Student Welding Report'.

WORKSHOP TASK 6.2

Flame Cutting Holes

Using a properly lit and adjusted cutting torch, wearing appropriate PPE, cut 12mm and 25mm holes in 6 – 12mm thick low carbon steel plate.

- Mark out circles using dividers and French chalk and pop-mark for greater visibility.
- Using the technique described for piercing a hole, start in the centre and make an outward spiral until the desired hole diameter is achieved.
- The hole must be within 2mm of being round and + or – 5° of being square.
- The hole may have dross on the bottom.
- Repeat this process until small and large holes within tolerance.
- Turn off equipment, re-instate the work area, cool down all hot components and dispose of scrap material in appropriate container.

  Complete a copy of the 'Student Welding Report'.

WORKSHOP TASK 6.3

Beveling a Plate

Using a properly lit and adjusted cutting torch, and wearing appropriate PPE make a 45° bevel cuts along a plate 150 long by 50mm wide in 6 – 12mm thick low carbon steel.

- Mark the bevel on the plate using French chalk and set the nozzle for beveling and cut using the procedure previously described.
- The bevel should be within 2mm of a straight line and + or – 5° of a 45° angle.
- There may be some soft dross, but no hard dross on the beveled plate.
- Repeat this process until the cuts are within tolerance.
- Turn off equipment, re-instate the work area, cool down all hot components and dispose of scrap material in appropriate container.

  Complete a copy of the 'Student Welding Report'.

WORKSHOP TASK 6.4

Cutting of Pipe

Using a properly lit and adjusted cutting torch and appropriate PPE, cut 13mm rings from a schedule 40 low carbon steel pipe with a 75mm diameter.

- Using a template and a piece of French chalk, mark several rings each 13mm wide and pop-mark these for better visibility, allowing for the kerf.
- Place the pipe horizontally on the cutting table. Start the cut at the top of the pipe using the proper piercing technique.
- Move the torch backward along the line and then forward, this will keep dross out of the cut.
- If the end of the cut closes in with dross, this will cause the oxygen to gouge the edge of the pipe when the cut.
- Keep the nozzle pointed straight down, and when the cut has gone as far as is comfortable quickly, flip the flame away from the pipe.
- Restart the cut at the top of the pipe and cut as far as possible in the other direction.
- Stop and turn the pipe so that the end of the cut on top and the cut can be continued around the pipe.
- When the cut is completed, the ring should fall free.
- To test the accuracy of the cut by place the ring on a flat surface and the sides of the ring should be within + or – 5° of the vertical and have no gaps greater than 2mm.
- Repeat the process until cuts are within tolerance.
- Turn off equipment and re-instate the work area. Cool down all hot components and place scrap material in appropriate containers.

  Complete a copy of the 'Student Welding Report'.

PLASMA ARC CUTTING

The plasma process was developed in the mid-1950s. Early experiments found that when the arc was restricted in a fast-flowing column of argon, *plasma* was formed, which was hot enough to rapidly melt any metal. The problem was that the fast-moving gas blew the molten metal away. Researchers could not find a way to control this scattering of the molten metal, so they decided to introduce this as a cutting process, not a welding process. Several years later, the problem was solved with the invention of the gas lens. Today the plasma arc can be used for plasma arc welding, spraying, cutting and arc gouging.

Plasma

In welding, the word plasma, refers to a state of matter that is found in the region of an electrical discharge. An electric arc is a discharge of current across a gap. The plasma created by an arc is an ionized gas that has electrons and positive ions whose charges are nearly equal.

Plasma is present in any electrical discharge and consists of charged particles that conduct the electrons across the gap between the work and an electrode. A plasma results when a gas is heated to a high enough temperature to convert into positive and negative ions, neutral atoms, and negative electrons. The temperature of an unrestricted arc is about 6000°C, but the temperature created when the arc is concentrated to form a plasma is about 23,800°C, This is hot enough to rapidly melt any metallic material the arc comes in contact with.

Plasma Torch

A plasma torch supplies electrical energy to a gas to change it into the high-energy state of a plasma. It may be designed for welding or cutting processes but both types of torch have the same basic parts and the plasma is created in essentially the same way.

The body of a manual torch is made of a special plastic that is resistant to high temperatures, ultraviolet light and impact, Figure 6.52. The torch body provides a good grip area and protects the cable and hose connections to the head. Various lengths and sizes are available. Generally, the longer, larger torches are used for the higher-capacity machines; however, sometimes a longer or larger torch will give better control or a longer reach.

The torch head is attached to the torch body where the cables and hoses attach to the electrode tip, nozzle tip and nozzle. The torch and head may be connected at various angles, e.g., 90°, 75° or 180° (straight), or the head can be flexible. The 75° and 90° angles are popular for manual operations and 180° straight torch heads are most often used for machine operations. Because of the heat produced in the head by the arc, some provisions for cooling the head and its internal parts must be made. For low-power torches this cooling may be either by air or by water. Higher-power torches must be liquid-cooled, Figure 6.53. It is possible to replace just the torch head on most torches if it becomes worn or damaged.

Most hand-held torches have a manual power switch that is used to start and stop the power source, gas and cooling water (if used). Usually, this is a thumb switch on the torch body, but it may be a foot control or be located on the panel for machine equipment. The thumb switch may

TIP ▶

On machine torches the body is often called a barrel and may come with a 'rack' attached to its side, which is a flat gear that allows the torch to be raised and lowered manually to the correct height above the work.

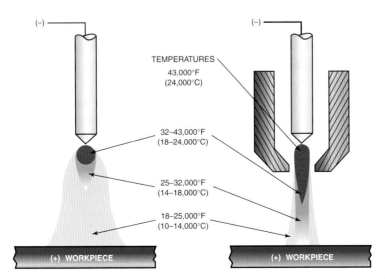

FIGURE 6.51 Approximate temperature differences between a standard arc and a plasma arc

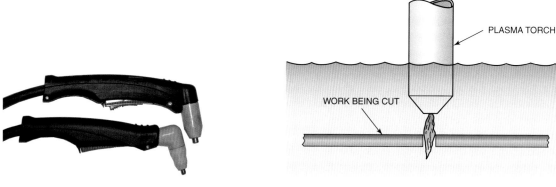

FIGURE 6.52 Hand-held torches

FIGURE 6.53 Water tables

be moulded into the torch body or attached to the torch body with a strap clamp. The foot control must be rugged enough to withstand the welding workshop environment. Some equipment has an automatic system that starts the plasma when the torch is brought close to the work.

The electrode tip, nozzle insulator, nozzle tip, nozzle guide and nozzle must be replaced periodically as they wear out or become damaged from use, Figure 6.54. The metallic parts are usually made out of copper, and they may be plated, which helps them stay spatter-free longer.

The electrode tip is often made of copper alloy with an embedded tungsten or hafnium tip. Copper/tungsten tips in newer torches have improved the quality of work. Copper allows the heat generated at the tip to be conducted away faster, lengthening the life of the tip and allowing better-quality cuts for a longer time. Some earlier torches required the welder to accurately grind the tungsten electrode into shape. If this grinding is necessary, there must be a guide or jig to ensure that the tungsten is properly prepared.

HEALTH & SAFETY

Improper use of the torch or assembly of torch parts may result in damage to the torch body and the need to replace parts frequently.

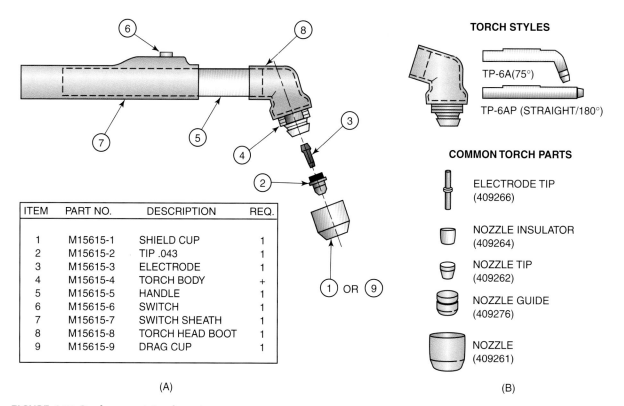

ITEM	PART NO.	DESCRIPTION	REQ.
1	M15615-1	SHIELD CUP	1
2	M15615-2	TIP .043	1
3	M15615-3	ELECTRODE	1
4	M15615-4	TORCH BODY	+
5	M15615-5	HANDLE	1
6	M15615-6	SWITCH	1
7	M15615-7	SWITCH SHEATH	1
8	M15615-8	TORCH HEAD BOOT	1
9	M15615-9	DRAG CUP	1

(A)

TORCH STYLES

TP-6A(75°)

TP-6AP (STRAIGHT/180°)

COMMON TORCH PARTS

ELECTRODE TIP (409266)

NOZZLE INSULATOR (409264)

NOZZLE TIP (409262)

NOZZLE GUIDE (409276)

NOZZLE (409261)

(B)

FIGURE 6.54 Replacement torch parts

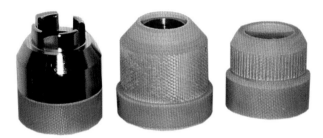

FIGURE 6.55 Nozzle tips

FIGURE 6.56 PAC torch replacement part kits

Other parts are:

- *Nozzle insulator*. This is situated between the electrode tip and the nozzle tip and provides the critical gap spacing and the electrical separation of the parts. The spacing between the electrode tip and the nozzle tip, called electrode setback, is critical to the proper operation of the system.

- *Nozzle tip*. This has a small, cone-shaped, constricting orifice in the centre. The electrode setback space, between the electrode tip and the nozzle tip, is where the electric current forms the plasma. The preset, close-fitting parts restrict the gas in the presence of the electric current so the plasma can be generated, Figure 6.55. Changes in the diameter of the orifice affect the action of the plasma arc. When the setback distance is changed, the arc voltage and current flow change.

- *Nozzle*. Sometimes called the 'cup', this is made of ceramic or any other high-temperature-resistant substance. It helps prevent the internal electrical parts from accidental shorting and provides control of the shielding gas or water injection if they are used, Figure 6.56.

- *Water shroud*. A water shroud nozzle may be attached to some torches. The water surrounding the nozzle tip is used to control the potential hazards of light, fumes, noise or other pollutants produced by the process.

Power and Gas Cables

A number of power and control cables and gas and cooling-water hoses may be used to connect the power supply with the torch, Figure 6.57. The multi-part cable is usually covered to provide some protection to the cables and hoses inside and to make handling the cable easier. The covering is heat resistant but will not prevent damage to the cables and hoses inside if it comes in contact with any hot metallic material or is exposed directly to the cutting sparks.

The power cable must have a high-voltage-rated insulation. It is made of finely stranded copper wire to allow for maximum flexibility of the torch. For all torches, there are two power conductors, one positive (+) and one negative (-). The size and current-carrying capacity of this cable are controlling factors for the power range of the torch. As the capacity of the equipment increases, the cable must be made larger to carry the increased current. Larger cables are less flexible and more difficult to manipulate. To make the cable smaller on water-cooled torches, the cable is run inside the cooling-water return line, which allows a smaller cable to carry more current. The water prevents the cable from overheating.

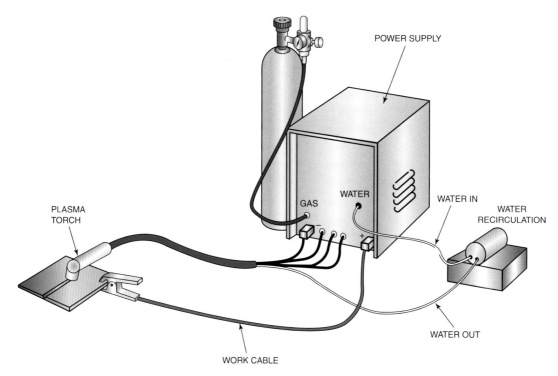

FIGURE 6.57 Typical manual plasma arc cutting setup

There may be two gas hoses running to the torch. One hose carries the gas used to produce the plasma, sometimes called the carrier gas and the other provides a shielding gas. On some small-amperage cutting torches there is only one gas line. The gas line is made of a special heat-resistant, ultraviolet-light-resistant plastic. If it is necessary to replace the tubing because it is damaged, be sure to use tubing provided by the manufacturer or a welding supplier. The tubing must be sized to allow the required gas flow rate within the pressure range of the torch, and it must be free from solvents and oils that might contaminate the gas. If the pressure of the gas supplied is excessive, the tubing may leak at the fittings or rupture.

The control wire is a two-conductor, low-voltage, stranded copper wire that connects the power switch to the power supply, allowing the welder to start and stop the plasma power and gas as needed.

Water tubing may also be used. Medium- and high-amperage torches may be water-cooled. The water for cooling early torch models had to be de-ionized to prevent them arcing out internally, which can destroy or damage the electrode tip and nozzle tip. Refer to the manufacturer's manual to check if a torch requires de-ionized water. If cooling water is required, it must be switched on and off at the same time as the plasma power.

Power Source Requirements

The production of the plasma requires a direct-current (DC), high-voltage, constant-current (drooping arc voltage) power supply. A constant-current machine allows for a rapid start of the plasma arc at the high open circuit voltage and a more controlled plasma arc as the voltage rapidly drops to the lower closed voltage level. The voltage required for most welding operations, such as shielded metal arc, gas metal arc, gas tungsten arc, and flux cored arc welding, ranges from 18 to 45 volts. The voltage for a plasma arc process ranges from 50 to 200 volts closed circuit and 150 to 400 **volts** open circuit. This higher electrical potential is required because the resistance of the gas increases as it is forced through a small orifice, and the gas is more difficult to ionize. The potential voltage of the power supplied must be high enough to overcome this resistance in the circuit in order for electrons to flow.

Although the voltage is higher, the current (amperage) flow is much lower than for most other welding processes. Some low-powered plasma systems will operate with as low as 10 amps of

> **TIP** ⊙
>
> Allowing the water to circulate continuously might result in corrosion in the torch which could cause internal arcing damage when the power is re-applied.

current flow. High-powered plasma cutting machines can have amperages as high as 200 amps, and some very large CNC cutting machines may have 1000 ampere capacities. The higher the amperage capacity, the faster the torch will cut and the thicker the material it will cut.

For example, a 3.25mm diameter, rutile electrode will operate at 18 volts and 90 amperes. The total joules (energy) used would be:

$$J = V \times A$$
$$J = 18 \times 90$$
$$= 1620 \text{ Joules}$$
$$\text{Or } 1.62 \text{ KJ}$$

A low-power plasma system torch operating with only 20 amperes and 85 volts would be using a total of:

$$J = V \times A$$
$$J = 85 \times 20$$
$$= 1700 \text{ Joules}$$
$$\text{Or } 1.7 \text{ KJ}$$

Heat Input

Although the total power used by plasma and non-plasma processes is similar, the actual energy input into the work per linear metre is less with plasma. The very high temperatures of the plasma process allow much higher travelling rates, so the same amount of heat input is spread over a much larger area. This has the effect of lowering the joules per mm of heat the weld or cut will receive. Figure 6.58 shows the cutting performance of a typical plasma arc system. Note the relationship between amperage, cutting speed and metal thickness. The lower the amperage, the slower the cutting speed or the thinner the metal can be cut. It should also be noted that the arc being constricted concentrates the arc in a much smaller cross-sectional area than an open arc.

A high travel speed with plasma cutting will result in a heat input that is much lower than that of the oxy-fuel cutting process. A steel plate cut using the plasma process may show only a slight increase in temperature after the cut. It is often possible to pick up a part only moments after it has been cut using plasma and find that it is cool to the touch.

Distortion

When a metallic material is heated in a localized zone or spot it expands in that area, and after the metal cools, it is no longer straight or flat, Figure 6.59. If a piece of metal, especially a thin sheet, is cut, there will be localized heating along the edge of the cut. Unless special care is taken,

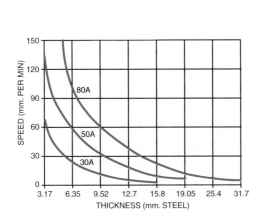

FIGURE 6.58 Plasma arc cutting parameters

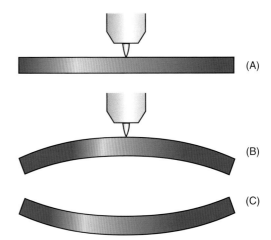

FIGURE 6.59 Heat distortion

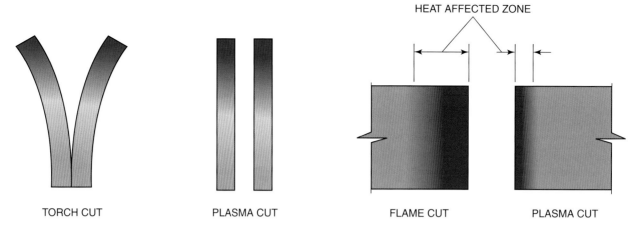

HEAT AFFECTED ZONE

| TORCH CUT | PLASMA CUT | FLAME CUT | PLASMA CUT |

FIGURE 6.60 Minimal bending with a plasma cut   **FIGURE 6.61 Heat-affected zone**

distortion will make the part unusable, Figure 6.60. A plasma cutter allows a body repair worker to cut the thin, low-alloy sheet metal of a damaged car with little problem from distortion.

On thicker sections, with the plasma process, the hardness zone along the edge of a cut will be so small that it is not a problem. In oxy-fuel cutting of thick plate, especially higher-alloyed materials, this hardness zone can cause cracking and failure if the metal is shaped after cutting, Figure 6.61. Often the plates must be pre-heated before they are cut using oxy-fuel to reduce the heat-affected zone, adding greatly to the time and costs of fabrication.

Applications

Early plasma arc cutting systems required that either helium or argon be used as the plasma and shielding gases. As the process improved, it was possible to start the arc using argon or helium and then switch to less expensive nitrogen, which greatly reduced costs. Originally, high operating expenses limited plasma cutting to metallic materials not easily cut using oxy-fuel, such as aluminum, stainless steel and copper. As the process developed, less expensive gases and even dry compressed air could be used, and the torches and power supplies improved. By the early 1980s, the plasma arc process had advanced to a point where it was used for cutting all but the thicker sections of mild steel.

High cutting speeds are possible, approximately four times faster than oxy-fuel cutting. Early high speed cutting machines could not reliably make cuts as fast as the manual plasma cutting systems. That problem has been resolved, and the new machines and robots can operate at the upper limits of the plasma system capacity. These are capable of automatically maintaining the optimum torch stand-off distance from the work and some cutting systems even follow the irregular surfaces of preformed part blanks, Figure 6.62.

FIGURE 6.62 Plasma arc machine cut in 2-in.-thick mild steel

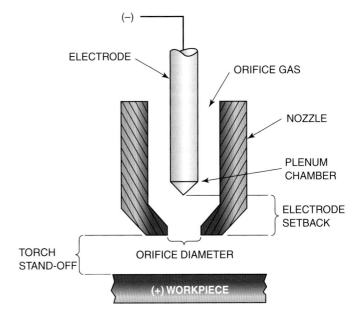

FIGURE 6.64 **Conventional plasma arc terminology**

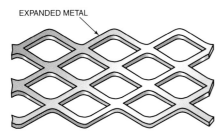

FIGURE 6.63 **Cutting expanded metal**

Any material that is electrically conductive can be cut using the plasma arc process. In a few applications non-conductive materials can be coated with conductive material so that they can also be cut. Although it is possible to make cuts in plate as thick as 175mm thick, it is not cost-effective to do so. The most popular materials cut are carbon steel up to 25mm stainless steel up to 100mm and aluminum up to 150mm. Beyond these limits other cutting processes may be less expensive. Other materials commonly cut using plasma arc are copper, nickel alloys, high-strength, low-alloy steels and clad materials. Plasma arc is also used to cut expanded metals, screens, and other items that would require frequent starts and stops if the oxy-fuel process were used, Figure 6.63.

The **stand-off distance** is the distance from the nozzle tip to the work, Figure 6.64. This distance is critical to producing quality plasma arc cuts. As the distance increases, the arc force is diminished and tends to spread out. This causes the kerf to be wider, the top edge of the plate to become rounded, and more dross to form on the bottom edge. If this distance is too small, the working life of the nozzle tip is reduced. In some cases an arc can form between the nozzle tip and the material that instantly destroys the tip.

On some torches, it is possible to drag the nozzle tip or guide along the surface of the work without shorting it out. This is a great help for work on plate or section in position or on thin sheet metal. Before using the torch in this manner, check the manufacturer's manual to see whether it will operate in contact with the work. A castle nozzle tip, like that shown in Figure 6.65, can be used to allow the torch to be dragged across the surface. This technique causes the nozzle tip orifice to become contaminated more quickly.

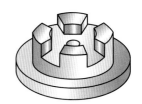

FIGURE 6.65 **Castle nozzle tip**

Starting Methods

Plasma cutters use a number of methods to start the arc. In some units, the arc is created by putting the torch in contact with the work piece. Some cutters use a high voltage, high frequency circuit to start the arc. This method has a number of disadvantages, including risk of electrocution, difficulty of repair, spark gap maintenance and the large amount of radio frequency emissions. Plasma cutters working near sensitive electronics, such as CNC hardware or computers, start the pilot arc by other means. The nozzle and electrode are in contact. The nozzle is the cathode, and the electrode is the anode. When the plasma gas begins to flow, the nozzle is blown forward. A less common method is capacitive discharge into the primary circuit via a silicon controlled rectifier.

Kerf – this is the gap left in the work piece as the metal is removed during a cut. A plasma arc cutting kerf is often smaller than an oxy-fuel cut kerf. Several factors affect the width of the kerf:

- *Stand-off distance.* The closer the torch nozzle tip is to the work, the narrower the kerf will be, Figure 6.66.
- *Orifice (hole) diameter.* Keeping the diameter of the nozzle orifice as small as possible will keep the kerf smaller.
- *Current setting.* Too high or too low a power setting will cause an increase in the kerf width.
- *Travel speed.* As the travel speed is increased, the kerf width will decrease; however, the bevel on the sides and the dross formation will increase if the speeds are excessive.
- *Gas.* The type of gas or gas mixture will alter the kerf width, as the choice of gas affects travel speed, power, concentration of the plasma stream and other factors.
- *Electrode and nozzle tip.* As these parts begin to wear out from use or are damaged by the plasma arc cutter, the quality and kerf width will be adversely affected.
- *Swirling of the plasma gas.* On some torches, the gas is directed in a circular motion around the electrode before it enters the nozzle tip orifice. This swirling produces a plasma stream that is more laminar with straighter sides, resulting in improved cut quality, and a narrow kerf, Figure 6.67.
- *Water injection.* The injection of water into the plasma stream as it leaves the nozzle tip is not the same as the use of a water shroud. Water injection into the plasma stream will increase the swirl and further concentrate the plasma. This improves the cutting quality, lengthens the life of the nozzle tip, and makes a squarer, narrower kerf, Figure 6.68.

Table 6.2 lists some standard kerf widths for several material thicknesses. Use these as a guide for nesting of parts on a plate to maximize the material used and minimize scrap. Your plasma arc system may vary and you should make test cuts to verify the size of the kerf before starting any large production cuts.

Because the sides of the plasma stream are not parallel as they leave the nozzle tip, a bevel is left on the sides of all plasma cuts. This bevel angle is from 0.5 to 3° depending on material thickness, torch speed, type of gas, stand-off distance, nozzle tip condition and other factors.

FIGURE 6.66 Torch height

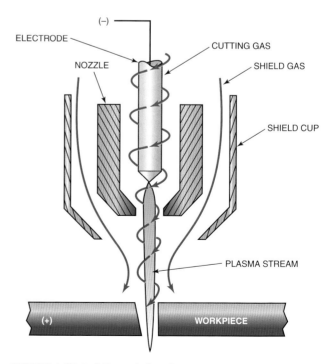

FIGURE 6.67 Swirling of the plasma gas

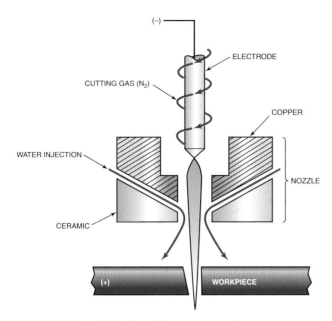

FIGURE 6.68 Water injection plasma arc cutting

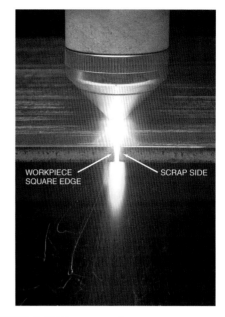

FIGURE 6.69 Plasma cutting

TABLE 6.2 Standard kerf widths for several metal thicknesses

Plate Thickness	Kerf Allowance
mm	mm
3.2 to 25.4	+2.4
25.4 to 51.0	+4.8
51.0 to 127.0	+8.0

On thin sheet, this bevel is undetectable and offers no problem. The use of a plasma swirling torch and the direction in which the cut is made can cause one side of the cut to be square and the scrap side (if there is one) to have all of the bevel, Figure 6.69.

Gases

Almost any gas or gas mixture can be used today for the plasma arc process. Changing the gas or gas mixture is one of the methods of controlling the plasma cut. The cutting factors affected by the choice of gas include:

- *Force*. This is the amount of mechanical impact on the material being cut, the density of the gas and its ability to disperse the molten metal. When using compressed air it is essential to monitor the bar pressure of the power unit as any decrease will affect the quality of the cut. Also check that a moisture trap is fitted and monitored.

- *Central concentration*. Some gases will have a more compact plasma stream. This factor will greatly affect the kerf width and cutting speed.

- *Heat content*. A rise in the electrical resistance of a gas or gas mixture increases the heat content of the plasma produced.

- *Kerf width*. A plasma that remains in a tightly compact stream will produce a deeper cut with less of a bevel on the sides.

- *Dross formation.* The dross that may be attached along the bottom edge of the cut can be controlled or eliminated.
- *Top edge rounding.* The rounding of the top edge of the plate can often be eliminated by selecting the correct gas or gas mixture.
- *Material type.* Because of the formation of undesirable compounds on the cut surface as the metal reacts to elements in the plasma, some metals should not be cut with a specific gas or gases.

Table 6.3 lists some of the popular gases and gas mixtures used for various materials in the plasma arc cutting process. The choice for a specific operation must be tested with the equipment and setup being used. With constant improvements in the plasma arc cutting system, new gases and gas mixtures are continuously being added to the list. It is also important to have the correct gas flow rate for the tip size, metal type and thickness. Too low a flow will result in a cut with excessive dross and sharply beveled sides, Figure 6.70. Too high a flow will produce a poor cut because of turbulence in the plasma stream and waste gas. A flow measuring kit can be used to test the flow at the plasma torch for more accurate adjustments.

TABLE 6.3 Gases for plasma arc cutting and gouging

Metal	Gas
Carbon and low alloy steel	Nitrogen Argon with 0 to 35% hydrogen air
Stainless steel	Nitrogen Argon with 0 to 35% hydrogen
Aluminum and aluminum alloys	Nitrogen Argon with 0 to 35% hydrogen
All plasma arc gouging	Argon with 35 to 40% hydrogen

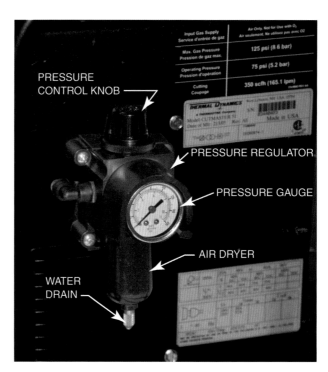

FIGURE 6.70 **Controlling pressure**

TIP

Controlling the pressure is one way of controlling gas flow. Some portable plasma arc cutting machines have their own air pressure regulator and dryer. Air must be dried to provide a stable plasma arc.

Stack Cutting

When sheets are stack-cut with oxy-fuel, it is important that there are no air gaps between layers to interrupt the cut. The plasma arc cutting process does not have these limitations and thin sheets can be stacked and cut efficiently. The sheets should be held together for cutting, using standard G-clamps. The clamping needs to be tight because, if the space between layers is too great, the sheets may stick together. Because of the kerf bevel, the parts near the bottom might be slightly larger if the stack is very thick. This problem can be controlled using the same techniques as described for making the kerf square.

Dross is the metal compound that re-solidifies and attaches itself to the bottom of a cut. It is made up mostly of un-oxidized metal, metal oxides, and nitrides. It is possible to make cuts dross-free if the plasma arc cutting equipment is in good operating condition and the metal is not too thick for the size of the torch being used. The thickness that a dross-free cut can be made is affected by the gas or gas mixture used, travel speed, standoff distance, nozzle tip orifice diameter, wear condition of the electrode tip and nozzle tip, gas velocity and plasma stream swirl.

Machine Cutting

Almost any plasma torch can be attached to some type of semi-automatic or CNC device to allow it to make machine cuts. The simplest devices are oxy-fuel portable flame-cutting machines that run on tracks, Figure 6.71. These are good for mostly straight or circular cuts. Complex shapes can be cut with a pattern cutter that uses a magnetic tracing system to follow the template's shape, Figure 6.72. High-powered plasma arc cutting (PAC) machines may have amperages of up to 1000 amps and must be used with some semi-automatic or automatic cutting system. The heat, light, and other potential hazards make these machines unsafe for manual operation. Large, dedicated, computer-controlled cutting machines have been built specifically for plasma arc cutting systems. These have the high travel speeds required to produce good-quality cuts and have a high volume of production. The operator can input specific cutting instructions such as speed, current, gas flow, location and shape of the part to be cut, and the machine will make the cut with a high degree of accuracy once or any number of times.

Robotic cutters are also available to perform high-quality, high-volume plasma arc cutting. See Figure 6.73 on page 154. The advantage of using a robot is that, in most cases, it can be set up for multi-tasking.

TIP

Stainless steel and aluminum are easy to cut dross-free but carbon steel, copper and nickel-copper alloys are much more difficult.

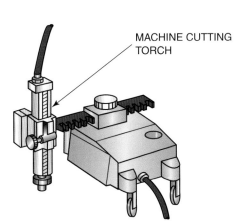

MACHINE CUTTING TORCH

FIGURE 6.71 Machine cutting tool

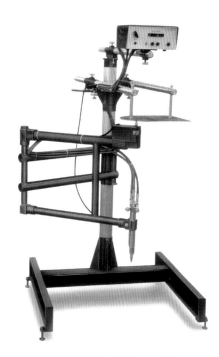

FIGURE 6.72 Portable pattern cutter

FIGURE 6.73 Automated cutting

Machine cutting lends itself to the use of water cutting tables, although they can also be used with most hand torches. The **water table** is used to reduce the noise level, control the plasma light, trap the sparks, eliminate most of the fume hazard and reduce distortion. Water tables either support the metal just above the surface of the water or submerge the metal about 75mm below the surface. All water tables must have some method of removing the cut parts, scrap, and dross that build up in the bottom. Often the surface-type tables have the plasma arc cutting torch connected to a water shroud nozzle. This offers the same advantages to the process as the submerged table offers. In most cases, the manufacturers of this type of equipment have made provisions for a special dye to be added to the water. This helps control the harmful light produced by the system.

HEALTH & SAFETY ⊙

Check with the equipment's manufacturer for limitations and application of the use of dyes.

Signpost Functional Skills

Manual Cutting

Manual plasma arc cutting is the most versatile of the plasma arc cutting processes. It can be used in all positions, on almost any surface, and on most metals. For safety reasons, it is limited to low-power (100 amperes or less) plasma machines, but even these machines can cut up to 37mm-thick metals. The higher-powered machines have extremely dangerous open circuit voltages that can kill a person if accidentally touched.

The setup is similar for most plasma equipment, but never attempt to set up a system without the owner's manual for the specific equipment. Be sure all of the connections are tight and that there are no gaps in the insulation on any of the cables. Check the water and gas lines for leaks. Visually inspect the complete system for possible problems. Before touching the nozzle tip, be sure that the main power supply is off. The open circuit voltage on even low-powered plasma machines is high enough to kill a person. Replace all parts to the torch before the power is restored to the machine.

Plasma Arc Gouging

Plasma arc gouging is a recent addition to the plasma arc cutting processes. It is similar to air carbon arc gouging in that a U-groove can be cut into the metal's surface. This process makes it easy to remove metal along a joint before welding, to remove a defect for repair. See Figure 6.74.

The torch is set up with a less-concentrated plasma stream. This allows the washing away of the molten metal instead of thrusting it out to form a cut. The torch is held at approximately a 30° angle to the metal surface. Once the groove is started it can be controlled by the rate of travel, torch angle and torch movement. Plasma arc gouging is effective on most metallic materials. Stainless steel and aluminum are especially good metals to gouge because there is almost no clean up. The groove is clean, bright and ready to be welded. Plasma arc gouging is especially beneficial with these metals because the only other process that can leave the metal ready to weld is to have the groove machined, which is slow and expensive. It is important not to try to remove too much material in one pass. If a deeper groove is required, multiple gouging passes can be made.

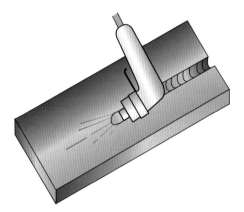

FIGURE 6.74 **Plasma arc gouging a U-groove in a plate**

Safety

Plasma arc cutting raises many of the same safety concerns as other electric arc welding or cutting processes. Some concerns are specific to this process:

- *Electrical shock.* Because the open circuit voltage is much higher for this process than for any other, extra caution must be taken. The chance of a fatal shock from this equipment is much higher than in the case of any other welding equipment.

- *Moisture.* Water is often used with plasma arc cutting torches to cool the torch, improve the cutting characteristic or as part of a water table. It is very important to avoid leaks or splashes. The chance of electrical shock is greatly increased if there is moisture on the floor, cables, or equipment. Moisture also tends to build up when compressed air is used and it is essential that a moisture trap is put in the line to prevent any moisture getting to the cutting nozzle.

- *Noise.* As the plasma stream passes through the nozzle orifice at high speed, a loud sound is produced, which increases as the power level rises. Even with low-power equipment the decibel (dB) level is above safety ranges. Some type of ear protection is required to prevent damage to the operator and other people in the area of the plasma arc cutting equipment. High levels of sound can have a cumulative effect on hearing over time unless proper precautions are taken. See the manufacturer's manual for recommendations for the equipment in use.

- *Light.* The plasma arc cutting process produces light radiation in all three spectrums. The large quantity of visible light will cause night blindness if the eyes are unprotected. The most dangerous of the lights is ultraviolet. As in other arc processes, this light can cause burns to the skin and eyes. Infrared light can be felt as heat, and it is not as much of a hazard. Some type of eye protection must be worn when any plasma arc cutting is in progress. Table 6.4 lists the recommended lens shade number for various power-level machines.

- *Fumes.* The plasma arc cutting process produces a large quantity of potentially hazardous fumes that must be removed from the work space. One solution is localized exhaust ventilation (LEV). A downdraft table is ideal for manual work, but some special hoods may be required for larger applications. The use of a water table or a water shroud nozzle, or both, will greatly help to control fumes. Often the fumes cannot be exhausted into the open air without first being filtered or treated to remove dangerous levels of contaminants. Before installing an exhaust system, you must check with local government and national guidelines to see whether specific safeguards are required.

- *Gases.* Some plasma gas mixtures include hydrogen; because this is a flammable gas, extra care must be taken to ensure that the system is leak-proof.

- *Sparks.* As with any process that produces sparks, the danger of an accidental fire is always present. This is a larger concern with plasma arc cutting because the sparks are often thrown some distance from the work area and the operator's vision is restricted by a welding head-shield. If there is any possibility that sparks will be thrown out of the immediate work area, a fire watch must be present. This is a person whose sole job is to watch for the possible starting of a fire, who knows how to sound the alarm and has appropriate firefighting equipment handy. Never cut in the presence of combustible materials.

- *Operator competence.* Never operate any plasma arc cutting equipment until you have read the manufacturer's, owner's and operator's manual for the specific equipment to be used. After reading the manual, it is a good idea to have someone who is familiar with the equipment to go through the operation.

TABLE 6.4 Recommended shade densities for filter lenses

Current Range A	Minimum Shade	Comfortable Shade
Less than 300	8	9
300 to 400	9	12
400 plus	10	14

WORKSHOP TASK 6.5

Flat, Straight Cuts in Thin Plate

Using a properly set-up and adjusted plasma arc cutting machine, proper safety protection, one or more pieces of mild steel, stainless steel, and aluminum 150mm long and 60mm wide, on 1.6mm sheet and 3mm thick plate and cut off 12mm wide strips, Figure 6.75.

- Mark off strips with French chalk.
- Starting at one end of the piece of metal that is 3mm thick, hold the torch as close as possible to a 90° angle.
- Lower the headshield and establish a plasma cutting stream.
- Move the torch in a straight line down the plate toward the other end, Figure 6.76.
- If the width of the kerf changes, speed up or slow down the travel rate to keep the kerf the same size for the entire length of the plate.
- Repeat the cut using both thicknesses of all three types of metals until consistently smooth cuts that are within ± 2mm of a straight line and ± 5° of being square.
- Turn off the plasma arc cutting equipment and clean up the work area when finished cutting.
- Repeat this procedure with a range of thicknesses to gain experience with the speed of travel.

  Complete a copy of the 'Student Welding Report'.

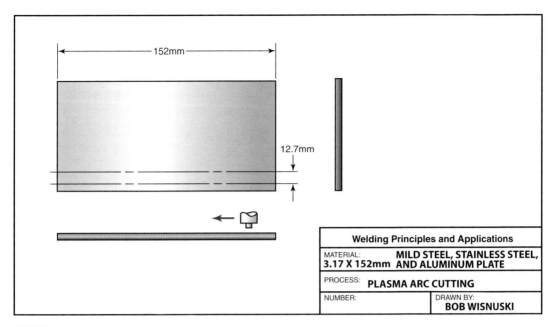

Welding Principles and Applications		
MATERIAL: 3.17 X 152mm	MILD STEEL, STAINLESS STEEL, AND ALUMINUM PLATE	
PROCESS:	PLASMA ARC CUTTING	
NUMBER:	DRAWN BY:	BOB WISNUSKI

FIGURE 6.75 Straight square plasma arc cutting

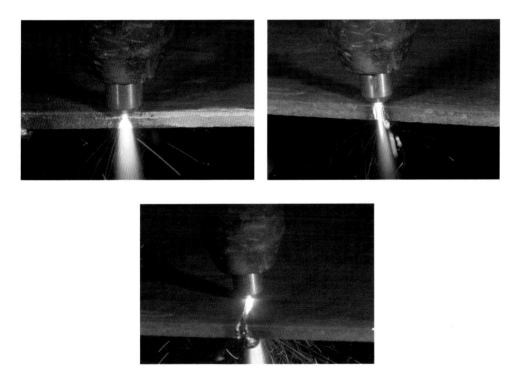

FIGURE 6.76 Flat, straight cut in thin plate

WORKSHOP TASK 6.6

Flat Cutting Holes

Using a properly set-up and adjusted plasma arc cutting machine, proper safety protection, one or more pieces of mild steel, stainless steel, and aluminum 1.6mm, 3mm, 6mm and 12mm thick, you will cut 12mm and 25mm holes.

- Mark out using dividers and French chalk.
- Starting with the piece of metal that is 3mm thick, hold the torch as close as possible to a 90° angle.
- Lower the headshield and establish a plasma cutting stream.
- Move the torch in an outward spiral until the hole is the desired size, Figure 6.77.
- Repeat the hole cutting process until both sizes of holes are made using all the thicknesses of all three types of materials and make consistently smooth cuts that are within ± 2mm of being round and ± 5° of being square.
- Turn off the plasma arc cutting equipment and clean up the work area when finished cutting.

  Complete a copy of the 'Student Welding Report'.

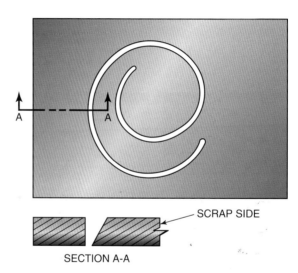

SECTION A-A

FIGURE 6.77 Flat cutting holes

TIP

When cutting a hole, make a test to see which direction to make the cut so that the beveled side is on the scrap piece.

WORKSHOP TASK 6.7

Beveling of a Plate

Using a properly set-up and adjusted plasma arc cutting machine, proper safety protection, one or more pieces of mild steel, stainless steel and aluminum 150mm long 60mm wide and in thickness ranges 6–12mm thick, cut a 45° bevel down the length of the plate.

- Mark out using French chalk.
- Starting at one end of the piece of metal that is 6mm thick, hold the torch as close as possible to a 45° angle.
- Lower the headshield and establish a plasma cutting stream.
- Move the torch in a straight line down the plate toward the other end.
- Repeat the cut using both thicknesses of all three types of metals until there are consistently smooth cuts that are within ± 2mm of a straight line and ±5° of a 45° angle.
- Turn off the plasma arc cutting equipment and clean up the work area when you are finished cutting.

  Complete a copy of the 'Student Welding Report'.

WORKSHOP TASK 6.8

Gouging a Plate

Using a properly set-up and adjusted plasma arc cutting machine, wearing appropriate PPE, practise cutting a gouge in 6–12mm plate, 150mm long in low carbon steel.

- Starting at one end of the piece of metal, hold the torch as close as possible to a 30° angle, Figure 6.78.
- Lower the headshield and establish a plasma cutting stream.
- Move the torch in a straight line down the plate towards the other end.
- If the width of the U-groove changes, speed up or slow down the travel rate to keep the groove the same width and depth for the entire length of the plate.
- Repeat this process in stainless steel and aluminium, until consistently smooth grooves that are within 2mm of a straight line and uniform in depth.
- Turn off the plasma arc cutting equipment and clean up the work area when you are finished cutting, depositing scrap metal in the appropriate bin.

  Complete a copy of the 'Student Welding Report'.

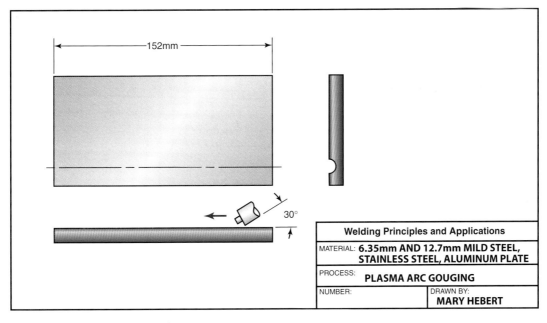

Welding Principles and Applications

MATERIAL: **6.35mm AND 12.7mm MILD STEEL, STAINLESS STEEL, ALUMINUM PLATE**

PROCESS: **PLASMA ARC GOUGING**

NUMBER: DRAWN BY: **MARY HEBERT**

FIGURE 6.78 Plasma arc gouging

SUMMARY

Used properly, oxy-fuel cutting can produce almost machine-cut quality requiring no post-cut clean up. Proper equipment setup and cutting nozzle cleaning are essential for the welder to produce quality oxy-fuel cuts.

Plasma arc cutting is quickly becoming one of the most popular cutting processes in the welding industry. Developments in the equipment and torch design have extended the effective life of the consumable torch parts. This has reduced costs and improved cut quality significantly. One of the most difficult parts of a plasma cutting process to be mastered by the beginner is the high cutting rate. Try to develop an eye and ear for the sights and sounds of the process as a high-quality cut is being produced, and your skills will increase.

With both processes a number of factors determine the travel speed for cutting, including plate thickness and the surface condition of the metal being cut. A good, clean, quality cut cannot be rushed. Learn to develop a sense for the cutting rate that produces the best quality cut.

OXY-FUEL REVIEW

1 State why acetylene is considered the most efficient oxy-fuel gas.

2 State one advantage of owning a combination welding–cutting torch as opposed to just having a cutting torch.

3 Describe the function of a mixing chamber and state where is it located.

4 Explain how an injector-type mixing chamber work.

5 State the benefits of using two oxygen regulators on a machine-cutting torch.

6 State why some copper alloy cutting tips are chrome-plated.

7 Explain the danger if an acetylene nozzle is used with propane.

8 If a cutting nozzle sticks in the cutting head, describe how it can be removed.

9 Explain the term exothermic reaction.

10 State the benefit of a slight forward torch angle for cutting.

11 Explain why drums, tanks or other sealed containers should not be opened with a cutting torch.

12 Explain why the torch nozzle should be raised as the cutting lever is depressed when cutting a hole.

13 List the materials that can be cut using the oxy-fuel gas process.

14 Explain the importance of having extra ventilation and/or a respirator when cutting some materials.

15 Name three features of a cut that can be read from the sides of the kerf after the cut is completed.

16 Describe the methods used to controlling distortion when making cuts.

PLASMA REVIEW

1 Describe an electrical plasma.

2 State why some copper parts are plated.

3 Explain how copper/tungsten tips have helped the plasma torch.

4 Explain how the gap between the electrode tip and the nozzle tip is controlled.

5 List the benefits of a water shroud nozzle for plasma cutting.

6 Explain how water is used to control power cable overheating.

7 Explain why the cooling water must be turned off when the plasma cutting torch is not being used.

8 State the form of current required for plasma cutting.

9 A plasma arc torch is operating with 90 volts (close circuit) and 25 amps. State the power output.

10 Explain why plasma cuts have little or no distortion.

11 List the materials that can be cut using the plasma arc cutting process.

12 List three things that will affect the quality of a plasma arc cutting kerf.

13 List four factors that may affect the choice of plasma arc cutting gas or gases.

14 Describe stack cutting.

15 Compare the amount of dross created between plasma arc and oxy-fuel gas cutting.

16 List the benefits of using a water table for plasma cutting.

Manual Metal Arc Welding

LEARNING OBJECTIVES

After completing this chapter, you should be able to:

- describe the process of manual metal arc welding (MMA)
- contrast constant current (CC) and constant voltage (CV) welding power supplies and which type the manual metal arc welding process requires
- define open circuit voltage, arc voltage and striking voltage
- identify arc blow and apply three different techniques to control it
- identify what the purpose of a welding transformer is and what kind of change occurs to the voltage and amperage with a step-down transformer
- identify the purpose of a rectifier
- read a welding machine duty cycle chart and explain its significance
- determine the proper welding cable size
- demonstrate safe MMA work practices
- strike an arc at a specific point
- select the correct diameter of welding electrode for a weld
- identify typical operating amperages for common sizes of electrodes
- compare a leading electrode angle to a trailing electrode angle
- identify weave patterns for weld beads
- match various MMA electrodes to the metal groups
- make a welded square butt joint in the flat position
- make an outside corner joint in the flat position
- make fillet welds in lap joints in the flat position
- make fillet welds in T-joints in the flat position
- prepare metal before welding
- make the root pass, filler weld and cover pass in the flat position
- make a visual inspection, and describe the appearance of an acceptable weld.

UNIT REFERENCES

NVQ units:

Complying with Statutory Regulations and Organizational Safety Requirements.
Joining Materials by the Manual Metal Arc Welding Process.
Producing Fillet Welded Joints using a Manual Welding Process.

VRQ units:

Engineering Environment Awareness.
Manual Welding Techniques.

KEY TERMS

back gouging a process of cutting a groove in the back side of a joint that has been welded.

ultrasonic inspection (UT) very short ultrasonic pulse-waves are launched into materials to detect internal flaws or to characterize materials somewhat similar to sonar detection.

welding code a document or specification governing aspects of process and procedures for the joining of materials by welding.

welding procedure specification (WPS) a document providing in detail the required variables for specific application to assure repeatability by properly trained welders and welding operators.

INTRODUCTION

Manual metal arc welding (MMA) is a welding process that uses a flux-covered metal electrode to carry an electrical current, Figure 7.1. The current forms an arc across the gap between the end of the electrode and the work, creating sufficient heat to melt both the electrode and the work. Molten metal from the electrode travels across the arc to the molten pool by electro-magnetic forces. The end of the electrode and molten pool of metal are surrounded, purified, and protected by a gaseous shield and a covering of molten flux is produced as the flux coating of the electrode burns or vaporizes. As the arc moves away, the mixture of molten electrode and base metal solidifies and becomes weld metal. At the same time, the molten flux solidifies forming a solid slag (silicate).

MMA is a widely used welding process because of its low cost, flexibility, portability and versatility. The machine and the electrodes are low cost. The machine itself can be as simple as a 110V step-down transformer. The electrodes are available from a large number of manufacturers in packages from 0.5–22kg.

The process is very flexible in terms of the metal thicknesses that can be welded and the variety of positions it can be used in. Metal as thin as 1.6mm (16 SWG standard wire gauge) can be welded using the same machine with different settings and sizes

of electrodes as thick plate. The flexibility of the process also allows metal to be welded in any position.

MMA is a very portable process because it is easy to move the equipment, and engine-driven generator-type welders are available. Also, the limited amount of equipment required for the process makes moving easy. The process is versatile, and it is used to weld almost any metal or alloy, including cast iron, aluminum, stainless steel and nickel.

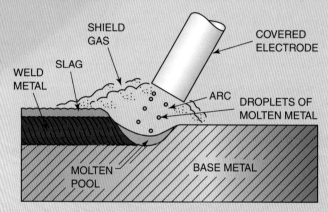

FIGURE 7.1 Shielded metal arc welding equipment

WELDING CURRENT

The source of heat for arc welding is an electric current, which is a flow of electrons. Electrons flow through a conductor from negative (–) to positive (+), Figure 7.2. Resistance to the flow of electrons produces heat. The greater the resistance, the greater the heat. Air has a high resistance to current flow. As the electrons jump the air gap between the end of the electrode and the work, a great deal of heat is produced. Electrons flowing across an air gap produce an arc.

> **TIP** ○
>
> The amount of arc energy being put into a weld per mm controls the width and depth of the weld bead.

Measurement

Three units are used to measure a welding current: voltage (V), amperage (A) and Arc energy (Joules). Voltage, or volts (V), is the measurement of electrical pressure. Voltage controls the maximum gap the electrons can jump to form the arc. A higher voltage can jump a larger gap. Amperage, or amperes (A), is the measurement of the total number of electrons flowing. Amperage controls the size of the arc. Arc energy (Joules) is measured in kj/mm and is calculated by multiplying volts (V) times amperes (A), Figure 7.3.

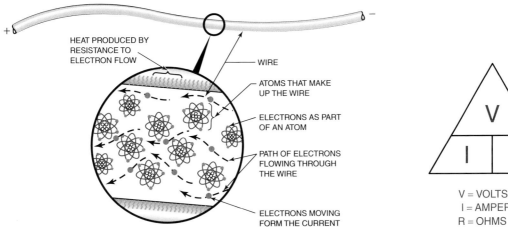

FIGURE 7.2 Electrons travelling along a conductor

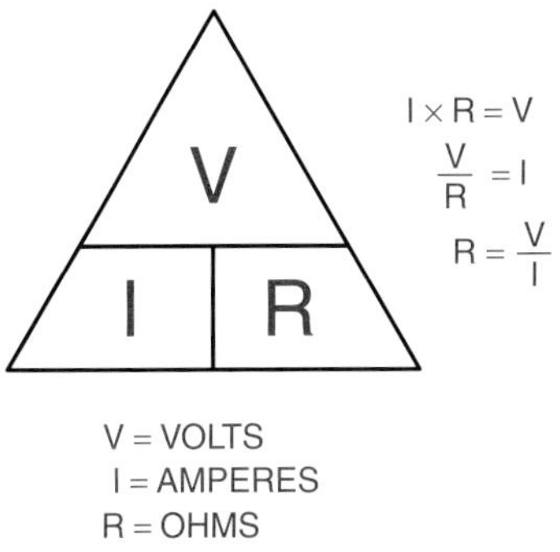

$$I \times R = V$$
$$\frac{V}{R} = I$$
$$R = \frac{V}{I}$$

V = VOLTS
I = AMPERES
R = OHMS

FIGURE 7.3 Ohm's Law

Temperature

The temperature of a welding arc exceeds 6000°C. The exact temperature depends on the resistance to the current flow, which is affected by the arc length and the chemical composition of the gases formed as the electrode covering burns. As the arc lengthens, the resistance increases, causing a rise in the arc voltage and temperature.

Most manual metal arc welding electrodes have chemicals added to their coverings to reduce the arc resistance and stabilize the arc. This also lowers the arc temperature. Other chemicals within the gaseous shield around the arc may raise or lower the resistance. The amount of heat produced is determined by the size of the electrode and the amperage setting. Not all of the heat produced by an arc reaches the weld; some is radiated away in the form of light and heat waves, Figure 7.4. Additional heat is carried away with the hot gases formed by the electrode covering and through conduction in the work. In total, about 50 per cent of all heat produced by an arc is missing from the weld.

The 50 per cent of the remaining heat the arc produces is not distributed evenly between both ends of the arc. This distribution depends on the composition of the electrode's coating, the type of welding current, and the polarity of the electrode's coating.

HEAT

FIGURE 7.4 Energy is lost from the weld in forms of radiation and convection

Currents

The three different types of current used for welding are alternating current (AC), direct current electrode negative (DCEN) and direct current electrode positive (DCEP). The terms DCEN and DCEP have replaced the former terms *direct current straight polarity (DCSP)* and *direct current reverse polarity (DCRP)*. DCEN and DCSP are the same currents, and DCEP and DCRP are the same currents. Some electrodes can be used with only one type of current. Others can be used with two or more types of current. Each welding current has a different effect on the weld.

- *DCEN.* In direct current electrode negative, the electrode is negative, and the work is positive, Figure 7.5. DCEN welding current produces a high electrode melting rate.

- *DCEP.* In direct current electrode positive, the electrode is positive, and the work is negative, Figure 7.6. DCEP current produces the deepest penetrating welding arc characteristics.

- *AC.* In alternating current, the electrons change direction 100 times per second or 50Hz so that the electrode and work alternate from *anode to cathode*, Figure 7.7. The positive side of an electrode arc is called the anode, and the negative side is the cathode. The rapid reversal of the current flow causes the welding heat to be evenly distributed on both the work and the electrode, which gives the weld bead a balance between penetration and build up.

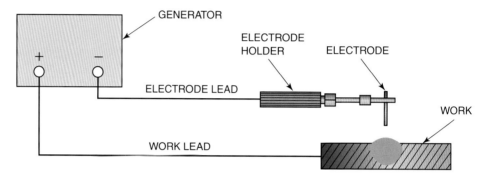

FIGURE 7.5 Electrode negative, straight polarity

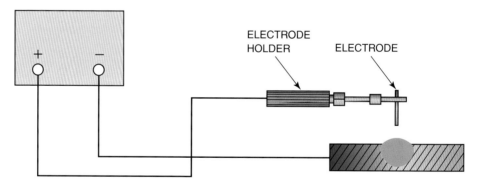

FIGURE 7.6 Electrode positive, reverse polarity

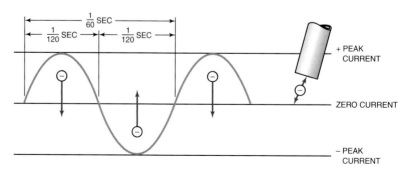

FIGURE 7.7 Alternating current sine wave

TYPES OF WELDING POWER

Welding power can be supplied as:

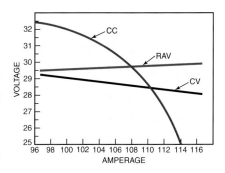

- *Constant voltage (CV)*. The arc voltage remains constant at the selected setting even if the arc length and amperage increase or decrease.
- *Constant current (CC)*. The total welding current remains the same. This type of power is also called 'drooping-arc' voltage, because the arc voltage drops once the arc has been established.

FIGURE 7.8 Constant voltage, rising arc voltage and constant current

The manual metal arc process requires a constant current arc voltage characteristic, illustrated by the CC line in Figure 7.8. The manual metal arc welding machine's voltage output decreases as current increases. This output power supply provides a reasonably high open circuit voltage before the arc is struck. The high open circuit voltage quickly stabilizes the arc. The arc voltage rapidly drops to the lower closed circuit level after the arc is struck. Following this short starting surge, the power remains almost constant despite the changes in arc length. With a constant voltage output, small changes in arc length would cause the power to make large swings and the welder would lose control of the weld.

Open Circuit Voltage

Open circuit voltage is the voltage at the electrode before striking an arc (with no current being drawn). This voltage is usually between 50V and 100V. The higher the open circuit voltage, the easier it is to strike an arc; however, the higher voltage also increases the chance of electrical shock.

Arc Voltage

Arc voltage is the voltage at the arc during welding and is measured from the tip of the electrode to the work piece. This voltage will vary with arc length, type of electrode being used, type of current and polarity. The arc voltage will be between 17-25V.

Striking Voltage

Striking voltage is the minimum voltage required to initiate particular electrode. Some types of electrodes require a high amperage range and therefore may not be suitable for use with DC.

Arc Blow

When electrons flow, they create lines of magnetic force that circle around the line of flow, Figure 7.9. Lines of magnetic force are referred to as 'magnetic flux lines', and space themselves evenly along a current-carrying wire. If the wire is bent, the flux lines on one side are compressed together, and those on the other side are stretched out, Figure 7.10. The unevenly spaced flux lines try to straighten the wire so that they can be evenly spaced again. The force that they place on the wire is usually small but when welding with very high amperages in excess of 250A the force may cause the wire to move.

The welding current flowing through a plate or any residual magnetic fields in the plate will result in uneven flux lines which can cause a movement of the arc called 'arc blow'. Arc blow makes the arc drift as a string would drift in the wind. It is more noticeable in corners, at the ends of plates, and when the work lead is connected to only one side of a plate, Figure 7.11. It can be controlled by connecting the work lead to the end of the weld joint and making the weld in the direction toward the work lead, Figure 7.12. Another solution is to use two work leads, one on each side of the weld. The best method to eliminate arc blow is to use alternating current. AC usually does not allow the flux lines to build long enough to bend the arc before the current changes direction. If none of these solutions are possible, a very short arc length can help control arc blow, as can a small tack weld or a change in the electrode angle.

If you are learning to weld with a steel welding table, arc blow may not be a problem. However, if you are using a pipe stand to hold your welding practice plates it can cause difficulties. In this case, try re-clamping your practice plates or switch to alternating current.

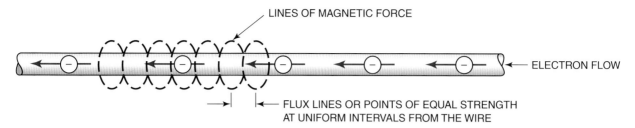

FIGURE 7.9 **Magnetic force around a wire**

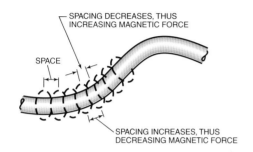

FIGURE 7.10 **Magnetic forces around bends in wires**

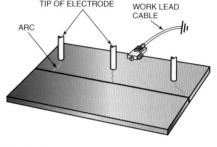

FIGURE 7.11 **Arc blow**

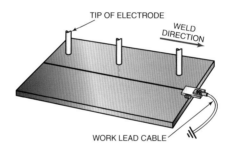

FIGURE 7.12 **Correct current connections to control arc blow**

TYPES OF POWER SOURCES

Two types of electrical devices can be used to produce the low-voltage, high-amperage current combination that arc welding requires. One type uses electric motors or internal combustion engines to drive alternators or generators. The other type uses step-down transformers. Because transformer-type welding machines are quieter, more energy efficient, less expensive and require less maintenance, they are now the industry standard. However, engine-powered generators are still widely used for portable welding.

Transformers

A welding transformer uses the alternating current (AC) supplied to the welding shop at a high voltage to produce the low-voltage welding power. As electrons flow through a wire they produce a magnetic field around it. If the wire is wound into a coil this field is concentrated to produce a much stronger central magnetic force. Because the current being used is alternating or reversing 100 times per second (50Hz), the magnetic field is constantly being built and allowed to collapse. By placing a second (secondary) winding of wire in the magnetic field produced by the first (primary) winding, a current will be induced in the secondary winding. The placing of an iron core in the centre of these coils will increase the concentration of the magnetic field, see Figure 7.13.

A transformer with more turns of wire in the primary winding than in the secondary winding is known as a step-down transformer. It takes a high-voltage, low-amperage current and changes it into a low-voltage, high-amperage current. Except for some power lost by heat within a transformer, the power into a transformer equals the power (joules) out because the volts and amperes are mutually increased and decreased. A transformer welder is a step-down transformer. It takes the high line voltage (110V, 240V, 415V, etc.) and low-amperage current (30A, 50A, 60A, etc.) and changes it into 17V to 45V at 190A to 590A.

Welding machines can be classified by the method through which they control or adjust the welding current. The major classifications are multiple-coil, movable-coil or movable-core, Figure 7.14, and inverter-type machines.

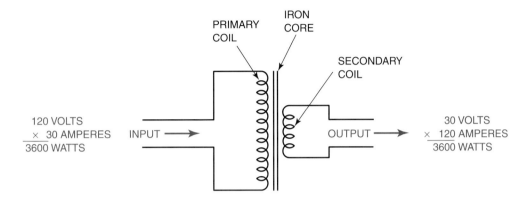

FIGURE 7.13 **Diagram of a step-down transformer**

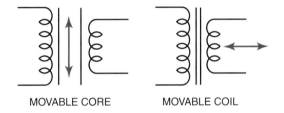

FIGURE 7.14 **Major types of adjustable welding transformers**

Multiple-coil Machines

The multiple-coil machine, or tap-type machine, allows the selection of different current settings by tapping into the secondary coil at a different turn value. The greater the number of turns, the higher the amperage is induced. These machines may have a large number of fixed amperes, or they may have two or more amperages that can be adjusted further with a fine adjusting knob, which may be marked in amperes or in steps.

Calculating the Amperage Setting

With the machine set on the medium range, from 50A to 250A, first subtract the low amperage from the high amperage to get the amperage spread (250–50 = 200). Now divide the amperage spread by the number of units shown on the fine adjusting knob (200 ÷ 10 = 20). Each unit is equal to a 20A increase, Table 7.1. When the knob points to 0, the amperage is 50; when the knob points to 1, the amperage is 70; and at 2, the amperage is 90, Figure 7.15. There are 100 small units on the fine adjusting knob. Dividing the amperage spread by the number of small units gives the amperage value for each unit (200 ÷ 100 = 2). Therefore, if the knob points to 6.1, the amperage is set at a value of 50 + 120 + 2 = 172A. This method provides a good starting place for the current setting, but if the welding is to be made in accordance with a welding procedure's specific amperage setting it will be necessary to use a calibrated meter to make the correct setting.

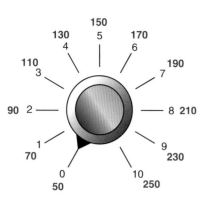

FIGURE 7.15 Fine adjusting knob

TABLE 7.1 Table to calculate amperage setting

Setting	Value in Amperes
0 = 50 + 0	or 50 A
1 = 50 + 20	or 70 A
2 = 50 + 40	or 90 A
3 = 50 + 60	or 110 A
4 = 50 + 80	or 130 A
5 = 50 + 100	or 150 A
6 = 50 + 120	or 170 A
7 = 50 + 140	or 190 A
8 = 50 + 160	or 210 A
9 = 50 + 180	or 230 A
10 = 50 + 200	or 250 A

Movable-coil or Movable-core Machines

These machines are adjusted by turning a hand wheel, or moving a lever, that moves the internal parts closer together or farther apart, Figure 7.17. These machines may have a high and low range, but they do not have a fine adjusting knob. The closer the primary and secondary coils are, the greater the induced current, Figure 7.18. Moving the core in concentrates more of the magnetic force on the secondary coil, thus increasing the current. Moving the core out allows the field to disperse, and the current is reduced, Figure 7.19.

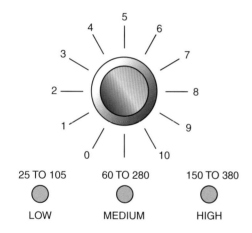

FIGURE 7.16 Activity

Activity: Using the amperage ranges given in table 7.1, calculate the amperage when the knob in Figure 7.16 is at 1, 4, 7, 9 and 8.5 settings.

CURRENT RANGE FOR RUTILE ELECTRODES

Diameter of Rod	Current in Amperes		
mm	Min	Max	Average
1.6	25	45	40
2.5	50	90	90
3.2	60	130	115
4.0	100	180	140
5.0	150	250	200
6.0	200	310	280

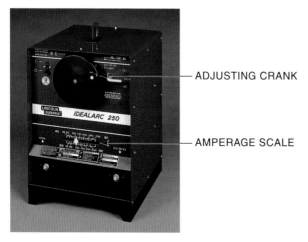

FIGURE 7.17 Movable core type welding machine

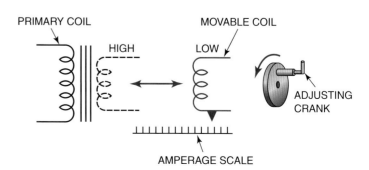

FIGURE 7.18 Movable coil

Inverter Machines

Inverter welding machines are much smaller than other types of machines of the same amperage range. This makes them much more portable and increases their energy efficiency. In a standard welding transformer, the iron core used to concentrate the magnetic field in the coils must be a specific size, determined by the length of time it takes for the magnetic field to build and collapse. By using solid-state electronic parts, the incoming power in an inverter welder is changed from 60 cycles a second to several thousand cycles a second. This higher frequency allows the use of a transformer that is light, but can still do the work of a standard transformer. Additional electronic parts remove the high frequency for the output welding power.

The electronics in the inverter-type welder allow it to produce any desired type of welding power. Previously each type of welding required a separate machine. Now a single welding machine can produce the specific type of current needed for manual metal arc welding, tungsten inert gas shielded, metal inert gas shielded and plasma arc cutting. Because the machine can be light enough to be carried closer to the work, shorter welding cables can be used. The welder does not have to walk as far to adjust the machine. Some manufacturers produce machines that can be stacked so that when you need a larger machine all you have to do is add another unit, Figure 7.20.

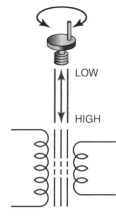

FIGURE 7.19 Movable core

FIGURE 7.20 Inverter type power supply

Generators and Alternators

Generators and alternators both produce welding electricity from a mechanical power source. Both devices have an armature that rotates and a stationary stator. As a wire moves through a magnetic force field, electrons in the wire move, producing electricity.

In an alternator, magnetic lines of force rotate inside a coil of wire, Figure 7.21. An alternator can produce AC only. In a generator, a coil of wire rotates inside a magnetic field. A generator produces DC. It is possible for alternators to use diodes to change the AC to DC for welding. In generators, the welding current is produced on the armature and is picked up with brushes, Figure 7.22. In alternators, the welding current is produced on the stator, and only the small current for the electromagnetic force field goes across the brushes. Therefore, the brushes in an alternator are smaller and last longer.

Some engine-driven generators and alternators run at welding speed all the time. Some can reduce their speed to an idle when welding stops, which saves fuel, reduces wear on the welding machine and is safer for the operator. To strike an arc when using this type of welder, touch the electrode to the work for a second. When you hear the welding machine pick up speed, remove the electrode from the work and strike an arc. In general, the voltage and amperage are too low to start a weld, so shorting the electrode to the work should not cause the electrode to stick. A

TIP

Alternators can be smaller in size and lighter in weight than generators and still produce the same amount of power.

timer can control the length of time that the welder maintains speed after the arc is broken and should be set long enough to change electrodes without losing speed.

Portable welders often have 110V or 240V plug outlets, which can be used to run grinders, drills, lights, and other equipment. The power provided may be AC or DC. If DC is provided, only equipment with brush-type motors or tungsten light bulbs can be used. If the plug is not specifically labelled 110V AC, check the owner's manual before using it for such devices as radios or other electronic equipment. A typical portable welder is shown in Figure 7.23.

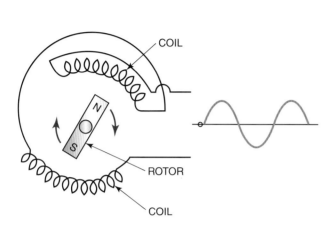

FIGURE 7.21 **Schematic diagram of an alternator**

FIGURE 7.22 **Diagram of a generator**

(A)

(B)

FIGURE 7.23 **Portable engine generator welder**

Rectifiers

Alternating-welding current can be converted to direct current by using a series of rectifiers. A rectifier allows current to flow in one direction only, Figure 7.24. If one rectifier is added, the welding power appears as shown in Figure 7.25. It would be difficult to weld with pulsating power such as this. A series of rectifiers, known as a bridge rectifier, can modify the alternating current so that it appears as shown in Figure 7.26.

Rectifiers become hot as they change AC to DC and must be attached to a heat sink and cooled by having air blown over them. The heat produced by a rectifier reduces the power efficiency of the welding machine. Figure 7.27 shows the amperage dial of a typical machine. Notice that at the same dial settings for AC and DC, the DC is at a lower amperage. The difference in amperage is due to heat lost in the rectifiers. The loss in power makes operation with AC more efficient and less expensive compared to DC. A DC adapter for small AC machines is available from manufacturers. For some types of welding, AC does not work properly.

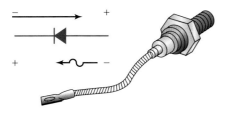

FIGURE 7.24 Rectifier

FIGURE 7.25 Rectifier in a welding power supply pulsating power

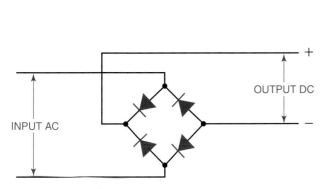

FIGURE 7.26 Bridge rectifier

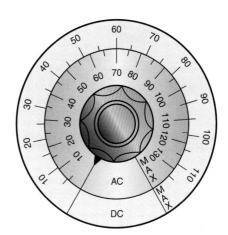

FIGURE 7.27 Typical dial on ac-dc transformer rectifier welder

DUTY CYCLE

Welding machines produce internal heat at the same time that they produce the welding current. Except for automatic welding machines, welders are rarely used continuously for long periods of time. The welder must take time to change electrodes, change positions, or change parts. The duty cycle is the percentage of time a welding machine can be used continuously. A 60 per cent duty cycle means that out of any ten minutes, the machine can be used for a total of six minutes at the maximum rated current and must be cooled off for four minutes. The duty cycle increases as the amperage is lowered and decreases for higher amperages, Figure 7.28. Most welding machines weld at a 60 per cent rate or less. Therefore, most manufacturers list the amperage rating for a 60 per cent duty cycle on the nameplate that is attached to the machine. Other duty cycles are given on a graph in the owner's manual.

The manufacturing cost of power supplies increases in proportion to their rated output and duty cycle. For this reason, some DIY welding machines may have duty cycles as low as 20 per cent even at a low welding setting of 90A to 100A. The duty cycle on these machines should never be exceeded because a build up of the internal temperature can cause the transformer insulation to break down, damaging the power supply.

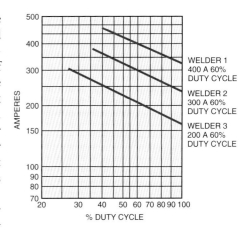

FIGURE 7.28 Duty cycle of metal arc welding machine

WELDING CABLES AND LEADS

The terms welding cables and welding leads are interchangeable. Cables to be used for welding must be flexible, well insulated and the correct size for the job. Most are made from stranded copper wire. Some manufacturers sell a newer type of cable made from aluminum wires which

FIGURE 7.29 Power lug protection provided by insulators

are lighter and less expensive. However, aluminum is not as good a conductor as copper, and aluminum wire should be one size larger than would be required for copper.

The insulation on welding cables is exposed to hot sparks, flames, grease, oils, sharp edges, impact and other types of wear. To withstand this, only specially manufactured insulation should be used for welding cable. Several new types of insulation are available that give longer service against these adverse conditions.

As electricity flows through a cable, the resistance to the flow causes the cable to heat up and increase the voltage drop. To minimize the loss of power and prevent overheating, the electrode cable and work cable must be the correct size. Large welding lead sizes make electrode manipulation difficult. Smaller cable can be spliced to the electrode end of a large cable to make it more flexible, although it must not be more than 13m long. Splices and end lugs are available from suppliers. Be sure that a good electrical connection is made because a poor electrical connection will result in heat build up and voltage drop. Splices and end lugs must be well insulated to avoid electrical shorting, Figure 7.29.

ELECTRODE HOLDERS

The electrode holder should be of the proper amperage rating and in good repair. Electrode holders are designed to be used at their maximum amperage rating or less. Higher amperage values will cause the holder to overheat and burn up. If the holder is too large for the amperage range being used, manipulation is hard, and operator fatigue increases. Make sure that the correct amperage holder is chosen, Figure 7.30. A correctly sized electrode holder can overheat if the jaws are dirty or loose, or if the cable is loose. This reduces welding power and is uncomfortable to work with. Replacement springs, jaws, insulators, handles, screws and other parts are available to keep the holder in good working order. To prevent excessive damage to the holder, welding electrodes should not be burned too short. A 50mm electrode stub is short enough to minimize electrode waste and save the holder.

200-AMP CAPACITY

FIGURE 7.30 Amperage capacity of electrode holder

WORK CLAMPS

The work clamp must be the correct size for the current being used, and it must clamp tightly to the material. Heat can build up in the work clamp, reducing welding efficiency. Power losses in the work clamp are often overlooked. The clamp should be carefully touched occasionally to check if it is getting hot. A loose clamp may also cause arcing, which can damage a part. Improper work clamp placement may also contribute to arc blow. To avoid this, keep the work clamp situated as close to the weld as practical, and weld away from it. If the part is to be moved during welding, a swivel-type work clamp may be needed, Figure 7.31. It may be necessary to weld a tab to thick parts so that the work lead can be clamped to the tab, Figure 7.32.

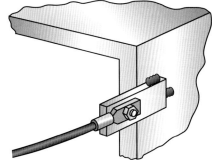

FIGURE 7.31 **A work clamp may be attached to workpiece**

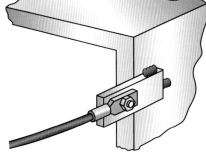

FIGURE 7.32 **Tack welded ground to part**

SETUP

Arc welding machines should be located near the welding site, but far enough away so that they are not covered with spark showers. The machines may be stacked to save space, but there must be enough room between them to ensure that air can circulate to keep them from overheating. The air that is circulated through the machine should be as free as possible of dust, oil and metal filings. Even in a good location, the power should be turned off periodically and the machine blown out with compressed air, Figure 7.33. The welding machine should be located away from cleaning tanks and any other sources of corrosive fumes. Water leaks must be fixed and puddles cleaned up before a machine is used. Power to the machine must be fused, and a power shut-off switch provided. The switch must be located so that it can be reached in an emergency without touching either the machine or the welding station. The machine case or frame must be earthed. The welding cables should be sufficiently long to reach the work station but not so long that they must always be coiled. Cables should not be placed on the floor in aisles or walkways. If cables must cross a walkway, they must be installed overhead, or protected by a ramp, Figure 7.34. The welding machine and its main power switch should be off during installation or work on the cables.

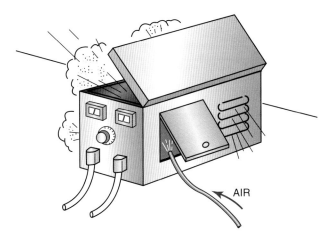

FIGURE 7.33 **Slag, chips from grinding, and dust must be blown out occasionally**

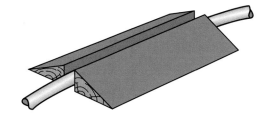

FIGURE 7.34 **To prevent tripping**

The work station must be free of combustible materials, and any portable tools removed from the work table as these will provide an alternative earth path. This will not only damage the tool but also the power supply to the socket. Screens or curtains should be provided to protect other workers from the arc light.

Never wrap the welding cable around your arms, shoulders, waist or any other part of your body. If the cable is caught by any moving equipment, such as a forklift, you could be pulled off balance or more seriously injured. If it is necessary to hold the weight off the cable, use a free hand. Hold the cable so that, if it is pulled, it can be easily released. Check the surroundings before starting to weld. If heavy materials are being moved in the area around you, there should be a safety watch. A safety watch can warn a person of danger while that person is welding.

STRIKING THE ARC

Using a properly set-up and adjusted arc welding machine, the proper safety protection and E6103 welding electrodes having a 3.25mm core wire diameter measured at the bare end of the electrode, and one piece of mild steel plate, 6mm thick, you will practise striking an arc, Figure 7.35.

1 With the electrode held over the plate, lower your helmet. Scratch the electrode across the plate (like striking a large match), Figure 7.36.

2 As the arc is established, slightly raise the electrode to the desired arc length.

3 Hold the arc in one place until the molten weld pool builds to the desired size.

4 Slowly lower the electrode as it burns off, and move it to start the bead.

5 If the electrode sticks to the plate, quickly squeeze the electrode holder lever to release the electrode or alternatively give it a twist to break the bond with the surface.

6 Do not touch the electrode without gloves because it will still be hot.

7 If the flux breaks away from the end of the electrode, restarting the arc will be very difficult, if you do not burn off on scrap metal to re-establish the flux coating, Figure 7.37.

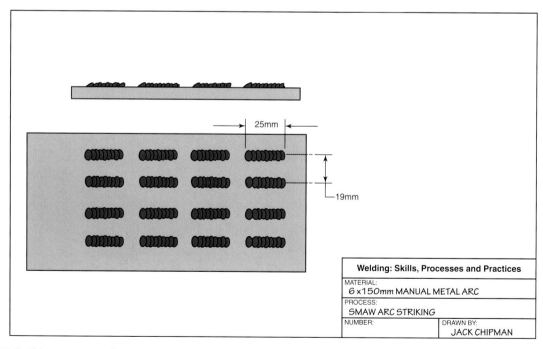

FIGURE 7.35 Striking an arc and running short beads

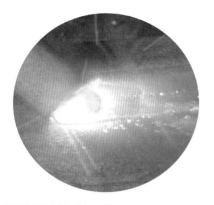

FIGURE 7.36 Striking the arc

FIGURE 7.37 If the flux is broken off

TIP

If the flux is broken off the end completely or on one side, the arc can be erratic or forced to the side.

8 Once you are able to easily strike an arc and make a weld, try to strike the arc where it will be re-melted by the weld you are making.

9 Arc strikes on the metal's surface that are not covered up by the weld are considered to be weld defects by most codes.

10 Break the arc by rapidly raising the electrode after completing a 25mm weld bead.

11 Restart the arc by striking in front and looping back to the previous weld as you did before, and make another short weld.

12 Repeat this process until you can easily start the arc each time.

13 Turn off the welding machine and re-instate your work area when you are finished welding. Place all stubs and surplus materials in the appropriate containers.

Effect of Too-High or Too-Low Current Settings

Each welding electrode must be operated in a specified current (amperage) range, Table 7.2. Too low a current results in poor fusion and poor arc stability, Figure 7.38. The weld may have slag or gas inclusions because the molten weld pool was not fluid long enough for the flux to react. There may also be little or no penetration of the weld into the base plate. With the current set too low, the arc length is very short, resulting in frequent shorting and sticking of the electrode.

The core wire of the welding electrode is limited in the amount of current it can carry. As the current is increased, the wire heats up. This preheating causes some of the chemicals in the covering to be burned out too early, Figure 7.39. The loss of the proper balance of elements causes poor arc stability, leading to spatter, porosity and slag inclusions.

Longer arc lengths can also cause an increase in spatter. The weld bead made at a high amperage setting is wide and flat with deep penetration. The spatter is excessive and is mostly 'hard', meaning that it fuses to the base plate and is difficult to remove, Figure 7.40. The electrode covering is discoloured for more than 3–6mm from the end of the electrode. Extremely high settings may also cause the electrode to discolour, crack, glow red or burn.

TABLE 7.2 Welding amperage range for common electrode types

Electrode	Classification					
Size	E6010	E6011	E6012	E6013	E7016	E7018
2.4mm	40–80	50–70	40–90	40–85	75–105	70–110
3.2mm	70–130	85–125	75–130	70–120	100–150	90–165
4mm	110–165	130–160	120–200	130–160	140–190	125–220

FIGURE 7.38 Welding with amperage too low

FIGURE 7.39 Welding with amperage too high

 TIP

Hard weld splatter is fused to base metal and is difficult to remove.

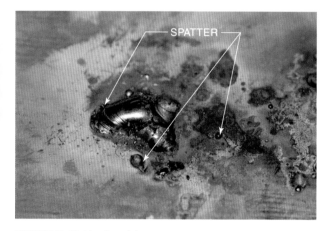

FIGURE 7.40 Hard weld spatter

Electrode Size and Heat

The correct size of welding electrode for a weld is determined by: the skill of the welder, the thickness of the metal to be welded, the size of the metal and welding codes or standards. Using small diameter electrodes requires less skill because the deposition rate, the rate at which the weld metal is added to the weld, is slower. Small diameter electrodes will make acceptable welds on thick plate, but more time is required to make the weld.

Large diameter electrodes may overheat the metal if they are used with thin or small pieces of metal. To determine if a weld is too hot, watch the shape of the trailing edge of the molten weld pool, Figure 7.41. Rounded ripples indicate the weld is cooling uniformly and that the heat is not excessive. If the ripples are pointed, the weld is cooling too slowly because of excessive heat. Extreme overheating can cause a burn-through, which is hard to repair. To correct an overheating problem, turn down the amperage, use a shorter arc, travel at a faster rate, use a chill plate (a large piece of metal used to absorb excessive heat) or use a smaller electrode at a lower current setting.

AMOUNT OF HEAT DIRECTED AT WELD	WELD POOL
TOO LOW	
CORRECT	
TOO HOT	

FIGURE 7.41 Effect on the shape of the molten weld pool caused by heat input

Arc Length

The arc length is the distance the arc must jump from the end of the electrode to the plate or weld pool surface. As the weld progresses, the electrode becomes shorter as it is consumed. To maintain a constant arc length, the electrode must be lowered continuously.

As the arc length is shortened, metal transferring across the gap may short out the electrode, causing it to stick to the plate. The weld that results from a short arc is narrow and has a high build up, Figure 7.42.

Long arc lengths produce more spatter because the metal being transferred may drop outside of the molten weld pool. The weld is wider and has little build up, Figure 7.43. There is a narrow range for the arc length in which stability is maintained, metal transfer is smooth, spatter is minimized, and the bead shape is controlled. Factors affecting the length are the type of electrode, joint design, metal thickness and current setting.

Some welding electrodes, such as E7024, have a thick flux covering. The rate at which the covering melts is slow enough to permit the electrode coating to be rested against the plate. The arc burns back inside the covering as the electrode is dragged along touching the joint, Figure 7.44. For this type of welding electrode, the arc length is maintained by the electrode covering.

An arc will jump to the closest metal conductor. On joints that are deep or narrow, the arc is pulled to one side and not to the root, Figure 7.45. As a result, the root fusion is reduced or may be non-existent. If a very short arc is used, it is forced into the root for better fusion. Because shorter arcs produce less heat and penetration, they are best suited for use on thin metal or thin-to-thick metal joints. To maintain a short arc that gives good fusion with a minimum of slag inclusions, higher amperage settings are required, but these must be within the amperage range for the specific electrode.

Finding the correct arc length often requires some trial and adjustment. Most welding jobs require an arc length of 1.5 × core wire diameter.

FIGURE 7.42 Welding with too short an arc length

FIGURE 7.43 Welding with too long an arc length

FIGURE 7.44 Welding with drag technique

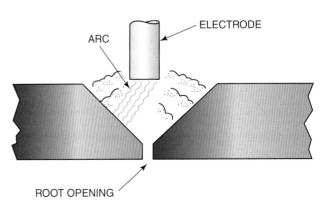

FIGURE 7.45 The arc may jump to the closest metal

Electrode Angle

The electrode angle is measured from the electrode to the surface of the metal and is described as leading or trailing, relative to the direction of travel, Figure 7.46.

A leading electrode angle pushes molten metal and slag ahead of the weld, Figure 7.47. When welding in the flat position, care must be taken to prevent overlap and slag inclusions. The solid metal ahead of the weld cools and solidifies the molten filler metal and slag. This rapid cooling prevents the metals from fusing together, Figure 7.48. As the weld passes over this area, heat from the arc may not melt it and some overlap and slag inclusions are left.

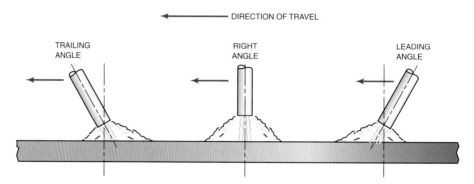

FIGURE 7.46 **Direction of travel and electrode angle**

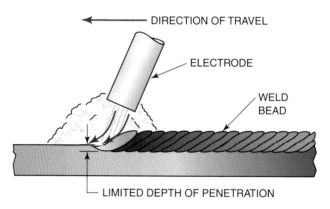

FIGURE 7.47 **Leading, lag or pushing electrode angle**

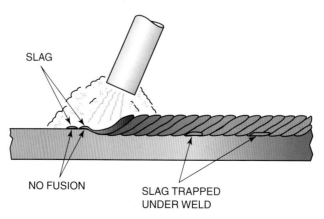

FIGURE 7.48 **Some electrodes, such as E7018, may not remove the deposits ahead of the molten weld pool, resulting in discontinuities within the weld**

The following are suggestions for preventing cold lap and slag inclusions:

- Use as little leading angle as possible.
- Ensure that the arc melts the base metal completely, Figure 7.49.
- Use a penetrating-type electrode that causes little build up.
- Move the arc back and forth across the molten weld pool to fuse both edges.

A leading angle can be used to minimize penetration or to help hold molten metal and is often the preferred choice for vertical welds, Figure 7.50.

TIP ⊙

Metal that is melted ahead of the molten weld pool helps to ensure good weld fusion.

FIGURE 7.49 Trailing drag electrode

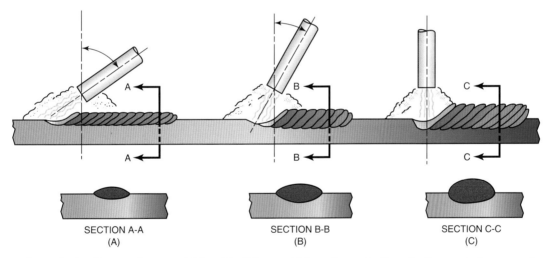

| SECTION A-A | SECTION B-B | SECTION C-C |
| (A) | (B) | (C) |

FIGURE 7.50 Effect of a leading angle on weld bead build up, width and penetration. As the angle increases toward the vertical position (C), penetration increases

A trailing electrode angle pushes the molten metal away from the leading edge of the molten weld pool toward the back where it solidifies. As the molten metal is forced away from the bottom of the weld, the arc melts more of the base metal, which results in deeper penetration. The molten metal pushed to the back of the weld solidifies and forms reinforcement for the weld and is often the preferred choice when welding in the flat position.

Electrode Manipulation

The movement or weaving of the welding electrode can control the following characteristics of the weld bead: penetration, build up, width, porosity, undercut, overlap and slag inclusions. The exact weave pattern for each weld is often the personal choice of the welder. However, some patterns are especially helpful for specific welding situations. The pattern selected for a flat (PA) butt joint is not as critical as is the pattern selection for other joints and other positions. Many weave patterns are available for the welder to use, Figure 7.51.

The circular pattern is often used for flat position welds on butt, T outside corner joints and for build up or surfacing applications. The circle can be made wider or longer to change the bead width or penetration, Figure 7.52.

The 'C' and square patterns are both good for most PA (flat) welds, and can also be used for vertical PF / PG positions. These patterns can also be used if there is a large gap to be filled when both pieces of metal are nearly the same size and thickness.

The 'J' pattern works well on flat PA lap joints, all vertical PF / PG joints, and horizontal PC butt and lap welds. This pattern allows the heat to be concentrated on the thicker plate, Figure 7.53. It also allows the reinforcement to be built up on the metal deposited during the first part of the pattern. As a result, a uniform bead contour is maintained during out-of-position welds.

The inverted 'T' pattern works well with fillet welds in the vertical PG/PF and overhead PE positions. It also can be used for deep butt welds for the hot pass. The top of the 'T' can be used to fill in the toe of the weld to prevent undercutting.

The 'straight step' pattern can be used for stringer beads, root pass welds and multiple pass welds in all positions. For this pattern, the smallest quantity of metal is molten at one time, allowing easy control of the weld. As the electrode is stepped forward, the arc length is increased so that no metal is deposited ahead of the molten weld pool. This action allows the molten weld pool to cool to a controllable size.

The 'figure-eight' pattern and the zig-zag pattern are used as cover passes in the flat and vertical positions. Do not weave more than three times the width of the electrode. These patterns deposit a large quantity of metal at one time. A shelf can be used to support the molten weld pool when making vertical welds using either of these patterns, Figure 7.54.

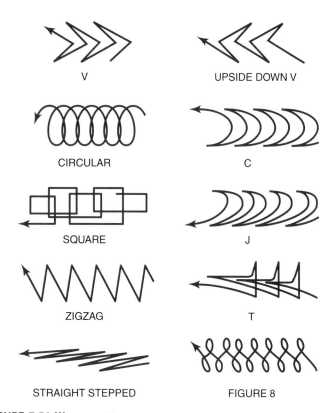

FIGURE 7.51 Weave patterns

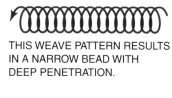

THIS WEAVE PATTERN RESULTS IN A NARROW BEAD WITH DEEP PENETRATION.

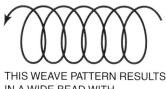

THIS WEAVE PATTERN RESULTS IN A WIDE BEAD WITH SHALLOW PENETRATION.

FIGURE 7.52 Changing the weave patterns

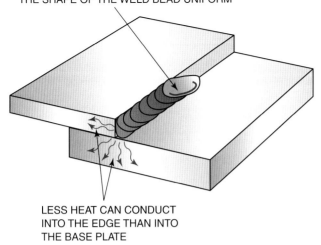

SHELF SUPPORTS MOLTEN WELD POOL, MAKING
THE SHAPE OF THE WELD BEAD UNIFORM

LESS HEAT CAN CONDUCT
INTO THE EDGE THAN INTO
THE BASE PLATE

FIGURE 7.53 The 'J' pattern allows heat to be concentrated
on the thicker plate

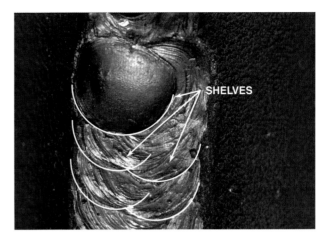

SHELVES

FIGURE 7.54 Using the shelf to support the molten pool
for vertical welds

Positioning of the Welder and the Plate

You should be in a relaxed, comfortable position before starting to weld. Welding in an awkward position can cause fatigue, which leads to poor coordination and poor-quality welds. You must have enough freedom of movement so that you do not need to change position during a weld.

When your welding helmet is down, you are blind to your surroundings. Due to the arc, your field of vision also very limited. These factors may cause you to sway. To avoid this, lean against or brace yourself against a stable object.

Welding is easier if you can find the most comfortable angle. With the welding machine turned off and with an electrode in place in the electrode holder, draw a straight line along the plate to be welded and position yourself. Turn the plate to several different angles, and find which angle and position is most comfortable for welding, Figure 7.55.

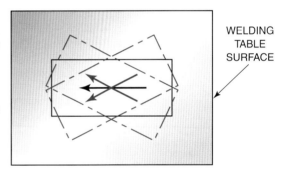

WELDING
TABLE
SURFACE

FIGURE 7.55 Change the plate angle to find
the most comfortable welding position

PRACTICE WELDS

Practice welds are grouped according to the type of joint and the type of welding electrode. The welder or instructor should select the order in which they are made. The stringer beads should be practiced first in each position before the welder tries the different joints in each position. Some time can be saved by starting with the stringer beads. If this is done, it is not necessary to cut or tack the plate together, and a number of beads can be made on the same plate.

Starting with the flat position allows the welder to build skills slowly. Practice in making the basic joints will give you the confidence to tackle other welding positions. Horizontal T and lap welds are almost as easy to make as the flat welds. Overhead welds are as simple to make as vertical welds, but they are harder to position.

TIP

Horizontal butt welds are more difficult to perform than most other welds.

Electrodes

Electrodes are often characterized by their flux composition and this determines what applications they are suitable for and typical operating characteristics. The flux coating has a number of functions:

● It provides a gas shield which prevents atmospheric contamination of the molten metal.

● It helps to ionize the arc gap and set up the conditions for transfer of molten droplets to the weld pool.

- It provides a protective coating which helps slow the cooling rate and refine the grain structure.
- It acts as a mould to contain the molten metal in positional welding.
- It can be a source of alloying in the weld zone.
- It can increase the deposition rate as in iron powder electrodes.
- It can contain elements which increase the arc voltage (as in cellulosic electrodes).

- *Iron Powder Electrodes.* E7024 and E7028 electrodes are produced with iron-powder-based fluxes that are substantially thicker than the other groups. They have very high deposition rates and produce welds with medium to low penetration, but are restricted to butt welds in the flat position and fillet welds in the flat or horizontal positions.
- *Rutile Electrodes.* E6012 and E6013 electrodes have rutile-based fluxes, giving a smooth, easy arc with a thick slag left on the weld bead. They may be used in all positions and do not require special rod ovens. These electrodes are popular because the slag is easily removed, and they may be used with all polarities to produce welds with medium to low penetration.
- *Cellulosic Electrodes.* E6010 and E6011 electrodes have cellulose-based fluxes. They have a forceful deep penetrating arc (hydrogen releasing) with little slag left on the weld bead. They do not require a rod oven, and are often chosen when surface conditions on the base metal are less than optimal. They are also commonly used for open root welds on plate and pipe.
- *Low Hydrogen Electrodes.* E7016 and E7018 electrodes have a mineral-based flux. The resulting medium penetration arc is smooth and easy, with a very heavy slag left on the weld bead. These electrodes are also referred to as low hydrogen or low-hi electrodes. They require special handling and storage in a rod oven after being removed from their factory packaging. Refer to manufacturer's requirements or the applicable welding code for specific handling directions. They are also available in vacuum packs, which prevents moisture absorption especially when working on site.

The cellulose and rutile-based groups of electrodes are the best for starting specific welds. Electrodes with cellulose-based fluxes do not have heavy slags that may interfere with the welder's view of the weld, an advantage for flat T and lap joints. Electrodes with the rutile-based fluxes (giving an easy arc with low spatter) are easier to control and are used for flat stringer beads and butt joints.

Unless a specific electrode has been required by a welding procedure specification (WPS), welders can select the best electrode for a specific weld. An accomplished welder should be capable of making defect-free welds on all types of joints using all types of electrodes in any weld position.

Stringer Beads

A straight weld bead on the surface of a plate, with little or no side-to-side electrode movement, is known as a stringer bead. Stringer beads are used by students to practise maintaining arc length and electrode angle so that their welds will be straight, uniform and free from defects. Stringer beads, Figure 7.56, are also used to set the machine amperage and for build up or surfacing applications and are the most commonly used type of bead for vertical, horizontal and overhead welds.

FIGURE 7.56 **Stringer bead**

A straight weld is easily made once the welder develops the ability to view the entire welding zone. At first, the welder sees only the arc. With practice, the welder begins to see parts of the molten weld pool. After much practice, the welder will see the molten weld pool (front, back and both sides), slag, build up and the surrounding plate. Often, at this skill level, the welder may not even notice the arc.

After making practice stringer beads, a variety of weave bead patterns should be practised to gain the ability to control the molten weld pool when welding out of position.

Restarting a Weld Bead

On all but short welds, the welding bead will need to be restarted after a welder stops to change electrodes. Because the metal cools as a welder changes electrodes and chips slag when restarting, the penetration and build up may be adversely affected.

When a weld bead is nearing completion, it should be tapered by increasing the travel rate just before welding stops. A 6mm taper is all that is required. The taper allows the new weld to be started and the depth of penetration re-established without having excessive build up, Figure 7.57.

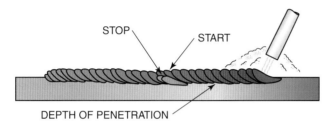

FIGURE 7.57 **Tapering the size of the weld**

> **TIP** ⊙
>
> Tapering the size of the weld bead helps keep the depth of penetration uniform.

The slag should always be chipped and the weld crater cleaned before restarting the weld. This is important to prevent slag inclusions at the start of the weld. Restart the arc in the joint ahead of the weld. Allow the electrodes to heat up so that the arc is stabilized and a shielding gas is re-established to protect the weld. Hold a long arc as the electrode heats up so that metal is not deposited. Slowly bring the electrode downward and toward the weld bead until the arc is directly on the deepest part of the crater where the crater meets the plate in the joint, Figure 7.58. The

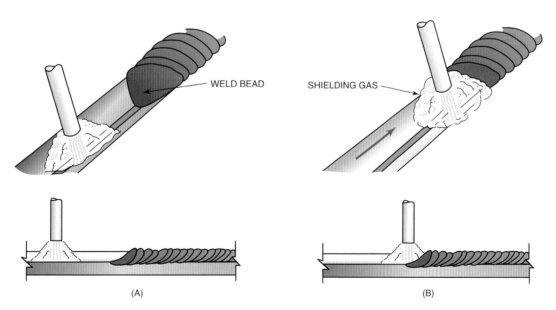

FIGURE 7.58 **When restarting the arc, strike the arc ahead of the weld in the joint (A). Hold a long arc and allow time for the electrode to heat up, forming the protective gas envelope. Move the electrode so that the arc is focused directly on the leading edge (root) of the previous weld crater (B)**

electrode should be low enough to start transferring metal. Next, move the electrode in a semicircle near the back edge of the weld crater. Watch the build up and match your speed to the deposit rate so that the weld is built up evenly, Figure 7.59. Move the electrode ahead and continue with the same weave pattern that was being used previously.

The movement to the root of the weld and back up on the bead helps build up the weld and reheat the metal so that the depth of penetration will remain the same. If the weld bead is started too quickly, penetration is reduced and build up is high and narrow. Avoid starting and stopping weld beads in corners, where tapering and restarting are especially difficult, often resulting in defects, Figure 7.60.

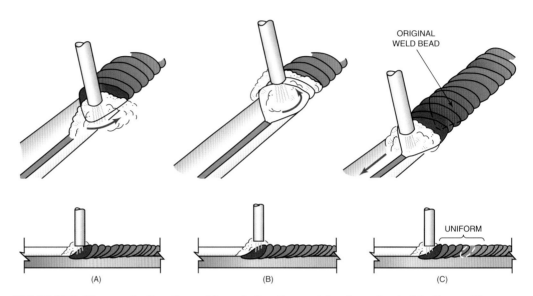

FIGURE 7.59 **When restarting the weld pool after the root has been heated to the melting temperature, move the electrode upward along one side of the crater (A). Move the electrode along the top edge, depositing new weld metal (B). When the weld is built up uniformly with the previous weld, continue along the joint (C)**

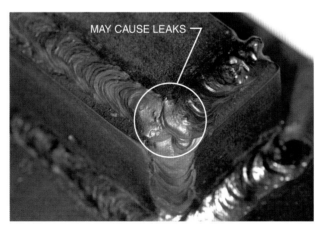

FIGURE 7.60 **Incorrect method of welding through a corner**

TABLE 7.3 Pre-heat temperatures for arc welding on low-carbon steels*

Plate Thickness (mm)	Minimum Temperature °C
Up to 13mm	21
13mm to 25mm	38
25mm to 51mm	95
Over 51mm	150

*Metal should be above the dew point.

Allow 1 hour for each inch in order to provide uniform heating or for localized preheating. Check to ensure that preheat temperatures are sufficient to melt thermal crayons a minimum of 76 mm in all directions from the area to be welded prior to welding arc application.

WORKSHOP TASK 7.1

Single Layer Pad in the Flat Position

Using a properly set-up and adjusted arc welding machine, and wearing appropriate PPE and 3.25mm electrodes you will deposit a single layer pad on a piece of low carbon steel 150mm × 100mm × 6mm thick in the flat position.

- Mark a distance 15mm in from all edges to establish a box in which the single layer pad will be contained in.

- Strike and establish the arc and move along the outside line to deposit the first run along the 120mm length, and terminate using the correct technique.

- Strike the arc to the side of the first run and continue the second run by overlapping the first run to maintain a consistent height along the run of weld.

- Continue with further runs until the box is completed, each run overlapping with the previous run and maintaining consistency on height.

- Practise pick-ups as well during this exercise, so that you will gain the confidence you will need to ensure you get the maximum deposition from the electrodes and therefore save on costing.

- Turn off machine and re-instate the work area when you are finished welding. Place all stubs and surplus materials in the appropriate containers.

- Cut across the centre of the plate, file and polish to a smooth finish and then apply an 'etch-ant' (acid based fluid) to reveal the detail (Macro-examination) such as penetration into base material and fusion of each run.

Complete a copy of the 'Student Welding Report' listed in Appendix I or provided by your instructor.

TIP

To make horizontal stringer beads use the same setup, material and electrodes as in workshop task 7.1. When practising recline the angle slightly, Figure 7.61. This placement allows the welder to build the required skill by practising the correct techniques successfully.

Square Butt Joint

The square butt joint is made by tack welding two flat pieces of plate together, Figure 7.62. The space between the plates is called the root opening or root gap. Changes in the root opening will affect penetration. As the space increases, the weld penetration increases. The root opening for most butt welds will vary from 1.0 – 2.0mm. Excessively large openings can cause burn-through or a cold lap at the weld root, Figure 7.63. If the plate strips are no longer flat after the weld has been cut off, they can be tack welded together and flattened with a hammer, Figure 7.64.

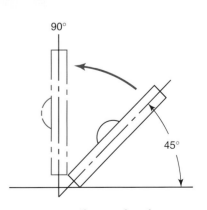

FIGURE 7.61 Change the plate angle as welding skill improves

FIGURE 7.62 Tack weld should be small and uniform to minimize effect on final weld

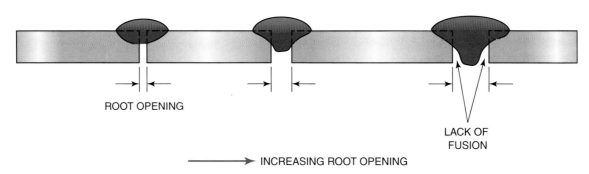

ROOT OPENING

LACK OF
FUSION

INCREASING ROOT OPENING

FIGURE 7.63 Effect of root opening on weld penetration

WORKSHOP PRACTICE 7.2

Square Butt Joint in the Flat Position

Using a properly set-up and adjusted arc welding machine, appropriate PPE, 2.5mm and 3.25mm electrodes, and two pieces of mild steel plate, 150mm × 50mm × 6mm thick, low carbon steel plate you will make a welded butt joint in the flat position, Figure 7.66.

● Tack weld the plates together and place them flat on the welding table. Starting at one end, establish a molten weld pool on both plates.

● Hold the electrode in the molten weld pool until it flows together, Figure 7.67. After the gap is bridged by the molten weld pool, start weaving the electrode slowly back and forth across the joint. Moving the electrode too quickly from side to side may result in slag being trapped in the joint, Figure 7.65.

● Continue the weld along the 150mm length of the joint and bring the electrode up to the vertical position followed by a circular action to ensure correct termination of the weld.

● Cool, chip and inspect the weld for uniformity and soundness.

● Repeat the welds as needed to master all groups of electrodes in this position.

● Turn off the welding machine and re-instate the work area when you are finished welding. Place all stub ends and surplus materials in appropriate containers.

Complete a copy of the 'Student Welding Report'.

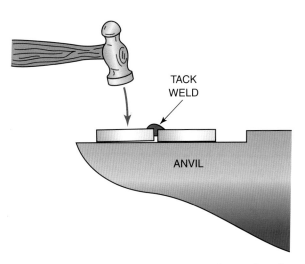

FIGURE 7.64 After plates are tack welded together they can be forced into alignment by striking with a hammer

FIGURE 7.65 Moving the electrode from side to side too quickly

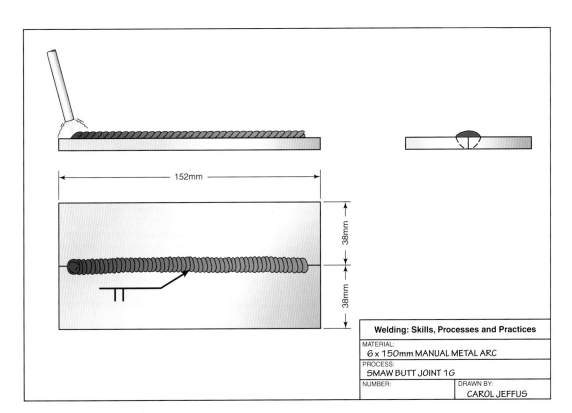

152mm

38mm

38mm

Welding: Skills, Processes and Practices

MATERIAL:
6 x 150mm MANUAL METAL ARC

PROCESS:
SMAW BUTT JOINT 1G

NUMBER:	DRAWN BY:
	CAROL JEFFUS

FIGURE 7.66 Square butt joint in flat position

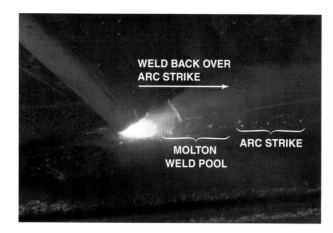
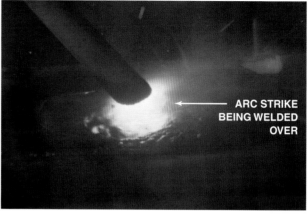

FIGURE 7.67 After the arc is established, hold the area long enough to establish the molten weld pool desired then weld back over the arc strike to melt into the weld

Outside Corner Joint

An outside corner joint is made by placing the plates at a 90° angle to each other, with the edges forming a V-groove, Figure 7.68. There may or may not be a slight root opening left between the plate edges. Small tack welds should be made approximately 12mm from both ends of the joint. The weld bead should completely fill the V-groove formed by the plates and may have a slightly convex surface build up. The back side of an outside corner joint can be used to practise fillet welds, or four plates can be made into a box tube shape, Figure 7.69.

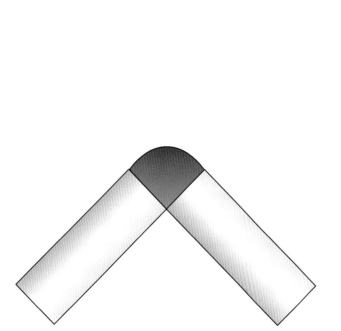
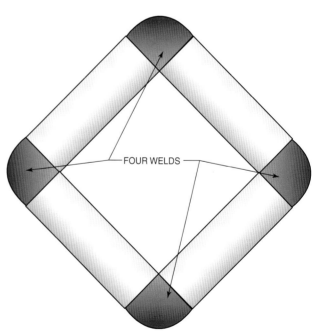

FIGURE 7.68 V formed by an outside corner joint

FIGURE 7.69 Box tube from four outside corner joint welds

WORKSHOP TASK 7.3

Outside Corner in the Flat Position

Using a properly set-up and adjusted arc welding machine, appropriate PPE and 2.5mm and 3.25mm diameter arc welding electrodes with two pieces of mild steel plate, 150mm × 50mm × 6mm thick low carbon steel plate, you will make a weld on an outside corner joint.

- The plates should be tack welded and checked for angular accuracy.
- Starting at one end of the plate, make a straight weld the full length of the plate.
- Watch the molten weld pool at this point, not the end of the electrode. As you become more skillful, it is easier to watch the molten weld pool.
- Cool, chip and inspect the weld for uniformity and defects.
- Repeat the welds as needed with all groups of electrodes until you can consistently make welds free of defects.
- Turn off the welding machine and re-instate your work area when you are finished welding. Place all welding stubs and surplus materials in the appropriate containers.

  Complete a copy of the 'Student Welding Report'.

Lap Joint

A lap joint is made by overlapping the edges of the two plates, Figure 7.70. The joint can be welded on one side or both sides with a fillet weld. As the fillet weld is made on the lap joint, the build up should equal the thickness of the plate, Figure 7.71. A good weld will have a smooth transition from the plate surface to the weld. If this transition is abrupt, it can cause stresses that will weaken the joint. Penetration for lap joints does not improve their strength; complete fusion is required. The root of fillet welds must be melted to ensure a completely fused joint. If the molten weld pool shows a notch during the weld, Figure 7.72, this is an indication that the root is not being fused together. The weave pattern will help prevent this problem, Figure 7.73.

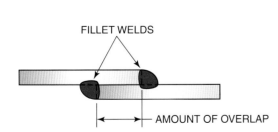

FILLET WELDS

AMOUNT OF OVERLAP

FIGURE 7.70 Lap joint

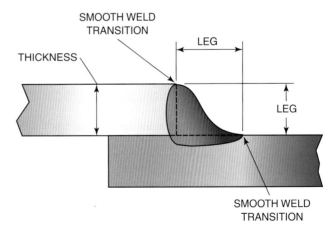

SMOOTH WELD TRANSITION

THICKNESS

LEG

LEG

SMOOTH WELD TRANSITION

FIGURE 7.71 Legs of a fillet weld generally should be equal to thickness of the base metal

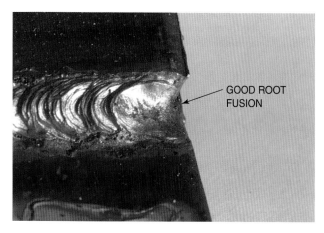

FIGURE 7.72 Watch the root of the weld bead to be sure there is a complete fusion

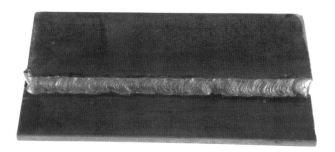

FIGURE 7.73 Lap joint

WORKSHOP TASK 7.4

Lap Fillet in the Flat Position

Using a properly set-up and adjusted arc welding machine, wearing appropriate PPE and with 2.5mm and 3.25mm electrodes, and two pieces of low carbon steel plate 150mm × 50mm × 6mm, you will make a welded lap joint in the flat position, Figure 7.74.

- Hold the plates together tightly with an overlap of no more than 10mm. Tack weld the plates together. A small tack weld may be added in the centre to prevent distortion during welding, Figure 7.75. Chip the tacks before you start to weld.
- The 'J,' 'C' or zigzag weave pattern works well on this joint. Strike the arc and establish a molten pool directly in the joint.
- Move the electrode out on the bottom plate and then onto the weld to the top edge of the top plate, Figure 7.76. Follow the surface of the plates with the arc.
- Do not follow the trailing edge of the weld bead. Following the molten weld pool will not allow for good root fusion and will cause slag to collect in the root.
- If slag does collect, a good weld is not possible. Stop the weld and chip the slag to remove it before the weld is completed.
- Cool, chip and inspect the weld for uniformity and defects.
- Repeat the welds with all groups of electrodes until you can consistently make welds free of defects.
- Turn off the welding machine and re-instate your work area when you are finished welding. Place all stubs and surplus materials in appropriate containers.

Complete a copy of the 'Student Welding Report'.

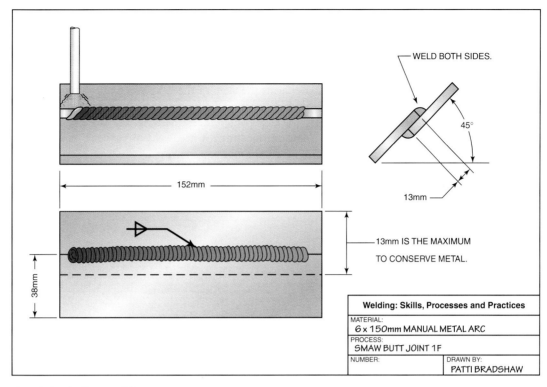

FIGURE 7.74 **Lap joint in flat position**

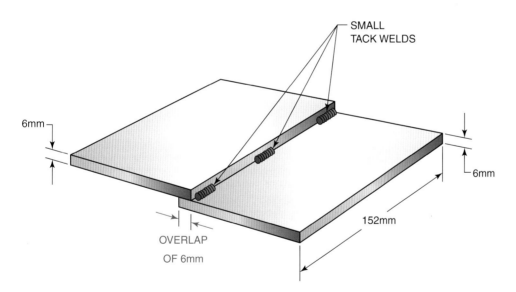

FIGURE 7.75 **Tack welding the plates together**

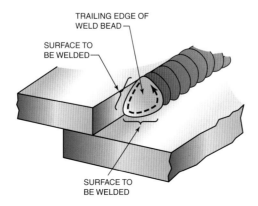

FIGURE 7.76 **Follow the surface of the plate to ensure good fusion**

T-Joint

The T-joint is made by tack welding one piece of metal on another piece of metal at a right angle, Figure 7.77. After the joint is tack welded together, the slag is chipped from the tack welds so that it is not included in the final weld. The heat is not distributed uniformly between both plates during a T-weld. Because the plate forms the stem of the T it can conduct heat away from the arc in only one direction, it will heat up faster than the base plate. When using a weave pattern, most of the heat should be directed to the base plate to keep the weld size more uniform and to help prevent undercut. A welded T-joint can be strong if it is welded on both sides, even without having deep penetration, Figure 7.78. The weld will be as strong as the base plate if the size of the two welds equals the total thickness of the base plate. The weld bead should have a flat or slightly concave appearance to ensure the greatest strength and efficiency, Figure 7.79.

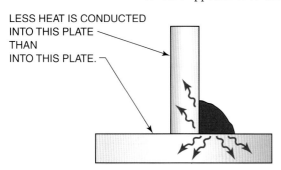

FIGURE 7.77 T-joint

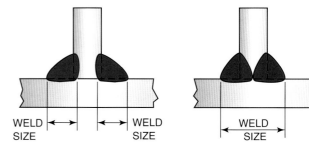

FIGURE 7.78 If total weld sizes are equal then both T-joints have equal strength

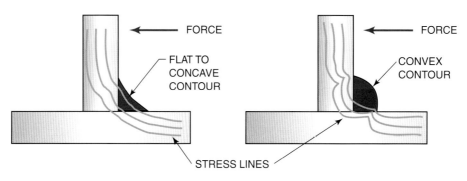

FIGURE 7.79 The stresses are distributed more uniformly through a flat or concave fillet weld

WORKSHOP TASK 7.5

T-fillet in Flat Position

Using a properly set-up and adjusted arc welding machine, wearing appropriate PPE and 2.5mm and 3.25mm electrodes, and two pieces of low carbon steel plate 150mm × 50mm × 6mm thick you will make a welded T-joint in the flat position, Figure 7.80.

● After the plates are tack welded together, place them on the welding table so the weld will be flat.

● Start at one end and establish a molten weld pool on both plates. Allow the molten weld pool to flow together before starting the bead.

● Any of the weave patterns will work well on this joint.

● To prevent slag inclusions, use a slightly higher than normal amperage setting. When the 150mm long weld is completed, cool, chip and inspect it for uniformity and soundness.

● Repeat the welds as needed for all groups of electrodes until you can consistently make welds free of defects.

● Turn off the welding machine and re-instate your work area when you are finished welding. Place all stubs and surplus materials in the appropriate containers.

Complete a copy of the 'Student Welding Report'.

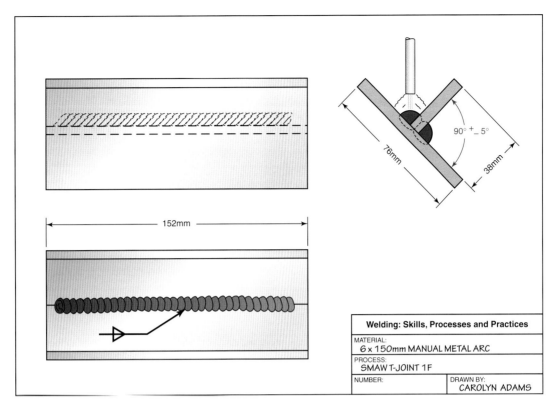

FIGURE 7.80 T-joint in the flat position

Root Pass

The root pass is the first weld bead of a multiple pass weld. It fuses the two parts together and establishes the depth of weld metal penetration. A good root pass is needed in order to obtain a sound weld. The root may be either open or closed, using a backing strip or backing ring, Figure 7.81.

The backing strip used in a closed root may remain as part of the weld, or it may be removed. Leaving the backing strip on a weld may cause it to fail due to concentrations of stresses, and removable back up tapes have been developed. Back up tapes are made of high-temperature ceramics, Figure 7.82, that can be used to increase penetration and prevent burn-through. The tape can be peeled off after the weld is completed. Most welds do not use backing strips.

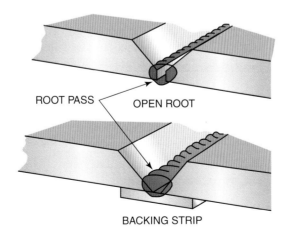

FIGURE 7.81 Root pass maximum deposit 6mm thick

(A) FIBREGLASS

(B) WELD ROOT PASS MADE
USING CERAMIC BACKING TAPE

FIGURE 7.82 **Welding backing tapes**

On plates that have the joints prepared on both sides, the root face may be ground or gouged clean before another pass is applied to both sides, Figure 7.83. This practice has been applied to some large diameter pipes. Welds that can be reached from only one side must be produced adequately, without the benefit of being able to clean and repair the back side.

The open root weld is widely used in plate and pipe designs. The face side of an open root weld is not so important as the root surface on the back or inside, Figure 7.84. The face may have some areas of poor uniformity in width, reinforcement, and build up, or other defects, such as undercut or overlap. As long as the root surface is correct, the front side can be ground, gouged, or burned out to produce a sound weld, Figure 7.85. For this reason, during the root pass practices, the weld will be evaluated from the root side only, as long as there are not too many defects on the face. To practise the open root welds, you will be using mild steel plate that is 3mm thick. The root face for most butt joints will be about the same size. This thin plate helps build skill without taking too much time beveling the plate. Two different methods are used to make a root pass. One method is used only on joints with little or no root gap and requires a high amperage and short arc length. The arc length is so short that the electrode flux may drag along on the edges of the joint. The setup for this method must be correct in order for it to work.

The other method can be used on joints with wide, narrow or varying root gaps. A stepping electrode manipulation and keyhole control the penetration. The electrode is moved in and out of the molten weld pool as the weld progresses along the joint. The edge of the metal is burned back slightly by the electrode just ahead of the molten weld pool, Figure 7.86. This is referred to as a 'keyhole', and metal flows through the keyhole to the root surface. The keyhole must be maintained to ensure 100 per cent penetration. This method requires more welder skill and can be used on a wide variety of joint conditions. The face of the bead resulting from this technique often is defect free.

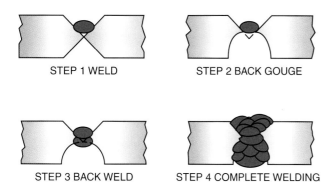

STEP 1 WELD

STEP 2 BACK GOUGE

STEP 3 BACK WELD

STEP 4 COMPLETE WELDING

FIGURE 7.83 **Using back gouging to ensure a sound weld root**

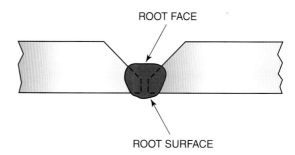

ROOT FACE

ROOT SURFACE

FIGURE 7.84 **Ideal bead shape for the root pass**

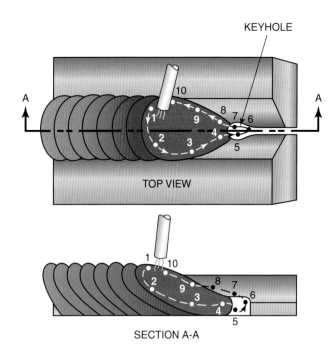

FIGURE 7.86 **Electrode movement to open and use a keyhole**

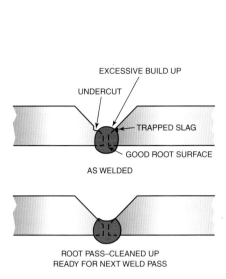

FIGURE 7.85 **Grinding back the root pass to ensure a sound pass**

WORKSHOP TASK 7.6

Root Pass on Plate With Backing Strip

Using properly set-up and adjusted welding machine, wearing appropriate PPE and E6013 electrodes 2.5mm and 3.25mm, and pieces of 150mm × 50mm × 3mm plate, Figure 7.87. Tack weld the plates together with a 1.0mm – 1.5mm root opening.

● Be sure there are no gaps between the backing strip and plates when the pieces are tacked together, Figure 7.88.

● If there is a small gap between the backing strip and the plates, it can be removed by placing the assembled test plates on an anvil and striking the tack weld with a hammer. This will close up the gap by compressing the tack welds, Figure 7.89.

● Use a straight step or T pattern for this root weld. Push the electrode into the root opening so that there is good fusion with the backing strip and bottom edge of the plates.

● Failure to push the penetration deep into the joint will result in a cold lap at the root, Figure 7.90.

● Watch the molten weld pool and keep its size as uniform as possible. As the molten weld pool increases in size, move the electrode out of the weld pool.

● When the weld pool begins to cool, bring the electrode back into the molten weld pool.

● Use these weld pool indications to determine how far to move the electrode and when to return to the molten weld pool.

● After completing the weld, cut the plate and inspect the cross section of the weld for good fusion at the edges.

● Repeat the welds as necessary until you can consistently make welds free of defects.

● Turn off the welding machine and clean up your work area when you are finished welding.

Complete a copy of the 'Student Welding Report'.

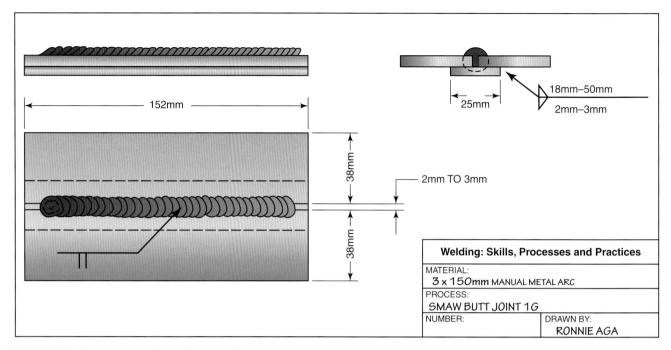

152mm

38mm

38mm

2mm TO 3mm

18mm–50mm

25mm

2mm–3mm

Welding: Skills, Processes and Practices

MATERIAL:
3 x 150mm MANUAL METAL ARC

PROCESS:
SMAW BUTT JOINT 1G

NUMBER:

DRAWN BY:
RONNIE AGA

FIGURE 7.87 Square butt joint with a backing strip

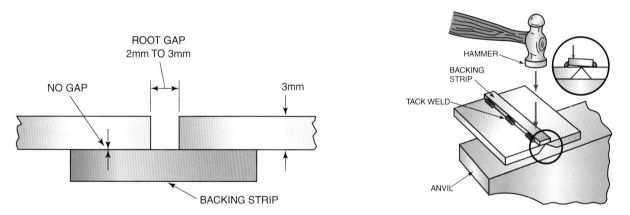

ROOT GAP
2mm TO 3mm

NO GAP

3mm

BACKING STRIP

FIGURE 7.88 Backing strip

HAMMER

BACKING
STRIP

TACK WELD

ANVIL

FIGURE 7.89 Using a hammer to align the backing strip and weld plates

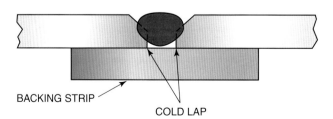

BACKING STRIP

COLD LAP

FIGURE 7.90 Incomplete root fusion

WORKSHOP TASK 7.7

TIP

Rocking the top of the electrode while keeping the end in the same place helps control the bead shape.

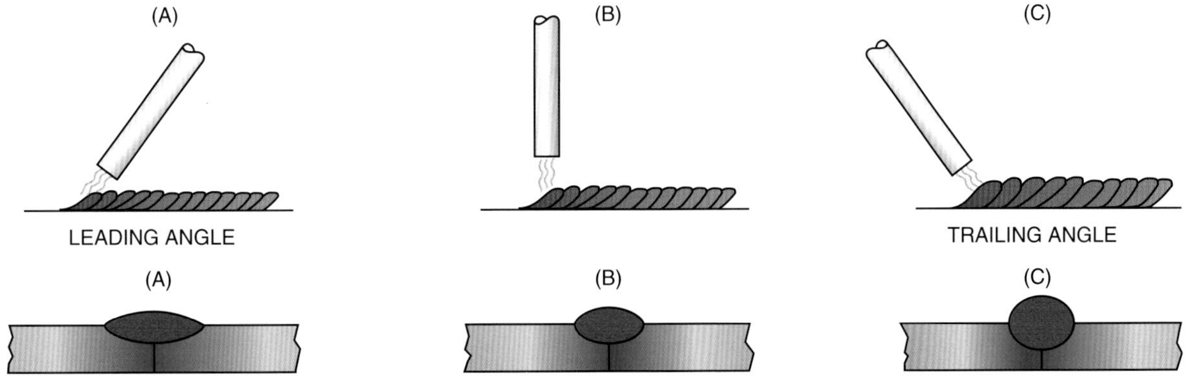

Root Pass on Plate With an Open Root

Using a properly set-up and adjusted arc welding machine, wearing appropriate PPE and E6013 arc welding electrodes with a 2.5mm and 3.25mm diameter, and two or more pieces of mild steel plate, 150mm × 50mm × 6mm.

● Tack weld the plates together with a root opening of 1.0mm – 2.5mm, use of a stub end helps to maintain the gap.

● Using a short arc length and correct amperage setting, make a weld along the joint.

● As the trailing angle is decreased, making the electrode flatter to the plate, penetration, depth and burn-through decrease, Figure 7.91, because both the arc force and heat are directed away from the bottom of the joint back toward the weld.

● Surface tension holds the metal in place, and the mass of the bead quickly cools the molten weld pool holding it in place.

● Increasing the electrode angle toward the perpendicular will increase penetration depth and possibly cause more burn-through. The arc force and heat focused on the gap between the plates will push the molten metal through the joint.

● The electrode holder can be slowly weaved from side to side while keeping the end of the electrode in the same spot on the joint, Figure 7.92. This will allow the arc force to fuse better in the sides of the root to the base metal.

● When a burn-through occurs, rapidly move the electrode back to a point on the weld surface just before the burn-through.

● Lower the electrode angle and continue welding. If the burn-through does not close, stop the weld, chip and wire brush the weld.

● Check the size of the burn-through. If it is larger than the diameter of the electrode, the root pass must be continued with the step method.

● If the burn-through is not too large, lower the amperage slightly and continue welding.

● Watch the colour of the slag behind the weld. If the weld metal is not fusing to one side, the slag will be brighter in colour on one side. The brighter colour is caused by the slower cooling of the slag because there is less fused metal to conduct the heat away quickly.

● After the weld is completed, cooled and chipped, check the back side of the plate for good root penetration.

● The root should have a small bead no more than 3mm that will look as though it was welded from the back side, Figure 7.93. The penetration must be completely free of any drips of metal from the root face.

● Repeat the welds as necessary until you can consistently make welds free of defects.

● Turn off the welding machine and re-instate work area when you are finished welding.

● Cool all components down and put any stubs or surplus materials in the appropriate containers.

Complete a copy of the 'Student Welding Report'.

(A)   (B)   (C)

LEADING ANGLE   TRAILING ANGLE

(A)   (B)   (C)

FIGURE 7.91 **Effect of rod angle on weld bead shape**

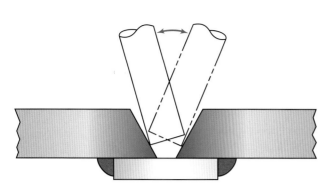

FIGURE 7.92 **Rocking the top of the electrode**

FIGURE 7.93 **The weld toes appear uniform**

Open Root Weld on Plate Using the Step Technique in All Positions

- Using the same setup, the electrode should be pushed deeply into the root to establish a keyhole that will be used to ensure 100 per cent root penetration. Once the keyhole is established, the electrode is moved out and back in the molten weld pool at a steady, rhythmic rate.

- Watch the molten weld pool and keyhole size to determine the rhythm and distance of electrode movement.

- If the molten weld pool size decreases, the keyhole will become smaller and may close completely.

- To increase the molten weld pool size and maintain the keyhole, slow the rate of electrode movement and shorten the distance the electrode is moved away from the molten weld pool.

- This will increase the molten weld pool size and penetration because of increased localized heating.

- If the molten weld pool becomes too large, metal may drip through the keyhole, forming on the back side of the plate. Extremely large molten weld pool sizes can cause a large hole to be formed or cause burn-through.

- Repairing large holes can require much time and skill. To keep the molten weld pool from becoming too large, increase the travel speed, decrease the angle, shorten the arc length or lower the amperage, Table 7.4.

- The distance the electrode is moved from the molten weld pool and the length of time in the molten weld pool are found by watching the molten weld pool.

- The molten weld pool size increases as you hold the arc in the molten weld pool until it reaches the desired size, about twice the electrode diameter, Figure 7.94.

- Move the electrode ahead of the molten weld pool, keeping the arc in the joint but being careful not to deposit any slag or metal ahead of the weld.

- To prevent metal and/or slag from transferring, raise the electrode to increase the arc length, Figure 7.95.

- Keep moving the electrode slowly forward as you watch the molten weld pool. The molten weld pool will suddenly start to solidify.

- At that time, move the electrode quickly back to the molten weld pool before it totally solidifies. Moving the electrode in a slight arc will raise the electrode ahead of the molten weld pool and automatically lower the electrode when it returns to the molten weld pool.

- Metal or slag deposited ahead of the molten weld pool may close the keyhole, reduce penetration and cause slag inclusions.

- Raising the end of the electrode too high or moving it too far ahead of the molten weld pool can cause all of the shielding gas to be blown away from the molten weld pool.
- If this happens, oxides can cause porosity. Keeping the electrode movement in balance takes concentration and practice.
- Changing from one welding position to another requires an adjustment in timing, amperage and electrode angle. The flat, horizontal and overhead positions use about the same rhythm, but the vertical position may require a shorter time cycle for electrode movement.
- The amperage for the vertical position can be lower than that for the flat or horizontal, but the overhead position uses nearly the same amperage as flat and horizontal.
- The electrode angle for the flat and horizontal positions is about the same. For the vertical position, the electrode uses a sharper leading angle than the overhead position, which is nearly perpendicular and may even be somewhat trailing.
- Cool, chip, wire brush and inspect both sides of the weld.
- The root surface should be slightly built up and look as though it was welded from that side (refer to Figure 7.93).
- Repeat the welds as necessary until you can consistently make welds free of defects.

TABLE 7.4 Changes affecting molten weld bead size

	Amperage	Travel Speed	Electrode Size	Electrode Angle
To decrease puddle size	Decrease	Increase	Decrease	Leading
To increase puddle size	Increase	Decrease	Increase	Trailing

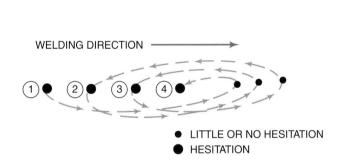

FIGURE 7.94 **Weave pattern**

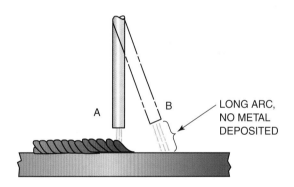

FIGURE 7.95 **A long arc prevents metal or slag from being deposited ahead of the weld bead**

Hot Pass

The surface of a root pass may contain some defects, depending upon the type of weld, its condition and the code or standards. The surface of a root pass can be cleaned by grinding or by using a hot pass. On critical, high-strength code welds, it is usually required that the root pass, as well as each filler pass, be ground (refer to Figure 7.85). This grinding eliminates weld imperfections caused by slag entrapments. It can also be used to remove most of the E60 series weld metal so that the stronger weld metal can make up most of the weld. When high-strength, low-alloy welding electrodes are used, grinding is important to remove most of the low-strength weld deposit, Figure 7.96.

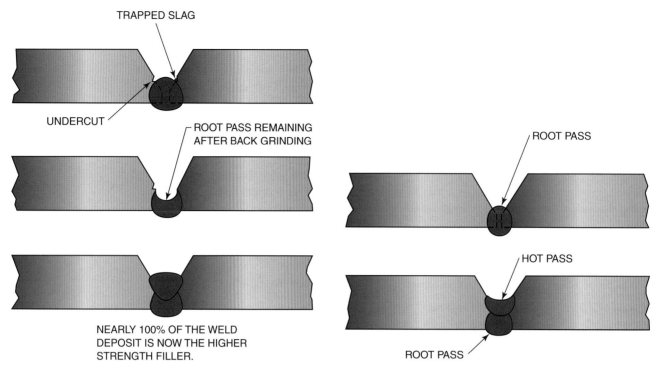

FIGURE 7.96 Back grinding to remove discontinuities

FIGURE 7.97 Using hot pass to clean up the face of the root pass

The fastest way to clean out trapped slag and make the root pass more uniform is to use a hot pass. This uses a higher than normal amperage setting and a fast travel rate to reshape the bead and burn out trapped slag. After chipping and wire brushing the root pass to remove all the slag possible, a welder is ready to make the hot pass. The best way to do this is to rapidly melt a large surface area, Figure 7.97. The slag, mostly silicon dioxide (SiO_2), may not melt itself, so the surrounding steel must be melted to enable it to float free. (The silicon dioxide melts at about 1700°C, which is more than 270°C hotter than the temperature at which the surrounding steel melts.) A very small amount of metal should be deposited during the hot pass so that the resulting weld is concave, which is more easily cleaned than a convex weld. Failure to clean the convex root weld will result in an imperfection showing up on an X-ray. Such imperfections are called 'wagon tracks' (slag traps), Figure 7.98.

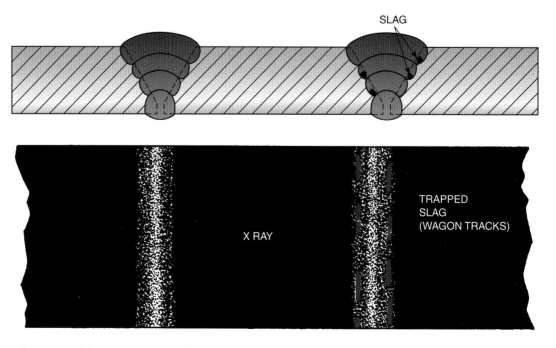

FIGURE 7.98 Slag trapped between passes will show on an X ray

The hot pass can also be used to repair or fill small spots of incomplete fusion or pinholes left in the root pass. The normal weave pattern for a hot pass is the straight step or 'T' pattern. The 'T' can be used to wash out stubborn trapped slag better than the straight step pattern. The frequency of electrode movement is dependent upon the time required for the molten weld pool to start cooling. As with the root pass, metal or slag should not be deposited ahead of the bead. Do not allow the molten weld pool to cool completely or let the shielding gas covering to be blown away from the molten weld pool. The hot pass technique can also be used to clean some welds that may first require grinding or gouging for a repair. The penetration of the molten weld pool must be deep enough to free all trapped slag and burn out all porosity.

Filler Pass

After the root pass is completed and has been cleaned, the groove is filled with weld metal. These filler passes are made with stringer beads or weave beads. More than one pass is often required and the weld beads must overlap along the edges so that the finished bead is smooth, Figure 7.99. Stringer beads usually overlap about 50 per cent, and weave beads overlap approximately 25 per cent. Each weld bead must be cleaned before the next is started. Slag left on the plate between welds cannot be completely burned out because filler welds should be made with a low amperage setting. Deep penetration will slow the rate of build up in the joint and deeply re-melting the previous weld metal may weaken the joint. All that is required of a filler weld is that it be completely fused to the base metal.

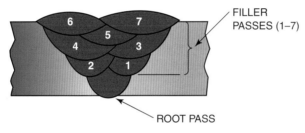

FIGURE 7.99 Burning out trapped slag

Chipping, wire brushing and grinding are the best ways to remove slag between filler weld passes. After the weld is completed, it can be checked by ultrasonic or radiographic non-destructive testing. As most colleges or training facilities are not equipped for this, a quick check for soundness can be made by destructive testing. One method to inspect the weld is to cut and cross-section it with an abrasive wheel.

Cover Pass

The last weld bead on a multiple pass weld is known as the cover pass. It may use a different electrode weave, or it may be the same as the filler beads. The cover pass should be uniform and neat looking. Most welds are not tested, and often appearance is the only factor used for accepting or rejecting welds. The cover pass should be free of any visual defects such as undercut, overlap, porosity or slag inclusions and be uniform in width and reinforcement. A cover pass should not be more than 3mm wider than the groove opening, Figure 7.100. Cover passes that are too wide do not add to the weld strength.

HEALTH & SAFETY

The hot pass technique is designed to be used on non-critical, non-coded welds only. It should not be used to cover bad welds or as a means of repairing the work of a welder who is less skilled.

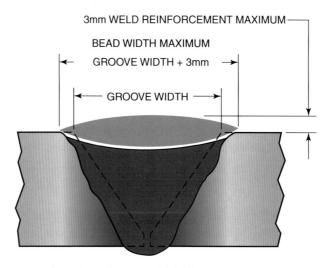

FIGURE 7.100 Cover pass should not be excessively large

Run-on and run-off tabs are tacked on at the front of the joint to establish the arc and ensure sufficient heat input at the start of the weld. They are also tacked at the end of the joint to act as a heat sink on the completion of the weld deposit. When the weld has been completed they are removed.

PLATE PREPARATION

When welding on thick plate, it is impossible to achieve 100 per cent penetration without preparing the plate for welding. This usually takes the form of a V, U or J preparation, which can be cut into one side or both sides of the plate, and into either just one plate or both plates of the joint, Figure 7.101. The type, depth, angle and location of the preparation are usually determined by a code standard that has been qualified for the specific job.

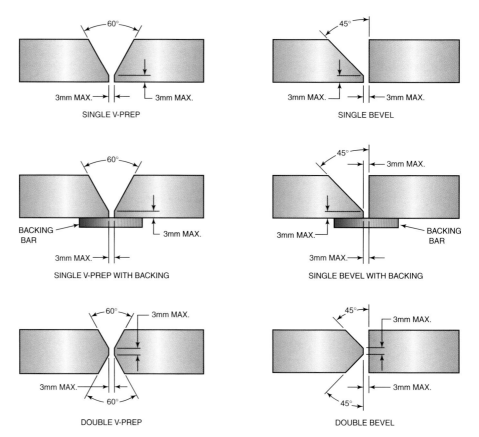

FIGURE 7.101 Typical butt joint preparation

For MMA welds on plate 6mm or thicker that need to have a weld with 100 per cent joint penetration, the plate must be prepared. The preparation may be ground, flame-cut, gouged or machined on the edge of the plate before assembly. Bevels and V-preparations are best if they are cut before the parts are assembled. J-grooves and U-grooves can be cut either before or after assembly, Figure 7.102. Plates that are thicker than 10mm can be prepared on both sides but may be prepared on only one side. This choice is determined by joint design, position and application. A T-joint in thick plate is easier to weld and will have less distortion if it is bevelled on both sides. Plate in the flat position is usually prepared on only one side unless it can be repositioned. Welds that must have little distortion or that are going to be loaded equally from both sides are usually prepared on both sides. Sometimes plates are either prepared and welded or just welded on one side, and then back gouged and welded, Figure 7.103. **Back gouging** is a process of cutting a groove in the back side of a joint that has been welded. It can ensure 100 per cent fusion at the root and remove discontinuities of the root pass, and also remove the root pass metal if its properties are not desirable to the finished weld. After back gouging, the groove is then welded.

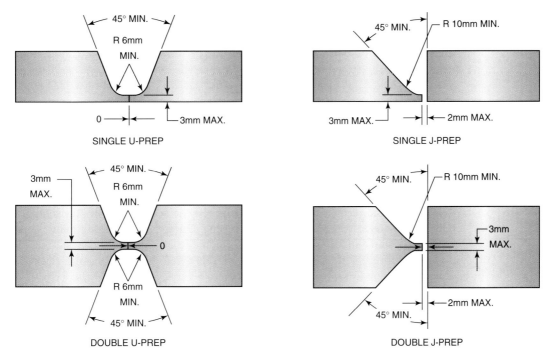

FIGURE 7.102 Typical butt joint preparation

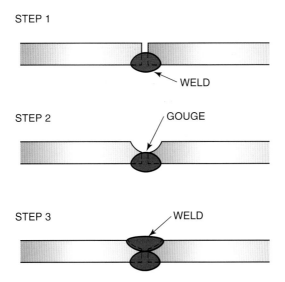

FIGURE 7.103 Back gouging sequence for a weld to ensure 100% joint penetration

Pre-Heating and Post-Heating

Pre-heating is the application of heat to the metal before it is welded. This process helps to reduce cracking, hardness, distortion and stresses by reducing the thermal shock from the weld and slowing the cool-down rate. It is most often required on large, thick plates, when the plate is very cold, on days when the temperature is very cold, when small diameter electrodes are used, on high-carbon or manganese steels, on complex shapes, or with fast welding speeds. It may also be used to reduce distortion on thick sections and to reduce hardness caused by the rapid cooling of the weld, which may result in weld failure. By slowing the weld cool rate, pre-heating produces a more ductile weld.

Post-heating is usually carried out in the range 590–760°C, to reduce internal stresses (residual stresses) and help to soften hardened areas in the HAZ which may well bring about failure if not addressed.

Poor Fitup

Ideally, all welding will be performed on joints that are properly fitted. Most welds produced to a code or standard are properly fitted. Repair, prototype and workshop welding, however, may not be cut and fitted properly. These welds must be performed under less than ideal conditions, but they still must be strong and have a good appearance. Making a good weld on a poorly fitted joint requires an experienced welder with some special skills. A skilled welder can read the molten weld pool correctly to make changes in amperage, current, electrode movement, electrode angle and timing before problems develop. The amperage setting may have to be adjusted up or down by only a few amperes to make the necessary changes in molten weld pool size. Adjusting the machine is often preferable to lengthening the weave pattern excessively. The current may be changed from AC to DCEN or DCEP to vary the amount of heat input to the molten weld pool. Some electrodes can operate better than other electrodes with lower amperages on some currents. The current will also alter the forcefulness of some electrodes.

The 'U', 'J' and straight step patterns are usually the best to use, but they should not be moved more than required to close the gap or opening. On some poor-fitting joints, it is necessary to break and restart the arc in order to keep the molten weld pool under control. This will result in a weld with porosity, slag inclusions and other defects. Changing the electrode angle from leading to trailing improves poor fit. Sometimes a very flat angle will also help. The timing used to move the electrode into and out of the molten weld pool is critical. Returning to a molten weld pool too often or too soon can cause it to drop out of the joint. In most cases, a welder should return to the molten weld pool only after it has started to cool.

On some joints, it is possible to make stringer beads on both sides of the joint until the gap is closed, Figure 7.104. Note that the beads are made alternately from the edges of the joint to the centre. Welds made in this manner can have good weld soundness and strength, but they require more time to complete.

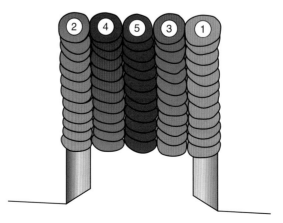

FIGURE 7.104 Multiple stringer beads used to close a large gap

SAFETY RULES FOR PORTABLE ELECTRIC TOOLS

The following safety precautions apply to all power tools. They should be strictly obeyed to avoid injury to the operator and damage to the power tool.

- Know the tool. Learn the tool's applications and limitations as well as its specific potential hazards by reading the manufacturer's literature.
- Earth portable power tools unless they are double-insulated.
- Never weld with a portable tool on the bench plugged in to a power supply, as the cable will offer less resistance than the return cable, which will expose the tool cable to higher current values, damaging the tool, the socket outlet and the circuit breaker.
- Do not expose a power tool to water or rain. Do not use a power tool in wet locations.
- Keep the work area well lit. Avoid chemical or corrosive environments.
- Always check the on-off switch is working correctly before use.
- Electric tools spark, so portable electric tools should never be started or operated in the presence of propane, natural gas, gasoline, paint thinner, acetylene or other flammable vapours that could cause a fire or explosion.
- Do not force a cutting tool to cut faster. It will do the job better and more safely if operated at the cutting rate for which it was designed.
- Use the right tool for the job. Never use a tool for any purpose other than that for which it was designed.
- Wear eye protectors. Safety glasses or goggles will protect the eyes while you operate power tools.
- Wear a face or dust mask if the operation creates dust.
- Many power tools operate above the 80dB rate, therefore ear defenders should be worn at all times.
- Take care of the power lead. Never carry a tool by its lead or yank it to disconnect it from the supply.
- Secure your work with clamps. It is safer than using your hands, and it allows both hands to operate the tool.
- Do not over-reach when operating a power tool. Keep proper footing and balance at all times.
- Maintain power tools. Follow the manufacturer's instructions for lubricating and changing accessories. Replace all worn, broken or lost parts immediately.
- Disconnect the tools from the power source when they are not in use.
- Form the habit of checking to see that any chuck keys or spanners are removed from the tool before turning it on.
- Avoid accidental starting. Do not carry a plugged-in tool with your finger on the switch. Be sure the switch is off when plugging in the tool.
- Be sure accessories and cutting bits are attached securely to the tool.
- Do not use tools with cracked or damaged housings.
- When operating a portable power tool, give it your full and undivided attention; avoid dangerous distractions.
- Never use a power tool with its safety features, guards removed or in-operable.
- Report all damaged tools and label not to be used.

HEALTH & SAFETY ◗

Some portable tools will exhibit vibration (hand and arm vibration (HAV)) and can lead to 'white finger syndrome'. You should limit the amount of time that you use this equipment. Many manufacturers now indicate the vibration value on their tools.

Grinders

Grinding, using a pedestal grinder or a portable grinder, is required to do many welding jobs correctly. Often it is necessary to grind a bevel, groove, remove rust or smooth a weld. Most abrasive wheels used are those found on hand-held machines. Only organic bonded wheels should be used. For wheels intended for cutting off purposes, some form of reinforcement should be incorporated usually fibreglass. Cutting off wheels should not be used for general grinding and vice versa. All machine guards must be secured in place and adjusted so that the guard is between the user and rotating wheels. Never remove the guard to allow oversized wheels to be mounted, as the wheel will be running at a speed in excess of the manufacturer's recommended maximum and there is a risk it could burst. All portable electric grinders should comply with BS EN 51044 Safety of Hand Held Electric Motor Operated Tools Part 2 Section 3 and carry the CE marking

Grinding stones have their maximum revolutions per minute (RPM) listed on the paper blotter. They must never be used on a machine with a higher rated RPM. If grinding stones are turned too fast, they can explode.

In the UK only approved personnel can fit or remove abrasive wheels, and anyone who does not comply with this ruling could be prosecuted if an accident were to occur.

Before a grinding stone is put on the machine, it should be tested for cracks. This is done by tapping the stone in four places and listening for a sharp ring, which indicates that it is good. A dull sound indicates that the grinding stone is cracked and should not be used. Once a stone has been installed and has been used, it may need to be trued and balanced using a special tool designed for that purpose. Truing keeps the stone face flat and sharp for better results.

Each grinding stone is made for grinding specific types of metal. Most stones are for ferrous metals (iron, cast iron, steel and stainless steel, etc). Some stones are made for non-ferrous metals such as aluminum, copper and brass. If a ferrous stone is used to grind non-ferrous metal, it will become glazed (the surface clogs with metal) and may explode as a result of frictional heat building up on the surface. If a non-ferrous stone is used to grind ferrous metal, the stone will be quickly worn away.

When the stone wears down, keep the tool rest adjusted to within 1mm, so that the metal being ground cannot be pulled between the tool rest and the stone surface. Stones should not be used when they are worn down to the size of the paper blotter. All permanent grinding wheels should have a means of extraction of dust particles. Never wear gloves when grinding. If a glove gets caught in a stone, the whole hand may be drawn in. The sparks from grinding should be directed down and away from other people or equipment.

Drills

Holes should be centre punched before they are drilled to help stop the drill bit from wandering. If the bit gets caught, stop the motor before trying to remove it. All metal being drilled on a drill machine should be securely clamped to the table. The sharp metal shavings should be avoided as they come out of the hole. If they start to become long, stop the downward pressure until the shaving breaks. Then the hole can be continued.

Metal-Cutting Machines

Mechanical metal-cutting machines used in the welding shop include shears, punches, cut-off machines and band saws. Their advantages over thermal cutting include little or no requirement for post-cutting clean up, the wide variety of metals that can be cut, and the fact that the metal is not heated.

Shears and Punches

These are often used in the fabrication of metal for welding. These machines can be operated either by hand or by powerful motors. Hand-operated equipment is usually limited to thin sheet stock or small bar stock. Powered equipment can be used on material 25mm or more in thickness and several millimeters wide, depending on its rating. Their power is a potential danger if these machines are not used correctly. Shears and punches are rated by the thickness, width and type of metal that they can be safely used to work. Failure to follow these limitations can result in injury, as well as damage to the equipment and to the metal being worked. Shears work like powerful scissors. The metal being cut should be placed as close to the pivot pin as possible. It must be securely held in place by the clamp on the shear before it is cut. If you are cutting a long piece of metal that is not being supported by the shear table, portable supports must be used. As the metal is being cut it may suddenly move or bounce around; if you are holding on to it, this can cause a serious injury.

Power punches are usually either hydraulic or flywheel operated. Both types move quickly, but usually only the hydraulic type can be stopped mid-stroke. Once the flywheel-type punch has been engaged it will make a complete cycle before it stops. Because punches move quickly or may not be stopped, it is very important that the operator's two hands be clear of the machine and that the metal is held firmly in place by the machine clamps before the punching operation is started.

Cut-off Machines

These may use abrasive wheels or special saw blades with tungsten carbide tips to make their cuts. Most abrasive cut-off wheels spin at high speeds (high RPMs) and are used dry (without coolant). Most saws operate much more slowly and with a liquid coolant. Both types of machines produce quality cuts which require little or no post-cut clean up in a variety of bar- or structural-shaped metals. Always wear eye protection when operating these machines. Before a cut is started, the metal must be clamped securely in the machine vice. Even the slightest movement of the metal can bind or break the wheel or blade. If the machine has a manual feed, the cutting force must be applied at a smooth and steady rate. Apply only enough force to make the cut without overloading the motor. Use only reinforced abrasive cut-off wheels that have an RPM rating equal to or higher than the machine-rated speed.

Band Saws

These can be purchased as vertical or horizontal, and some can be used in either position. Some band saws can be operated with a cooling liquid and are called 'wet saws'; most small saws operate dry. The blade guides must be adjusted as closely as possible to the metal being cut. The cutting speed and cutting pressure must be low enough to prevent the blade from overheating. When using a vertical band saw with a manual feed, keep your hands away from the front of the blade so that, if your hand slips, it will not strike the moving blade. If the blade breaks, sticks, or comes off the track, turn off the power, lock it off and wait for the band saw drive wheels to come to a complete stop before touching the blade.

HEALTH & SAFETY
Be careful of hot flying chips.

SUMMARY

Manual metal arc welding is a process which requires great dexterity and visual awareness. It is one of the last hand skills, in that the welder can actually feel the weld deposit going down and know whether the result is acceptable. To achieve this level of skill, you also need to be fully conversant with the equipment, the power supply and the advantages and disadvantages of choosing AC or DC. Selection of the correct electrode for the job depends on your knowledge of materials, operating characteristics and service conditions. For a new welder, it is often difficult to concentrate on anything other than the bright sparks and glow at the end of the electrode. As you develop your skills, your visual field will increase, allowing you to see a much larger welding zone. This skill comes with practice and nothing enhances your welding skills more than time spent actually welding, cleaning off the weld, inspecting it to identify any necessary corrections and trying to improve your technique. Finally, all welders need to make sure that their immediate work area is safe and kept in a clean condition for the benefit of all who work there.

ACTIVITY AND REVIEW

1 State how the welding heat is created in MMA.

2 Define what the following terms refer to:
   a Open circuit voltage.
   b Arc voltage.
   c Striking voltage.

3 Define what arc blow is, and how it can be controlled.

4 State how step down transformers work.

5 Define what is meant by the term 'Duty Cycle.

6 Indicate what the term 'drooping characteristic' refers to.

7 Name five power source machines for welding and state whether they are AC, DC or both.

8 State why is it important to strike the arc only in the weld joint.

9 Identify the potential problems with:
   a too low a current.
   b too high a current.

10 Explain what factors should be considered when selecting an electrode size.

11 Define what 'stringer' beads are.

12 Explain what effect changing the arc length has on the weld.

13 Sketch four edge preparations that may be used for welded joints.

14 Indicate why backing tapes are used on some joints.

15 Draw a typical edge preparation setup and indicate the following:
   a Root gap or opening.
   b Root face.
   c Included angle.

16 Explain what the term 'keyhole' refers to.

17 State the functions of the following:
   a Hot pass.
   b Filler pass.
   c Cover pass.

18 Sketch the correct technique for a pick up on previous weld deposit.

19 Indicate what advantage there is in back-gouging the root run.

20 Sketch the effect on the trailing edge of a weld deposit if:
   a the current is too high.
   b the current is too low.
   c correct current settings.

21 Explain why welds should never be terminated in a corner.

22 Describe which joint configuration requires more heat input and where the heat input should be directed.

8

Metal Inert Gas Shielded and Flux Cored Welding

LEARNING OBJECTIVES

After completing this chapter, you should be able to:

- demonstrate the proper use of personal protective equipment (PPE) for metal inert gas shielded welding (MIG)
- describe methods of metal transfer including the axial spray metal transfer process, globular transfer, pulsed-arc metal transfer and short-circuiting transfer dip transfer
- define *voltage, electrical potential, amperage* and *electrical current* as related to MIG welding
- describe and demonstrate the backhand and forehand welding techniques and their relationship to weld bead profile and penetration in the short-circuiting transfer mode
- set up a MIG welding work station
- thread the electrode wire on a MIG machine
- set the shielding gas flow rate on a MIG machine
- demonstrate the effect of changing the electrode extension on a weld
- describe the effects of changing the welding gun angle on the weld bead
- evaluate weld beads made with various shielding gas mixtures
- make MIG welds in butt joints, lap joints, and T-joints in the flat (PA) position that will pass a specified standard's visual or destructive examination criteria
- describe the flux cored arc welding (FCAW) process
- list the equipment required for an FCAW workstation
- tell how electrodes are manufactured and explain the purpose of the electrode cast and helix
- explain what each of the digits in a standard FCAW electrode identification number mean
- make root, filler and cover passes with the FCAW process
- make butt welds in the flat (PA) position
- make fillet welds in T-joints and lap joints in the flat (PA) position.

UNIT REFERENCES

NVQ units:

Complying with Statutory Regulations and Organizational Safety Requirements.
Joining Materials by the Manual MIG/MAG and Other Continuous Wire Processes.
Producing Fillet Welded Joints using a Manual Welding Process.

VRQ units:

Engineering Environment Awareness.
Manual Welding Techniques.

KEY TERMS

arc plasma a state of matter found in the region of an electrical discharge (arc). See also plasma.

mechanical testing tests carried out to the specimen's failure in order to understand its structural performance or material behaviour under different loads. These tests are generally much easier to carry out, give more information and are easier to interpret than non-destructive testing.

specification a specification is an explicit set of requirements to be satisfied by a material, product or service.

tack weld a weld made to hold the parts of a weldment in proper alignment until the final welds are made.

venturi a condition in which gas or fluid can be introduced or accelerated. This effect can be induced in welding operations by reducing the angle of the nozzle which reduces gas coverage and increases potential for atmospheric gases to be drawn into the welding zone.

weld a localized joining of metals or non-metals produced either by heating the materials to suitable temperatures, with or without the application of pressure, or by the application of pressure alone and with or without the use of the filler material.

INTRODUCTION

The majority of metal gas shielded welding on carbon steels uses *active* shielding gases, so the title of this chapter could refer to metal active gas shielding (MAGS) welding. However, the process was first designed for the welding of aluminum and its alloys, where we use inert gases, and it has always been referred to as metal inert gas shielded (MIG) welding. Performing a satisfactory MIG or FCAW weld requires more than just manipulative skill. The setup, voltage, amperage, electrode extension, welding angle, as well as other factors, can dramatically affect the result. The best welding conditions allow a welder to produce the largest quantity of successful welds in the shortest period of time with the highest productivity. Because these are semi-automatic or automatic processes, increased productivity may only require the welder to increase the travel speed and current. There has been a major development in the use of robotics for MIG welding. Many industrial robots now produce a high level of quality welds 24 hours a day, 365 days a year with very little maintenance. Robots have the advantage that they can access more restricted areas than manual operators and they are not affected by fume.

WELDING POWER SUPPLIES

In order to understand how the different welding power supplies are described, you need to know the following electrical terms:

- *Voltage or volts (V)*. A measurement of electrical pressure and the force that causes the current (amperage) to flow. Electrical potential means the same thing as voltage and is usually expressed by the term *potential (P)*. The terms *voltage*, *volts* and *potential* are all used to refer to electrical pressure.
- *Amperage or amps (A)*. The measurement of the total number of electrons flowing, in the same way.
- *Electrical current*. The same thing as amperage and usually expressed as *current (C)*. The terms *amperage*, *amps* and *current* used to refer to electrical flow.

MIG power supplies are constant-voltage, constant-potential (CV, CP) machines and are available as transformer-rectifiers, motor generators or inverters, Figure 8.1. Some newer machines use electronics, enabling them to supply both types of power at the flip of a switch; they may be referred to as CC/CV (constant current/constant voltage) power supplies.

The relationships between current and voltage with different combinations of arc length or wire-feed speeds are called volt-ampere characteristics. The volt-ampere characteristics of arcs in argon with constant arc lengths or constant wire-feed speeds are shown in Figure 8.2.

FIGURE 8.1 Transformer rectifier welding power supply

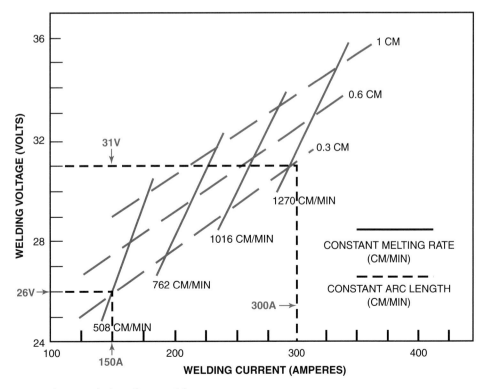

FIGURE 8.2 Volt-ampere characteristics of arcs with argon

To maintain a constant arc length while increasing current, the voltage is increased. For example, with a 3mm arc length, increasing current from 150 to 300 amperes requires increasing the voltage from about 26 to 31 volts. The current increase illustrated here results from increasing the wire-feed speed.

Speed of the Wire Electrode

TIP ⊙

A self-adjusting arc can also happen when the operator increases or reduces his arc length and this produces an increase or decrease in voltage.

The wire-feed speed, which is the speed at which the wire exits the contact tube, is generally recommended by the electrode manufacturer and is measured in metres per minute (m/min) The welder uses a control dial on the wire-feed unit to modify the speed and control the burn-off rate to match the welder's skill. Note the direct relationship between current (amps) and wire-feed speed; as wire-feed speed increases, the amperage increases. If wire-feed speed is reduced, amperage will decrease, and this is what is referred to as the self-adjusting arc, Table 8.1.

To accurately measure wire-feed m/min, snip off the wire at the contact tube. Wearing safety glasses and pointing the contact tube away from your face, work piece or table, squeeze the trigger for ten seconds; release and snip off the wire electrode. Measure the number of metres of wire that was fed out in the ten seconds. Now multiply this figure by six to find how many metres of wire are fed per minute.

TABLE 8.1 Typical amperages for carbon steel

Wire-feed Speed (m/min)	Wire Diameter Amperages			
	0.8mm	0.9mm	1.2mm	1.6mm
(2.5)	40	65	120	190
(5.0)	80	120	200	330
(7.6)	130	170	260	425
(10.2)	160	210	320	490
(12.7)	180	245	365	–
(15.2)	200	265	400	–
(17.8)	215	280	430	–

Power Supplies for Short-circuiting Transfer

Although the MIG power source is said to have a constant voltage (CV) or constant potential (CP), it is not perfectly constant. The graph in Figure 8.3 shows that there is a slight decrease in voltage as the amperage increases within the working range. The rate of decrease is known as *slope*. It is expressed as the voltage decrease per 100-ampere increase – for example, 10V/100A. For short-circuiting welding, some welding power supplies are equipped to allow changes in the slope by steps or continuous adjustment.

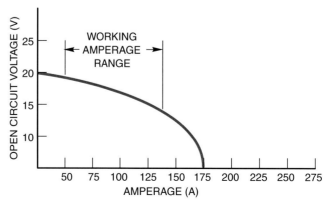

FIGURE 8.3 Constant potential welder slope

The slope, which is also known as the 'volt ampere curve', is often drawn as a straight line because it is fairly straight within the working range of the machine. Whether it is drawn as a curve or a straight line, the slope can be established by finding two points. The first point is the set voltage as read from the voltmeter when the gun switch is activated but no welding is being done. This is referred to as the open circuit voltage. The second point is the voltage and amperage as read during a weld. The voltage control is not adjusted during the test but the amperage can vary. The slope is found by subtracting the second voltage reading from the first. For settings over 100 amperes, it is easier to calculate the slope by adjusting the wire feed so that you are welding with 100 amperes, 200 amperes, 300 amperes and so on. In other words, the voltage difference can be simply divided by 1 for 100 amperes, 2 for 200 amperes and so forth.

The machine slope is affected by circuit resistance. This may result from a number of factors, including poor connections, long leads or a dirty contact tube. A higher resistance means a steeper slope. In short-circuiting machines, increasing the inductance increases the slope. This increase slows the current's rate of change during short circuiting and the arcing intervals, Figure 8.4. As the slope increases, both the short-circuit current and the *pinch effect* are reduced. A flat slope has both an increased short-circuit current and a greater pinch effect.

The machine slope affects the short-circuiting metal transfer mode more than it does the other modes. Too much current and pinch effect from a flat slope cause a violent short and arc restart cycle, which results in increased spatter. Too little current and pinch effect from a steep slope result in the short circuit not being cleared as the wire freezes in the molten pool and piles up on the work, Figure 8.5.

The slope should be adjusted so that a proper spatter-free metal transfer occurs. On machines that have adjustable slopes, this is easily set by varying the contact tube-to-work distance. The MIG filler wire is much too small to carry the welding current and heats up due to its resistance to the current flow. The greater the tube-to-work distance, the greater the circuit resistance and the steeper the slope. By increasing or decreasing this distance, a proper slope can be obtained so that the short circuiting is smoother with less spatter.

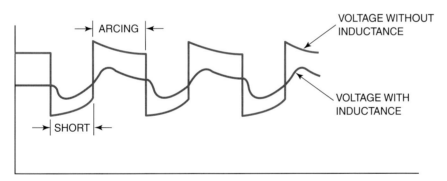

FIGURE 8.4 Voltage pattern with and without inductance

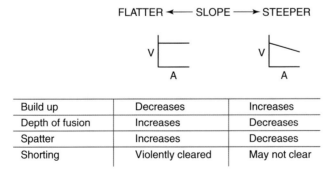

FLATTER ← SLOPE → STEEPER

	Flatter	Steeper
Build up	Decreases	Increases
Depth of fusion	Increases	Decreases
Spatter	Increases	Decreases
Shorting	Violently cleared	May not clear

FIGURE 8.5 Effects of slope

Inductance

Many newer MIG welding power supplies are supplied with an inductance control. These are used primarily in the MIG short-circuiting (dip) transfer mode, especially when open root joints are involved. Inductance affects the rise up to peak current during MIG short-circuiting transfer.

A low inductance control setting will provide a greater short-circuiting frequency, which may be beneficial in welding thin material where burn-through is an issue. A higher setting will reduce the frequency of short circuits, creating slightly longer arc periods, which allows greater current to go to the work. This can be used to increase penetration in thicker sections or where complete penetration is required on open root joints welded from one side.

In the case of the MIG welding arc, the rate of change in the amperage relative to the arc voltage determines how the metal droplet detaches from the end of the electrode. If the rate of change is too rapid (inductance too low), the droplet detaches violently and produces excessive spatter. If the rate of change is too slow (inductance too high) the metal droplet does not detach cleanly and the arc is unstable. A mid-range setting on the inductance control can be used for general-purpose short-circuit MIG welding and adjustments made for thick, thin, or open root conditions.

MOLTEN WELD POOL CONTROL

The MIG molten weld pool can be controlled by varying the following factors: shielding gas, power settings, gun manipulation, travel speed, electrode extension and gun angle.

Shielding Gas

The shielding gas selected for a weld affects the method of metal transfer, welding speed, weld contour, arc cleaning effect and fluidity of the molten weld pool. The selection of shielding gas also depends on the metal to be welded. Some metals must be welded with an inert gas such as argon or helium or mixtures of argon and helium. Other metals weld more favourably with reactive gases such as carbon dioxide or with mixtures of inert gases and reactive gases such as argon and oxygen or argon and carbon dioxide, Table 8.2. The most commonly used shielding gases are 80 per cent argon + 20 per cent CO_2, argon + 1 – 5 per cent oxygen and carbon dioxide, Figure 8.6.

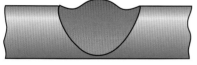
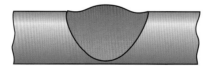

ARGON + OXYGEN   ARGON + CARBON DIOXIDE   CARBON DIOXIDE

FIGURE 8.6 **Effect of shielding gas on bead shape**

Argon

The atomic symbol for argon is *Ar*, and it is an inert gas. Inert gases do not react with other substances and argon is insoluble in molten metal. 100 per cent argon is used on non-ferrous metals such as aluminium, copper, magnesium, nickel and their alloys; but 100 per cent argon is not normally used for making welds on ferrous metals. Because argon is denser than air, it effectively shields welds by pushing the lighter air away. Argon is relatively easy to ionize, which means that it can carry long arcs at lower voltages. This makes it less sensitive to changes in arc length. Argon gas is naturally found in air and is collected in air separation plants.

Argon gas blends

Oxygen, carbon dioxide, helium and nitrogen can be blended with argon to change its welding characteristics. Adding reactive gases (oxidizing), such as oxygen or carbon dioxide, to argon tends to stabilize the arc, promote favourable metal transfer and minimize spatter. This improves the penetration pattern and undercutting is reduced or eliminated. Adding helium or nitrogen gases (non-reactive or inert) increases the arc heat for deeper penetration.

The amount of the reactive gases required to produce the desired effects is quite small. As little as a 0.5 per cent change in the amount of oxygen will produce a noticeable effect on the weld. Most of the time, blends containing 1 per cent to 5 per cent of oxygen are used. Carbon dioxide may be added to argon in the range of 5 per cent to 20 per cent. The most commonly used blend for short-circuiting transfer is 80 per cent argon and 20 per cent CO_2. When using oxidizing shielding gases with oxygen or carbon dioxide added, a suitable filler wire containing deoxidizers should be used to prevent porosity in the weld. The presence of oxygen in the shielding gas can also cause some loss of certain alloying elements, such as chromium, vanadium, aluminium, titanium, manganese and silicon.

Helium

The atomic symbol for helium is *He*, and it is an inert gas that is a byproduct of the natural gas industry. Helium is lighter than air; and its flow rates must be about twice as high as argon's for the gas stream to be able to push air away from the weld. Proper protection is difficult in draughts unless high flow rates are used. It requires a higher voltage to ionize, which produces a much hotter arc. This makes it easier to make welds on thick sections of aluminium and magnesium. Small quantities of helium are blended with heavier gases. These blends take advantage of the heat produced by the lightweight helium and weld coverage by the other heavier gas.

Carbon dioxide (CO_2)

In the short-circuiting transfer mode CO_2 allows higher welding speed, better penetration, good mechanical properties and costs less than the inert gas mixes. The chief drawback is the less-steady arc characteristics and a considerable increase in weld spatter. The spatter can be kept to a minimum by maintaining a very short, uniform arc length and strict attention to amperage and voltage parameters. CO_2 can produce sound, spatter-free welds of the highest quality, provided established procedures are followed and a filler wire with the proper deoxidizing additives is selected. Carbon dioxide is now used as part of a blended gas as this overcomes the spatter problem and produces a more tranquil arc.

TABLE 8.2(A) Metals matched with MIG shielding gases

MIG Metals, Shielding Gas, and Gas Blends

Gases: Argon (Ar), CO_2. Blends of Two Gases — Argon + Oxygen, Argon + Carbon Dioxide, Argon + Helium. Blends of Three Gases.

Metals	Argon (Ar)	CO_2	Ar + 1% O_2	Ar + 2% O_2	Ar + 5% CO_2	Ar + 5% CO_2	Ar + 10% CO_2	Ar + 25% CO_2	Ar + 25% He	Ar + 50% He	Ar + 75% He	Ar + CO_2 + O_2	Ar + CO_2 + Nitrogen	Ar + CO_2 + Helium
Aluminum	•								•	•	•			•
Copper Alloys	•								•	•	•			
Stainless Steel		•	•	•	•			•			•		•	•
Steel		•	•	•	•	•	•	•				•		
Magnesium	•								•	•	•			
Nickel Alloys	•								•	•	•			

TABLE 8.2(B) Metals matched with MIG shielding gases and welding processes

MIG Shielding Gas, Gas Blends, Metals, and Welding Process

Gases/Blend	Gas Reaction	Application	Remarks
Argon (Ar)	Inert	Nonferrous metals	Provides spray transfer
Helium (He)	Inert	Aluminum and magnesium	Very hot arc for welds on thick sections, usually used in gas blends to increase the arc temperature and penetration
Ar + 1% O_2	Oxidizing	Stainless steel	Oxygen provides arc stability
Ar + 2% O_2	Oxidizing	Stainless steel	Oxygen provides arc stability
Ar + 5% O_2	Oxidizing / Oxidizing	Mild and low-alloy steel / Low-alloy steel	Provides spray transfer
Ar + 5% CO_2	Oxidizing	Low-alloy steel	Pulse spray and short-circuit transfer in out-of-position welds
Ar + 10% CO_2	Oxidizing	Low-alloy steel	Same as above with a wider, more fluid weld pool
Ar + 25% CO_2	Oxidizing	Mild, low-alloy steels and stainless steel	Smooth weld surface, reduces penetration with short-circuiting transfer

Gas/Blend	Gas Reaction	Application	Remarks
CO_2	Oxidizing	Mild, low-alloy steels and stainless steel	Least expensive gas, deep penetration with shortcircuiting or globular transfer
Nitrogen	Almost inert	Copper and copper alloys	Has high heat input with globular transfer
Ar + 25% He	Inert	Al, Mg, copper, nickel, and their alloys	Higher heat input than Ar, for thicker metal
Ar + 50% He	Inert	Al, Mg, copper, nickel, and their alloys	Higher heat in arc use on heavier thickness with spray transfer
Ar + 75% He	Inert	Copper, nickel, and their alloys	Highest heat input
Ar + CO_2 + O_2	Oxidizing / Oxidizing	Low-alloy steel and some stainless steels	All metal transfer for automatic and robotic applications
Ar + CO_2 + N	Almost inert	Stainless steel	All metal transfer, excellent for thin gauge material
He + 7.5% Ar + 2.5% CO_2	Almost inert	Stainless steel and some low-alloy steels	Excellent toughness, arc stability, wetting characteristics, and bead contour, little spatter with short-circuiting transfer

Nitrogen

The atomic symbol for nitrogen is *N*. It is not an inert gas but is relatively non-reactive to the molten weld pool. It is often used in blended gases to increase the arc's heat and temperature and is an economical choice for gas purging of some austenitic stainless steel pipe welds.

Power Settings

As the power settings are adjusted, the weld bead is affected. Making an acceptable weld requires a balancing of voltage and amperage. If either or both are set too high or too low, the weld penetration can decrease. A MIG welding machine has no direct amperage settings. Instead, the amperage at the arc is adjusted by changing the wire-feed speed. As a result of the welding machine's maintaining a constant voltage when the wire-feed speed increases, more amperage flows across the arc. This higher amperage is required to melt the wire so that the same arc voltage can be maintained. The higher amperage is used to melt the filler wire and does not increase the penetration. In fact, the weld penetration may decrease significantly.

Gun Manipulation

The MIG welding process is greatly affected by the location of the electrode tip and molten weld pool. During the short-circuiting process if the arc is directed to the base metal and outside the molten weld pool, the welding process may stop. Without the resistance of the hot molten metal, high-amperage surges occur each time the electrode tip touches the base metal, resulting in a loud pop and a shower of sparks. This is something that occurs each time a new weld is started. When you make a weave pattern (a gun manipulation technique of moving side to side in order to produce a wider weld bead), you must keep the arc and electrode tip directed into the molten weld pool. Other than the sensitivity to arc location, most of the MMA weave patterns that keep the electrode wire at or near the leading edge of the weld pool can be used for short-circuiting MIG welds.

Travel Speed

Because the location of the arc inside the molten weld pool is important, the welding travel speed cannot exceed the ability of the arc to melt the base metal. Fusion between the base metal and filler metal can completely stop if the travel rate is too fast. If the travel rate is too slow and the weld pool size increases excessively, it can also restrict fusion to the base plate.

> **TIP**
>
> Too high a travel speed can result in over running of the weld pool and an uncontrollable arc.

Electrode Extension

The electrode extension (stickout) is the distance from the contact tube to the arc measured along the wire. Adjustments in this distance cause a change in the wire resistance and the resulting weld bead, Figure 8.7. MIG welding currents are relatively high for the wire sizes, even for the low current values used in short-circuiting arc metal transfer, Figure 8.8. As the length of wire extending from the contact tube to the work increases, the voltage, too, should increase. Since this is impossible with a constant-voltage power supply, the system compensates by reducing the current. In other words, by increasing the electrode extension and maintaining the same wire-feed speed, the current has to change to provide the same resistance drop. This situation leads to a reduction in weld heat, penetration, and fusion, and an increase in build up. On the other hand, as the electrode extension distance is shortened, the weld heats up, penetrates more, and builds up less, Figure 8.9.

> **TIP**
>
> Some nozzles can be extended to provide coverage. Others must be exchanged with the correct-length nozzle, Figure 8.10.

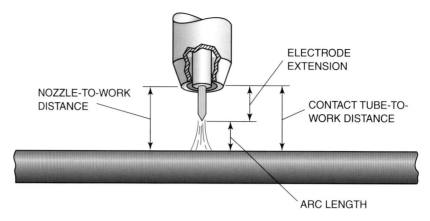

NOZZLE-TO-WORK DISTANCE

ELECTRODE EXTENSION

CONTACT TUBE-TO-WORK DISTANCE

ARC LENGTH

FIGURE 8.7 Electrode to work distance

> **TIP**
>
> Heat is built up due to extremely high current for the small conductor (electrode).

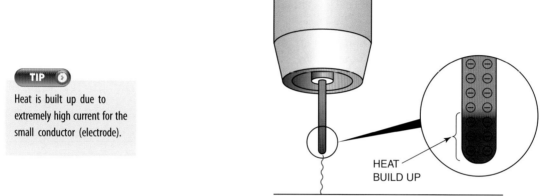

HEAT BUILD UP

FIGURE 8.8 Heat build up due to extremely high current for the small conductor (electrode)

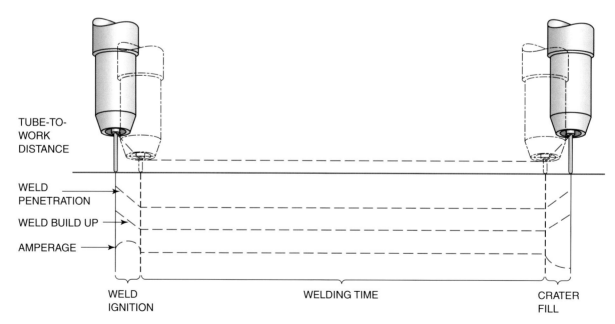

TUBE-TO-WORK DISTANCE

WELD PENETRATION

WELD BUILD UP

AMPERAGE

WELD IGNITION

WELDING TIME

CRATER FILL

FIGURE 8.9 Changing tube-to-work distance

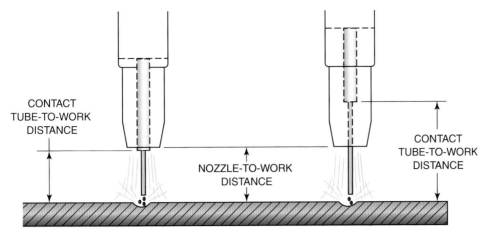

FIGURE 8.10 **Nozzle-to-work distance can differ from the tube to work distance**

Gun Angle

The MIG welding gun may be held so that the relative angle between the gun, work and welding bead is either vertical or has a drag angle or a push angle. Changes in this angle will affect the weld bead, particularly during the short-circuiting arc and globular transfer modes. Backhand welding uses a drag angle, Figure 8.11, and forehand welding uses a push angle. Figure 8.12.

FIGURE 8.11 **Backhand welding**

FIGURE 8.12 **Forehand welding**

Backhand Welding (drag technique)

This technique, although not the most popular, directs the arc force into the molten weld pool of metal. This forces the molten metal back onto the trailing edge of the molten weld pool and exposes more of the un-melted base metal, Figure 8.13. The digging action pushes the penetration deeper into the base metal while building up the weld cap. If the weld is sectioned, the profile of the bead is narrow and deeply penetrated, with a high build up. It is often used for vertical down welding of thin sheet.

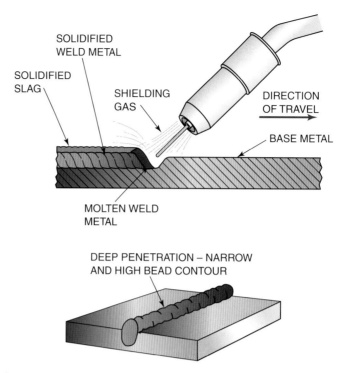

FIGURE 8.13 Backhand welding, or dragging angle

Forehand Welding (push technique)

This is the most widely used technique. The arc force pushes the weld metal forward and out of the molten weld pool onto the cooler metal ahead of the weld, Figure 8.14. The heat and metal are spread out over a wider area. The sectional profile of the bead is wide, showing good penetration with little build up. The greater the angle, the more defined the effect on the weld. As the angle approaches vertical, the effect is reduced. This allows the welder to change the weld bead as effectively as by adjusting the machine current settings.

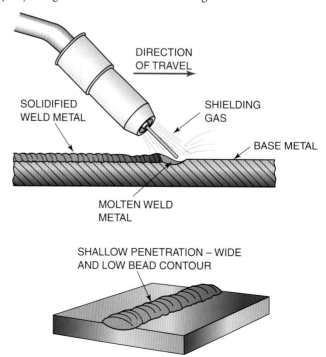

FIGURE 8.14 Forehand welding

EQUIPMENT

The basic MIG equipment consists of the gun, electrode (wire) feed unit, electrode (wire) supply, power source, shielding gas supply with flow meter and regulator, control circuit, and related hoses, liners and cables, Figure 8.15 and Figure 8.16. Larger, more complex systems may have water for cooling, solenoids for controlling gas flow and carriages for moving the work or the gun or both, Figure 8.17. The system may be stationary or portable, Figure 8.18. Most systems are meant to be used for only one process although some manufacturers do make multi-functional power sources that can be used for all arc processes.

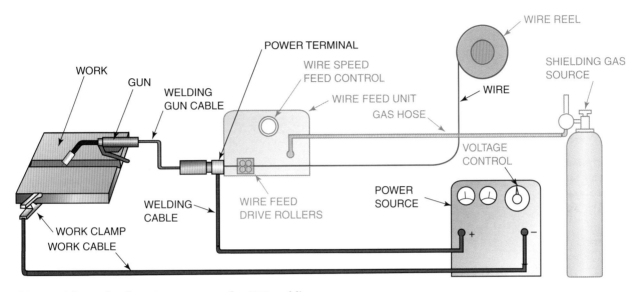

FIGURE 8.15 Schematic of equipment setup for MIG welding

FIGURE 8.16 Small 110V MIG welder

FIGURE 8.17 Robot welding

FIGURE 8.18 Portable water cooler for MIG plant welding equipment

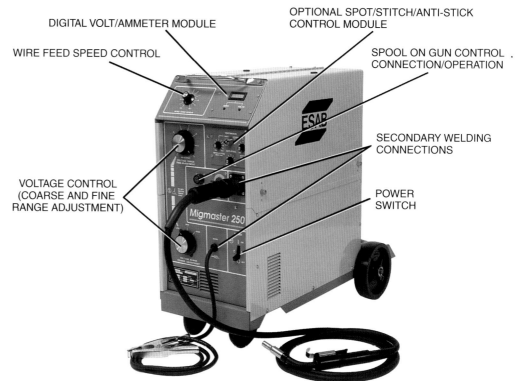

DIGITAL VOLT/AMMETER MODULE

OPTIONAL SPOT/STITCH/ANTI-STICK CONTROL MODULE

WIRE FEED SPEED CONTROL

SPOOL ON GUN CONTROL CONNECTION/OPERATION

VOLTAGE CONTROL (COARSE AND FINE RANGE ADJUSTMENT)

SECONDARY WELDING CONNECTIONS

POWER SWITCH

(A) 200-ampere constant-voltage power supply

(B) 650-ampere constant-voltage and constant-current power supply

FIGURE 8.19 Power supplies for multipurpose MIG applications

Power Source

The power source may be a transformer-rectifier, inverter or generator type. The transformer-rectifiers are stationary and usually require a three-phase power source. The inverter power sources are smaller, lighter, and may be designed to accept a variety of different electrical inputs, from 240 volts to 415 volts, single or three-phase. Engine generators are ideal for portable use or where sufficient power is not available.

Typical MIG welding machines produce a DC welding current ranging from 40 amperes to 600 amperes with 10 volts to 40 volts, depending upon the machine, Figure 8.19. Because many MIG power supplies are used in automation, it is not unusual for MIG welding machines to have a 100 per cent duty cycle. This allows the machine to be run continuously at its highest-rated output without damage.

Electrode (Wire) Feed Unit

The purpose of the electrode feeder is to provide a steady and reliable supply of wire to the weld. Slight changes in the rate at which the wire is fed have distinct effects on the weld.

The motor used in a feed unit is usually a DC type that can be continuously adjusted over the desired range. Figure 8.20 and Figure 8.21 show typical wire-feed units and accessories.

(A) A 90-ampere power supply and wire feeder for welding sheet steel with carbon dioxide shielding

(B) Modern wire feeder with digital preset and readout of wire-feed speed and closed-loop control

FIGURE 8.20 Examples of wire feeders

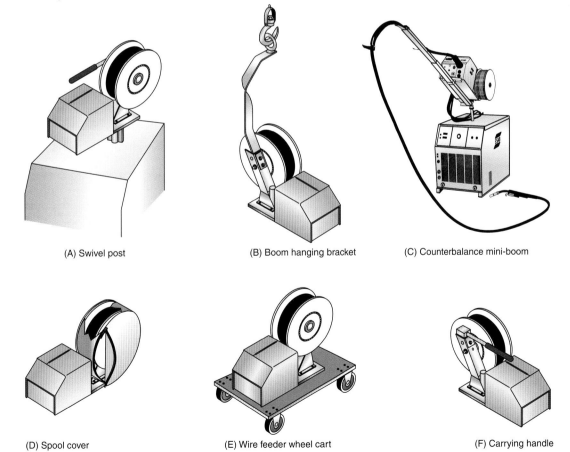

(A) Swivel post

(B) Boom hanging bracket

(C) Counterbalance mini-boom

(D) Spool cover

(E) Wire feeder wheel cart

(F) Carrying handle

FIGURE 8.21 Accessories for electrode feed systems

Push-type Feed System

The wire rollers are clamped securely against the wire to provide the necessary friction to push the wire through the conduit to the gun. The pressure applied on the wire can be adjusted. A groove in the roller aids alignment and lessens the chance of slippage. Most manufacturers provide rollers with smooth or knurled U-shaped or V-shaped grooves, Figure 8.22. Knurling (a series of ridges cut into the groove) helps grip larger-diameter wires so that they can be pushed along more easily. Soft wires, such as aluminium, can be damaged by knurled rollers and are best used with U-grooved rollers. V-grooved rollers are best suited for hard wires, such as mild and stainless steel. It is also important to use the correct-size grooves in the rollers.

Variations of the push-type electrode wire feeder include the pull type and push-pull type. The difference is in the size and location of the drive rollers. In the push-type system, the electrode must have enough strength to be pushed through the conduit without kinking. Mild steel and stainless steel can be readily pushed 4 to 6m, but aluminium is much harder to push more than 3m.

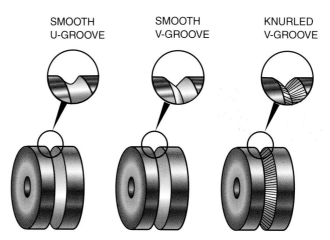

FIGURE 8.22 Feed rollers

Pull-type Feed System

In pull-type systems, a smaller but higher-speed motor is located in the gun to pull the wire through the conduit, making it possible to move even soft wire over great distances. The disadvantages are that the gun is heavier and more difficult to use, re-threading the wire takes more time, and the operating life of the motor is shorter.

Push-pull-type Feed System

Push-pull-type feed systems use a synchronized system with feed motors located at both ends of the electrode conduit, Figure 8.23. This system can be used to move any type of wire over long distances by periodically installing a feed roller into the electrode conduit. Compared to the pull-type system, this system can move wire over longer distances. Other advantages include faster re-threading, and increased motor life due to the reduced load. However, the system is more expensive.

Linear Electrode Feed System

Linear electrode feed systems use a different method to move the wire and change the feed speed. Standard systems use rollers that pinch the wire between them and a system of gears between the engine and the rollers to vary the roller speed. The linear feed system uses a small motor with a hollow armature shaft through which the wire is fed. The rollers are attached so that they move

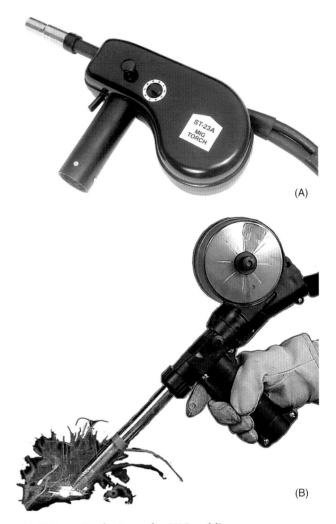

(A)

(B)

FIGURE 8.24 **Feeder/guns for MIG welding**

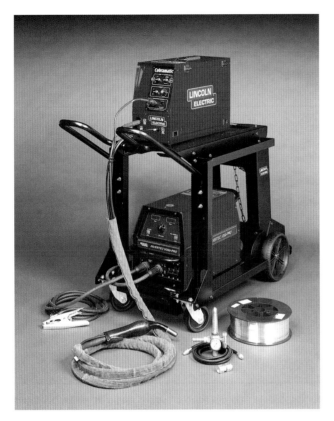

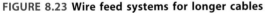

FIGURE 8.23 **Wire feed systems for longer cables**

around the wire. Changing the roller pitch (angle) changes the speed at which the wire is moved without changing the motor speed.

The advantage of a linear system is that the reduced size allows the system to be housed in the gun or within an enclosure in the cable. Several linear wire feeders can be synchronized to provide an extended operating range. The disadvantage is that the wire may become twisted as it is moved through the feeder.

Spool Gun

A spool gun is a compact, self-contained system consisting of a small drive system and a wire supply, Figure 8.24. It allows the welder to move freely around a job with only a power lead and shielding gas hose to manage. The major control system is usually mounted on the welder. The feed rollers and motor are found in the gun just behind the nozzle and contact tube, Figure 8.24B. Because of the short distance the wire must be moved, very soft wires (aluminium) can be used. A small spool of welding wire is located just behind the feed rollers. The spools of wire required in these guns are often very expensive. Although the guns are small, they feel heavy when being used.

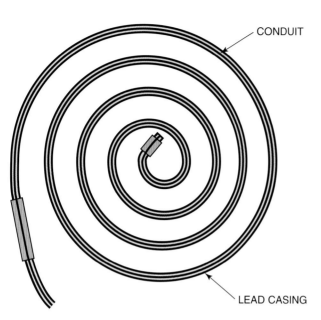

FIGURE 8.25 Tightly coiled lead casing will force the liner out of the gun

FIGURE 8.26 A typical MIG welding gun

Electrode Conduit

HEALTH & SAFETY ❯

Guns like that shown in Figure 8.26 are used for most welding processes with a head shield attached to protect the welder's gloved hand from intense heat generated when welding with high amperages.

The electrode conduit, or liner, guides the welding wire from the feed rollers to the gun. It may be encased in a lead that contains the shielding gas. Power cable and gun switch circuit wires are contained in a conduit that is made of a tightly wound coil with the required flexibility and strength. The steel conduit may have a nylon or Teflon liner to protect soft, easily scratched metals, such as aluminium, as they are fed. If the conduit is not an integral part of the lead, it must be firmly attached to both ends of the lead. Failure to attach the conduit can result in misalignment, which causes additional drag or makes the wire jam completely. If the conduit does not extend through the lead casing to make a connection, it can be drawn out by tightly coiling the lead, Figure 8.25. If the conduit is too long for the lead, it should be cut off and filed smooth. Too long a lead will bend and twist inside the conduit, which may cause feed problems.

Welding Gun

The welding gun attaches to the end of the power cable, electrode conduit, and shielding gas hose, Figure 8.26. It is used by the welder to produce the weld. A trigger switch is used to start and stop the weld cycle. The gun also has a contact tube, which is used to transfer welding current to the electrode moving through the gun, and a gas nozzle, which directs the shielding gas onto the weld, Figure 8.27.

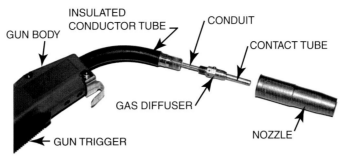

(A)

Accessories Selection Guide

To select correct accessories choose tip based on wire, follow chart to determine nozzle adaptor.

200 to 500 AMP MIG Guns

Contact Tips & Tubes
Standard Duty Contact Tips

Short (S)
Medium (M)
Long (L)

.058 cm. M or L   .132 cm. S or M
.076 cm. M or L   2.946 cm. S or M
.089 cm. M or L   14.326 cm. S or M
.114-.119 cm. S, M or L

Heavy Duty Contact Tips

Short (S)
Medium (M)

.114-.119 cm. M   .198 cm. S
.132 cm. M   .238 cm. S
.159 cm. M

Heavy Duty Contact Tubes

Short (S)
Medium (M)

.089 cm. M   .198 cm. S
.114-.119 cm. M   .238 cm. S
.132 cm. M
.159 cm. M

Nozzles
Slide-On Self Insulated Nozzles

Standard Duty   Heavy Duty   Extra Heavy Duty
#6 Tapered #12   #6   #12 Brass
#8   #12 Spot #10
#10   #12
                    #12 Spot

Extra Heavy Duty Threaded Nozzles

#6
#10
#12
#16 Spot

Adaptors
Tip Adaptor

Tip Adaptor
Tip Adaptor

Tip Adaptor
Tip Adaptor

.114 cm. - .159 cm.
.198 cm. - .238 cm.

Collet Body   Collet

Nozzle Adaptor

Nozzle Adaptor
Nozzle Adaptor

Extra Heavy Duty Adaptor

——————— Primary Use of accessories.

— — — Alternate use of accessories.

Note: *Using Heavy Duty contact tips or tubes and Extra Heavy Duty threaded Nozzles will greatly increase the rated current capacity of a MIG torch.*

(B)

FIGURE 8.27 MIG welding gun parts

SPOT WELDING

MIG can be used to make high-quality arc spot welds, using standard or specialized equipment, Figure 8.28. The MIG spot weld starts on one surface of one member and burns through to the other member, Figure 8.29. Fusion between the members occurs, and a small nugget is left on the metal surface. MIG spot welding has some advantages:

- Welds can be made in thin-to-thick materials.
- The weld can be made when only one side of the materials to be welded is accessible.
- The weld can be made when there is paint on the interfacing surfaces.

The arc spot weld can also be used to assemble parts for welding to be done at a later time. Thin metal can be attached to thicker sections using an arc spot weld. If a thin-to-thick butt, lap, or T-joint is to be welded with complete joint penetration, often the thin material will burn back, leaving a hole, or there will not be enough heat to melt the thick section. With an arc spot weld, the burning back of the thin material allows the thicker metal to be melted. As more metal is added to the weld, the burn-through is filled, Figure 8.29.

CAUTION

A welding shield must be worn to protect the eyes from flying sparks.

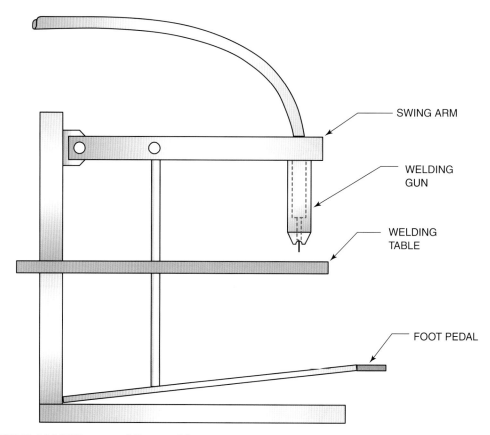

FIGURE 8.28 MIG spot welding machine

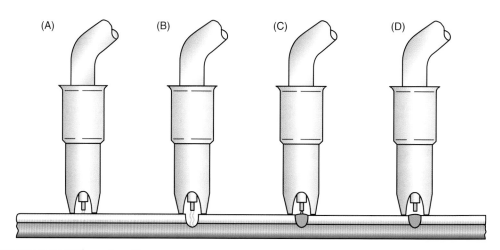

FIGURE 8.29 MIG spot weld

The MIG spot weld is produced from one side only. This means that it can be used on awkward shapes and in situations where the other side of the surface being welded should not be damaged. This makes it an excellent process for auto body repair. Also, because the metals are melted and the molten weld pool is agitated, thin films of paint between the members being joined need not be removed. This is an added benefit for auto body repair work.

Specially designed nozzles provide flash protection, part alignment and arc alignment, Figure 8.30. The optional control timer provides weld time and burn-back time. To make a weld, the amperage, voltage and length of welding time must be set correctly. The burn-back time is a short period at the end of the weld when the wire feed stops but the current continues. This allows the wire to be burned back so it does not stick in the weld, Figure 8.29.

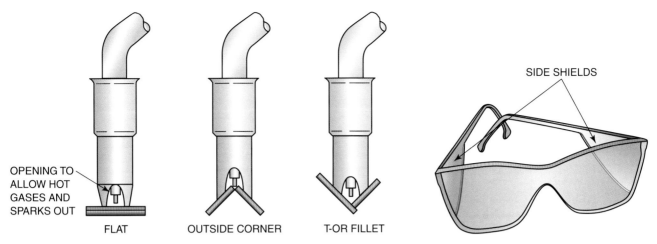

OPENING TO ALLOW HOT GASES AND SPARKS OUT

FLAT   OUTSIDE CORNER   T-OR FILLET

FIGURE 8.30 Specialized nozzles for MIG spot welding

SIDE SHIELDS

FIGURE 8.31 Safety glasses with side shields

FACE, EYE AND EAR PROTECTION

Face and Eye Protection

Eye protection must be worn in the workshop at all times. It can consist of safety glasses with side shields, Figure 8.31, goggles, or a full face shield, Figure 8.37. To give better protection when working in brightly lit areas or outdoors, some welders wear flash glasses, which are special, lightly tinted, safety glasses which provide protection from both flying debris and reflected light.

Suitable eye protection is important because eye damage caused by excessive exposure to arc light is not noticed at the time it happens. Therefore, welders must take appropriate precautions in selecting appropriate filters or goggles. Selecting the correct shade lens is important, because this process uses an opaque shielding gas and therefore requires a higher grade of shielding.

Any approved arc welding lenses will filter out the harmful ultraviolet light. Ultraviolet light can burn the eye in two ways. This light can injure either the white of the eye or the retina, which is the back of the eye, Figure 8.32. Anytime you receive an eye injury you should see a doctor.

Welding Helmets

Even with quality welding helmets, the welder must check for potential problems that may occur from accidents or daily use. Small, undetectable leaks of ultraviolet light in an arc welding helmet can cause a welder's eyes to itch or feel sore after a day of welding. To prevent these leaks, make sure the lens gasket is installed correctly, Figure 8.33. The outer and inner clear lenses must be plastic. As shown in Figure 8.34, you can check a lens for cracks by twisting it between your fingers. Worn or cracked spots on a helmet must be repaired.

Safety Glasses

Safety glasses with side shields are adequate for general use, but if heavy grinding, chipping or overhead work is being done, goggles or a full face shield should be worn in addition to safety glasses. Safety glasses must be worn under an arc welding helmet at all times.

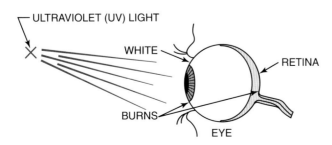

ULTRAVIOLET (UV) LIGHT

WHITE   RETINA

BURNS

EYE

FIGURE 8.32 The eye can be burned on the white or on the retina by ultraviolet light

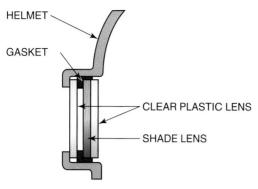

FIGURE 8.33 **Correct placement of the gasket around the shade lens**

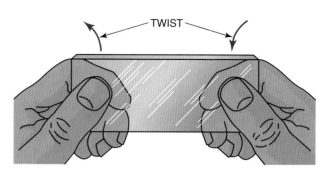

FIGURE 8.34 **To check the shade lens for possible cracks, gently twist it**

Ear Protection

The sound level in the welding environment is sometimes high enough to cause pain and some loss of hearing if the welder's ears are unprotected. Hot sparks can also drop into an open ear, causing severe burns.

Ear protection is available in several forms. Earmuffs cover the outer ear completely, Figure 8.35. Earplugs fit into the ear canal, Figure 8.36. Both of these protect a person's hearing, but only the earmuffs protect the outer ear from burns.

FIGURE 8.35 **Ear muffs**

FIGURE 8.36 **Ear plugs**

GENERAL WORK CLOTHING

As discussed in previous chapters, clothing should be chosen to match the working conditions. Above all, the use of flame retardant overalls is to be recommended. Fine chrome leather welding gauntlets are the preferred choice as these give the operator the most manipulative freedom while protecting the skin from the hazards of radiation.

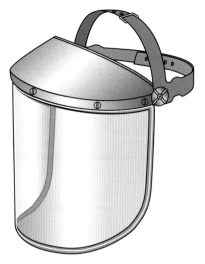

FIGURE 8.37 **Full face shield**

FIGURE 8.38 **Gas cylinder chained**

FIGURE 8.39 **Attaching flowmeter regulator**

FIGURE 8.40 **Wire label**

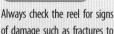

CAUTION

Always check the reel for signs of damage such as fractures to the casing, as these can affect the rate at which the wire is delivered. Also check the surface of the wire for any indication of moisture contamination. In all new reels there should be a desicant (silica-gel) to remove moisture from the wire. If this is not present, the reel may be old stock that could have been exposed to contamination. Do not use any reel that does not have an identifying label as you could be introducing the wrong composition to the weld zone.

SETUP

The same equipment may be used for semi-automatic MIG, MAGS and flux cored arc welding (FCAW). If the shielding gas supply is a cylinder, it must be chained securely in place before the valve protection collar is removed, Figure 8.38. Standing to one side of the cylinder and making sure no bystanders are in line with the valve, quickly crack the valve to blow out any dirt in the valve before the flow meter and regulator is attached, Figure 8.39. With the flow meter and regulator attached securely to the cylinder valve, attach the correct hose from the flow meter to the 'gas-in' connection on the electrode feed unit or machine.

Install the reel of electrode (welding wire) on the holder and secure it. When installing the spool wire, check the label to be sure that the wire is the correct type and size, Figure 8.40. Check the feed roller size to ensure that it matches the wire size, Figure 8.41. Check the conduit liner size to be sure that it is compatible with the wire size and connect the conduit to the feed unit. The conduit or an extension should be aligned with the groove in the roller and set as close to the roller as possible without touching, Figure 8.42. Misalignment at this point can contribute to a bird's nest, Figure 8.43, where the wire does not go through the outlet side of the conduit and the feed roller pushes it into a tangled ball.

FIGURE 8.41 **Checking feed roller size**

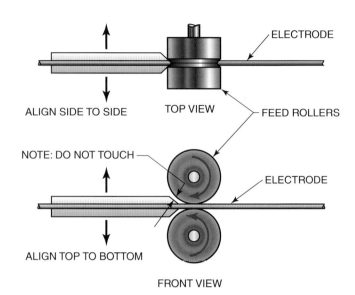

FIGURE 8.42 **Feed**

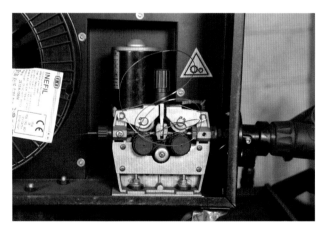

FIGURE 8.43 **'Bird's nest' in the filler wire at the feed rollers**

Be sure the power is off before attaching the welding cables, which should be attached to the proper terminals. The electrode lead should be attached to the terminal marked electrode or positive (+). If necessary, it is also attached to the power cable part of the gun lead. The work lead (return lead) should be attached to the terminal marked work or negative (−). The shielding 'gas-out' side of the solenoid is then also attached to the gun lead. If a separate splice is required from the gun switch circuit to the feed unit, it should be connected at this time, Figure 8.44. Check to see that the welding contactor circuit is connected from the feed unit to the power source. The welding gun should be securely attached to the main lead cable and conduit, Figure 8.45. There should be a gas diffuser attached to the end of the conduit liner to ensure proper alignment. A contact tip of the correct size to match the electrode wire size being used should be installed, Figure 8.46. A shielding gas nozzle is attached to complete the assembly.

Recheck all fittings and connections for tightness. Loose fittings can leak; loose connections can cause added resistance, reducing welding efficiency. Some manufacturers include detailed setup instructions with their equipment, Figure 8.47.

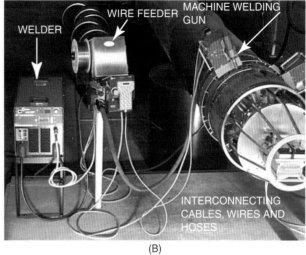

(A)   (B)

FIGURE 8.44 MIG station setup

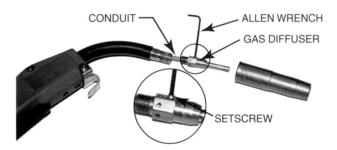

CONDUIT ——————   ————— ALLEN WRENCH

GAS DIFFUSER

SETSCREW

FIGURE 8.45 MIG welding gun assembly

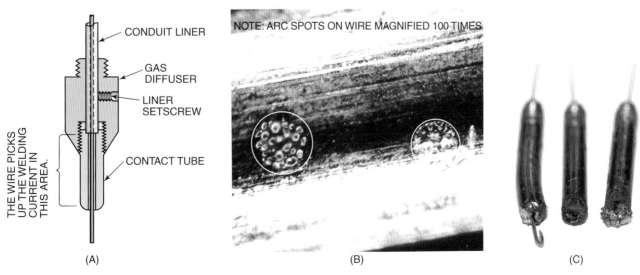

CONDUIT LINER

GAS DIFFUSER

LINER SETSCREW

CONTACT TUBE

THE WIRE PICKS UP THE WELDING CURRENT IN THIS AREA.

NOTE: ARC SPOTS ON WIRE MAGNIFIED 100 TIMES

(A)   (B)   (C)

FIGURE 8.46 The contact tube must be the correct size

Open the side cover.

With the gun trigger pressed, adjust the feed roller tension.

Remove the empty wire spool.

Check the setting guide inside the machine door.

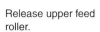
Release upper feed roller.

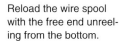

Set the voltage and wire feed for the metal you are going to be welding.

Reload the wire spool with the free end unreeling from the bottom.

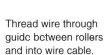

Attach work cable clamp to work to be welded.

Thread wire through guide between rollers and into wire cable.

Connect gas to coupling at rear of case and turn on shielding gas.

Set the polarity as DCEP from MIG welding.

ALWAYS WEAR PROPER SAFETY EQUIPMENT. Pull trigger and weld.

Turn the input switch on.

FIGURE 8.47 Manufacturer's setup instructions

Threading MIG Wire

Once the MIG machine has been properly assembled, thread the electrode wire through the system.

Cut the end of the electrode wire free. Hold it tightly so that it does not unwind. The wire has a natural curve known as its cast, which is measured by the diameter of the circle that the wire makes if it is loosely laid on a flat surface, Figure 8.48. The cast helps the wire make a good electrical contact as it passes through the contact tube, but can be a problem when threading the system. To make threading easier, straighten the end of the wire and cut off any kinks. Separate the wire feed rollers and push the wire first through the guides, Figure 8.49, then between the rollers, and finally into the conduit liner. Reset the rollers so there is a slight amount of compression on the wire, Figure 8.50. Set the wire-feed speed control to a slow speed. Hold the welding gun so that the electrode conduit and cable are as straight as possible. With safety glasses on and the gun pointed away from the welder's face, press the gun switch, or the cold feed switch if your wire feeder is equipped with one. The cold feed switch feeds wire without delivering current to the gun. The wire should start feeding into the liner. Watch to make certain that the wire feeds smoothly and release the switch as soon as the end comes through the contact tube.

> **CAUTION**
>
> If your machine does not have a cold start or inch control, remove the return lead when feeding the wire through the liner to prevent any accidental arcing on to the work table. If the wire stops feeding before it reaches the end of the contact tube, stop and check the system. If no obvious problem can be found, mark the wire with tape and remove it from the gun. You can then hold it next to the system to find the location of the problem.

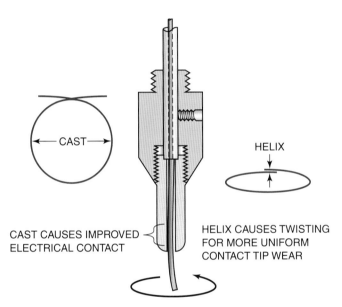

CAST

CAST CAUSES IMPROVED ELECTRICAL CONTACT

HELIX

HELIX CAUSES TWISTING FOR MORE UNIFORM CONTACT TIP WEAR

FIGURE 8.48 Cast of welding wire

FIGURE 8.49 Push wire through guides

FIGURE 8.50 Adjust wire feed tensioner

With the wire feed running, adjust the feed roller compression so that the wire reel can be stopped easily by a slight pressure. Too light a roller pressure will cause the wire to feed erratically. Too high a pressure can turn a minor problem into a major disaster. If the wire jams at a high roller pressure, the feed rollers keep feeding the wire, causing it to bird-nest and possibly short out. With a light pressure, the wire can stop. This is very important with soft wires.

With the feed running, adjust the spool drag so that the reel stops when the feed stops. The reel should not coast to a stop because the wire can be snagged easily. Also, when the feed restarts, a jolt occurs when the slack in the wire is taken up which can be enough to stop the wire momentarily and cause a discontinuity in the weld.

When the test runs are completed, the wire can either be rewound or cut off. Some wire-feed units have a button which retracts the wire automatically. To rewind the wire on units without this retraction feature, release the rollers and turn them backward by hand. If the machine will not allow the feed rollers to be released without upsetting the tension, you must cut the wire.

FILLER METAL SELECTION

MIG welding filler metals are available for a variety of base metals, Table 8.3. The most frequently used filler metals are A5.18 (AWS code) for carbon steel and A5.9 for stainless steel. Wire electrodes are produced in diameters of 0.6mm–1.6mm, and other, larger diameters are available for production work. Table 8.4 lists the most common sizes and the amperage ranges for these electrodes. The amperage will vary depending on the method of metal transfer, type of shielding gas and base metal thickness. Some steel wire electrodes have a thin copper coating. This provides some protection to the electrode from rusting and improves the electrical contact between the wire electrode and the contact tube. It also acts as a lubricant to help the wire move more smoothly through the liner and contact tube. These electrodes may look like copper wire, but the amount of copper is so small that it either burns off or is diluted into the weld pool with no significant effect on the weld deposit.

TABLE 8.3 Filler metal specifications for different base metals

Base Metal Type	AWS Filler Metal Specification
Aluminum and aluminum alloys	A5.10
Copper and copper alloys	A5.6
Magnesium alloys	A5.19
Nickel and nickel alloys	A5.14
Stainless steel (austenitic)	A5.9
Steel (carbon)	A5.18
Titanium and titanium alloys	A5.16

TABLE 8.4 Filler metal diameters and amperage ranges

Base Metal	Electrode Diameter (Millimeter)	Amperage Range
Carbon Steel	0.6	35–190
	0.8	40–220
	0.9	60–280
	1.2	125–380
	1.6	275–450
Stainless Steel	0.6	40–150
	0.8	60–160
	0.9	70–120
	1.2	140–310
	1.6	280–450

Solid Wire

The specification for carbon steel filler metals for gas-shielded welding wire is A5.18. Filler metal classified within this specification can be used for MIG, tungsten inert gas shielded welding (TIG), and plasma arc welding (PAW) processes. Because in MIG and PAW the wire does not carry the welding current, the letters *ER* are used as a prefix, followed by two numbers to indicate the minimum tensile strength of a good weld. The actual strength is obtained by adding three zeroes to the right of the number. For example, ER70S-X is 70,000 psi. The *S* located to the right of the tensile strength indicates that this is a solid wire. The last number – 2, 3, 4, 5, 6 or 7 – or the letter *G* is used to indicate the filler metal composition and the weld's mechanical properties, Figure 8.51.

Carbon Steel Wires and Electrodes Classification ER70S-2 (AWS A5.18, EN ISO 636 – A)

This is a deoxidized mild steel filler wire. The deoxidizers allow the wire to be used on metal that has light coverings of rust or oxides. There may be a slight reduction in the weld's physical properties in this case, but the weld will usually still pass the classification test standards. This is a general-purpose filler that can be used on killed, semi-killed and rimmed steels. Argon-oxygen, argon-CO_2, and CO_2 can be used as shielding gases. Welds can be made in all positions.

Classification ER70S-3 (AWS A5.18, EN ISO 14341 – A)

This is a popular filler wire. It can be used in single or multiple-pass welds in all positions. ER70S-3 does not have the deoxidizers required to weld over rust, over oxides or on rimmed steels. It produces high-quality welds on killed and semi-killed steels. Argon-oxygen, argon-CO_2 and CO_2 can be used as shielding gases. This is the low-carbon steel filler most commonly used to weld galvanized steel.

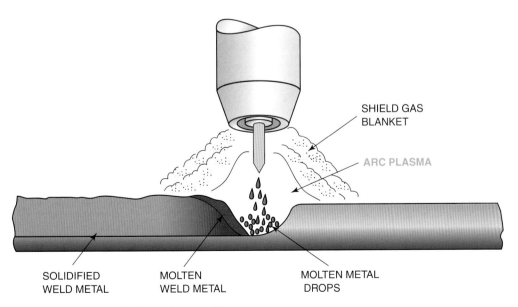

FIGURE 8.51 **Gas shielded metal arc welding**

Classification ER70S-6 (AWS. 18, EN ISO 14341 – A)

This is a good general-purpose filler wire. It has the highest levels of manganese and silicon. The wire can be used to make smooth welds on sheet metal or thicker sections. Welds over rust, oxides and other surface impurities will lower the mechanical properties, but not normally below the specifications of this classification. Argon-oxygen, argon-CO_2 and CO_2 can be used as shielding gases. Welds can be made in all positions. This filler wire is not suggested for galvanized steel as its higher silicon content may induce cracking.

Stainless Steels Wires and Electrodes

The specification for stainless steel–covered arc electrodes is A5.4 AWS (E308 – 17, EN 1600) or for stainless steel bare, cored, and stranded electrodes and welding rods is A5.9. AWS (ER 308L, ER309L, ER 316L, EN ISO 14343). Filler metal classified within A5.4 uses the letter E as its prefix, and the filler metal within A5.9 uses the letters ER as its prefix. Following the prefix, the three-digit stainless steel number from the American Iron and Steel Institute (AISI) is used. This number indicates the type of stainless steel in the filler metal, Table 8.5.

Aluminium and Aluminium Alloys

The International Alloy Designation System is the most widely accepted naming scheme for aluminium alloys. Each alloy has a four-digit number, in which the first digit indicates the major alloying elements. Heat-treatable alloys can be strengthened by furnace operations after they are produced, while non-heat-treatable alloys are work hardened by plastic deformation such as bending and rolling operations. Heat treatment of aluminium alloys is sometimes referred to as precipitation hardening, and work hardening is sometimes described as strain hardening.

- 1000 series are essentially pure aluminium with a minimum 99 per cent aluminium content by weight, and can be work hardened.
- 2000 series are alloyed with copper, and are heat treatable.
- 3000 series are alloyed with manganese, and can be work hardened.
- 4000 series are alloyed with silicon and can be work hardened.
- 5000 series are alloyed with magnesium, derive most of their strength from heat treatment, but can also be work hardened.
- 6000 series are alloyed with magnesium and silicon, are easy to machine, and are heat treatable, but not to the high strengths that 2000, 5000 and 7000 can reach.
- 7000 series are alloyed with zinc, and can be heat treated to the highest strengths of any aluminium alloy.
- 8000 series are a miscellaneous category where other elements not listed above may be used.

The specifications for aluminium and aluminium alloy filler metals are A5.3 (E4043, E4047) for covered arc welding electrodes and A5.10 for bare welding rods and electrodes. Filler metal classified within A5.3 uses the atomic symbol Al, and in A5.10 the prefix ER is used with the Aluminium Association number for the alloy, Table 8.6.

TABLE 8.5 Filler metal selection for joining different types of stainless

AISI TYPE NUMBER	442 / 446	430F / 430FSE	430 / 431	501 / 502	416 / 416SE	403 405 / 410 420 / 414	321 348 347	317	316L	316	314	310 / 310S	309 / 309S	304L	303 / 303SE	201 202 301 / 302 / 302B / 304 / 305 / 308	Mild Steel
201–202–301	310/312	310/312	310/312	310/312	309/310/312	309/310/312	308	308	308	308	308	308	308	308	308	308	312/310/309
302–3028–304	310/312	310/312	310/312	310/312	309/310/312	309/310/312	308	308	308	308	308	308	308	308	308	308	312/310/309
305–308	310/312	310/312	310/312	310/312	309/310/312	309/310/312	308	308	308	308	308	308	308	308	308-15	308	312/310/309
303	310	310	310	310	309/310	309/310	308	308	308-L	308	308	308	308	308-L	308	308	310/309/312
303SE	310	310	310	310	309/310	309/310	308	308	308	308	308	309	308	308	308	308	310/309/312
304L	310	310	310	310	309/310	309/310	308	316	316	316	309	310	309	308-L	308	308	310/309/312
309	310/312	310/312	310	310	309/310	309/310	309	309	309	309	309	309/310	309	309	309	309	310/309/312
309S	309/310	309/310	309/310	309/310	310	310	310	316	316	316	310	310	309/310	310	310	310	312
310	310	310	310	310	309/310	310/310	308	310	310	316	310/316	310	309/310	310	309/310	310	310/309/312
310S	309/312	309	309	309	309/310	309/310	308	309/310	308/309/316	309/310	309/310/316	309/316	310/316	310	310	310	312
314	310	410-15*	310	310	310	310-15	308	316/317/308	309/310	309/310	310-15	310	309/310	309/310	309/310	309/310	309/310/312
316	310	310	310	310	316	316	316	316/308	316	316	309/310/316	309/310/316	309/310	316	316	308/316	310/309
316L	310	310	310	310	316	316	316/308	316/317/308	316-L	316	309/310/316	310/309/316	309/310/316	316	316	308	310/309
317	310	310	310	310	309/310	308	308/347	317	316/308	316/308	316/317/308/309	317/316/309	317/316/309	308	308	308	309/310
321 / 348 / 347	310/312	310	310/312	310/312	309/310/312	309/310	347	308/347	347/308	347/308	309/310/317	347/308	347/308	308	308	308	310/309
403–405 / 410–420 / 414	310/309/312	309	310/309	310/309	410-15*/309**/310**	410*/309**	309/310	308/347	309/310	309/310	310-15/316	310/309	309/310	308	308	308	309/310/312
416 / 416SE	310/309/312	309	309	309/312	410-15*/310	410-15*/309*/310**	309/310	309/310	309/310	309/310	309/310	309/310	309/310	309/310	309/310	309/310	309/310/312
501 / 502	310	310	310	502*/310**	310	502*/310**	310/309	310/309	310/309	310/309	310/309	310/309	310/309	310/309	310/309	310/309	310/309/312
430 / 431	310/309	310/309	430-15*/310**/309**	310	309	410*/309**	310/309	310/309	310/309	310/309	310/309	310/309	310/309	310/309	310/309	310/309	310/309/312
430F / 430FSE	310/309/312	410-/15*	310/309	310/309	310/309	310/309	310/309/312	310/309	310/309	310/309	310/309/312	310/309	310/309/312	310/309	310/309/312	310/309/312	310/309/312
442 / 443	309/310/312	309/310/312	310/309/312	310/309/312	310/309/312	310/309/312	309/310/312	309/310/312	310/309/312	309/310/312	310/309/312	310/309/312	309/310/312	309/310/312	309/310/312	309/310/312	309/310/312

This choice can vary with specific applications and individual job requirements.

*Pre-heat.

**No pre-heat necessary.

Source: Courtesy of Thermacote Welco

TABLE 8.6 Filler metal selection for aluminium

Base Metal	319 355	43 356	214	6061 6063 6151	5456	5454	5154 5254	5086	5083	5052 5652	5005 5050	3004	1100 3003	1060
1060	4145 4043 4047	4043 4047 4145	4043 5183 4047	4043 4047	5356 4043	4043 5183 4047	4043 5183 4047	5356 4043	5356 4043	4043 4047	1100 4043	4043	1100 4043	1260 4043 1100
1100 3003	4145 4043 4047	4043 4047 4145	4043 5183 4047	4043 4047	5356 4043	4043 5183 4047	4043 5183 4047	5356 4043	5356 4043	4043 5183	4043 5183 5356	4043 5183 5356	1100 4043	
3004	4043 4047	4043 4047	5654 5183 5356	4043 5183 5356	5356 5183 5556	5654 5183 5356	5654 5183 5356	5356 5183 5556	5356 5183 5556	4043 5183 4047	4043 5183 5356	4043 5183 5356		
5005 5050	4043 4047	4043 4047	5654 5183 5356	4043 5183 5356	5356 5183 5556	5654 5183 5356	5654 5183 5356	5356 5183 5556	5356 5183 5556	4043 5183 4047	4043 5183 5356			
5052 5652	4043 4047	4043 5183	5654 5183 5356	4043 5356 5183	5356 5183 5556	5654 5183 5356	5654 5183 5356	5356 5183 5556	5356 5183 5556	5654 5183 4043				
5083	NR	4043 5183 5356	5356 5183 5556	5356 5183 5556	5356 5183 5556	5356 5183 5556	5356 5183 5556	5356 5183 5556	5356 5183 5556					
5086	NR	4043 5183	5356 5183 5556	5356 5183 5556	5356 5183 5556	5356 5183 5556	5356 5183 5556	5356 5183 5556						
5154 5254	NR	4043 5183	5654 5183 5356	4043 5356 5183	5356 5183 5556	5654 5183 5356	5654 5183 5356							
5454	4043 4047	4043 5183 4047	5654 5183 5356	4043 5356 5183	5356 5183 5554	5554 4043 5183								
5456	NR	4043 5183	5356 5183 5554	5356 5183 5556	5556 5183 5356									
6061 6063 6151	4145 4043 4047	4043 5183 4047	4043 5654 5183 5356	4043 5356 5183 4047										
214	NR	4043 5183	5654 5183 5356											
43 356	4043 4047	4043 4145 4043 4047												
319 355	4145 4043 4047													

Note: First filler alloy listed in each group is the all-purpose choice. NR means that these combinations of base metals are not recommended for welding.

Courtesy of Thermacote Welco

Aluminium Bare Welding Rods and Electrodes Classification ER1100 (AWS A5.10, EN ISO 18273)

1100 aluminium has the lowest percentage of alloy agents of all the aluminium alloys, and it melts at 650°C. The filler wire is also relatively pure. ER1100 produces welds that have good corrosion resistance and high ductility, with tensile strength. The weld deposit has a high resistance to cracking during welding. This wire can be used with oxy-fuel gas welding, TIG and MIG. Pre-heating to 140°C to 170°C is required for TIG welding on plate or pipe 10mm and thicker to ensure good fusion. Flux is required for 1100 aluminium, which is commonly used for items such as food containers, food-processing equipment, storage tanks, and heat exchangers. ER1100 can be used to weld 1100 and 3003 grade aluminium.

Classification ER4043 (AWS A5.10, EN ISO 18273)

ER4043 is a general-purpose welding filler metal. It has 4.5 per cent to 6.0 per cent silicon added, which lowers its melting temperature to 620°C, helping to promote a free-flowing molten weld pool. The welds have high ductility and a high resistance to cracking during welding. This wire can be used with TIG and MIG. Pre-heating to 140°C to 170°C is required for TIG welding on plate or pipe 10mm. and thicker to ensure good fusion. ER4043 can be used to weld on 2014, 3003, 3004, 4043, 5052, 6061, 6062 and 6063 and cast alloys 43, 355, 356 and 214.

Classification ER5356 (AWS A5.10, EN ISO 18273)

ER5356 has 4.5 per cent to 5.5 per cent magnesium added to improve the tensile strength. The weld has high ductility but only an average resistance to cracking during welding. This wire can be used for TIG and MIG. Pre-heating to 140°C to 170°C is required for TIG welding on plate or pipe 10mm and thicker to ensure good fusion. ER5356 can be used to weld on 5050, 5052, 5056, 5083, 5086, 5154, 5356, 5454 and 5456.

Classification ER5556 (AWS A5.10, EN ISO 18273)

ER5556 has 4.7 per cent to 5.5 per cent magnesium and 0.5 per cent to 1.0 per cent manganese added to produce a weld with high strength. The weld has high ductility and only average resistance to cracking during welding. This wire can be used for TIG and MIG. Pre-heating to 140°C to 170°C is required for TIG welding on plate or pipe 10mm and thicker to ensure good fusion. ER5556 can be used to weld on 5052, 5083, 5356, 5454 and 5456.

GAS DENSITY AND FLOW RATES

Density is the chief determinant of how effective a gas is for arc shielding. The lower the density of a gas, the higher is the flow rate required for equal arc protection. However, flow rates are not in proportion to the densities. Helium, with about one-tenth the density of argon, requires about twice the flow for equal protection.

The correct flow rate can be set by checking welding guides that are available from the welding equipment and filler metal manufacturers. These list the gas flow required for various nozzle sizes and welding amperage settings. Some welders feel that a higher gas flow will provide better weld coverage, but that is not always the case. High gas flow rates waste shielding gases and may lead to contamination due to turbulence in the gas. Air is drawn into the weld as envelope by the venturi effect around the edge of the nozzle. Air can also be drawn in under the nozzle if the torch is held at too sharp an angle to the metal. Using a wind screen, as shown in Figure 8.52, can help prevent the shielding gas from being blown away.

Standing to one side, turn on the shielding gas supply valve. If the supply is a cylinder, the valve is opened all the way. With the machine power on and the welding gun switch depressed, you are

CAUTION

If you need more shielding gas coverage in a windy or draughty area, use both a larger-diameter gas nozzle and a higher gas flow rate. With a larger nozzle size, a higher flow rate can be used without causing turbulence. Larger nozzle sizes may restrict your view of the weld. You might also consider setting up a wind barrier to protect your welding from the wind, Figure 8.53.

Make sure you are reading
the correct scale for the gas
being used.

ready to set the flow rate. Slowly turn in the adjusting screw and watch the float ball as it rises in a tube on a column of gas. The faster the gas flows, the higher the ball will float. A scale on the tube allows you to read the flow rate. Different scales are used with each type of gas being used. Since gases have different densities, the ball will float at varying levels even though the flow rates are the same, Figure 8.54. You may need to read the line corresponding to the flow rate as it compares to the top, centre, or bottom of the ball, depending upon the manufacturer's instructions. There should be some marking or instruction on the tube or regulator to tell how it should be read, Figure 8.55.

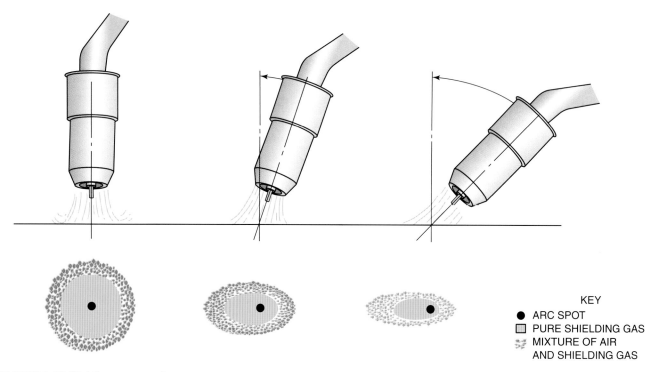

KEY
● ARC SPOT
▢ PURE SHIELDING GAS
▨ MIXTURE OF AIR
    AND SHIELDING GAS

FIGURE 8.52 **Welding gun angle**

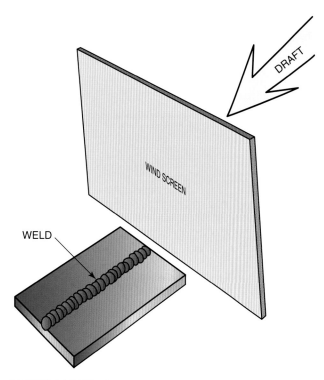

FIGURE 8.53 **Wind screen**

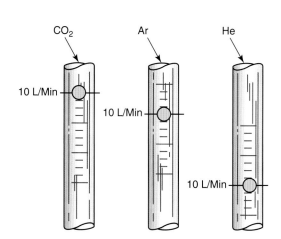

FIGURE 8.54 **Reading gas flow rates**

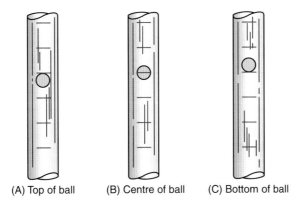

(A) Top of ball   (B) Centre of ball   (C) Bottom of ball

FIGURE 8.55 **Three methods of reading a flowmeter**

Release the welding gun switch, and the gas flow should stop. Turn off the power and spray the hose for leaks. When not in use, turn off the cylinder valve and de-load the regulator but leave the residual gas in the line to prevent atmospheric gases entering the hoses and gun assembly. This residual gas will be purged before the next operation of the welder.

ARC-VOLTAGE AND AMPERAGE CHARACTERISTICS

The arc-voltage and amperage characteristics of MIG welding are different from those for most other welding processes. The voltage is set on the welder, and the amperage is set by changing the wire-feed speed. More amperage is required to melt the wire the faster it is fed, Because changes in the wire-feed speed directly change the amperage, it is possible to set the amperage by using a chart and measuring the length of wire fed per minute, Table 8.7. The voltage and amperage required for a specific metal transfer method differ for various wire sizes, shielding gases and metals. The voltage and amperage setting will be specified for all welding done according to a welding procedure specification (WPS) or other codes and standards.

TABLE 8.7 **Welding amperage range for common electrode types**

Wire-feed Speed (m/min)	Wire Diameter			
	0.8mm	0.9mm	1.2mm	1.6mm
(2.5)	40	65	120	190
(5.0)	80	120	200	330
(7.6)	130	170	260	425
(10.2)	160	210	320	490
(12.7)	180	245	365	–
(15.2)	200	265	400	–
(17.8)	215	280	430	–

Setting the Voltage and Current Settings

On a scale of 0 to 10, set the wire-feed speed control dial at 5, or halfway between the low and high settings of the unit. The voltage is also set at a point halfway between the low and high settings. The shielding gas can be CO_2, argon, or a mixture. The gas flow should be adjusted to a rate of 12–14L/Min

METAL PREPARATION

All hot-rolled steel has a thin layer of dark grey or black iron oxide (mill scale), which is formed during the rolling process. It is possible to purchase some hot-rolled steels that have had this layer removed but most still has this layer because it offers some protection from rusting. Mill scale is not removed for non-code welding, because it does not prevent most welds from being suitable for service. For practice welds that will be visually inspected, mill scale can usually be left on the plate. Filler metals and fluxes usually have deoxidizers added to them so that the adverse effects of the mill scale are reduced or eliminated, Table 8.8. But with MIG welding wire it is difficult to add enough deoxidizers to remove all effects of mill scale. The porosity that mill scale causes is most often confined to the interior of the weld and is not visible on the surface, Figure 8.56. Because it is not visible, it usually goes unnoticed and the weld passes visual inspection. If the practice results are going to be destructively tested or if the work is of a critical nature, then all welding surfaces within the weld preparation and the surrounding surfaces within 25mm must be cleaned to bright metal, Figure 8.57. This may be done by grinding, filing, sanding or shot blasting.

TABLE 8.8 Shielding gas mixtures used for MIG welding

Shielding Gas	Chemical Behaviour	Uses and Usage Notes
1. Argon	Inert	Welding virtually all metals except steel
2. Helium	Inert	Al and Cu alloys for greater heat and to minimize porosity
3. Ar and He (20% to 80% to 50% to 50%)	Inert	Al and Cu alloys for greater heat and to minimize porosity but with quieter, more readily controlled arc action
4. N_2	Reducing	On Cu, very powerful arc
5. Ar + 25% to 30% N_2	Reducing	On Cu, powerful but smoother operating, more readily controlled arc than with N_2
6. Ar + 1% to 2% O_2	Oxidizing	Stainless and alloy steels, also for some deoxidized copper alloys
7. Ar + 3% to 5% O_2	Oxidizing	Plain carbon, alloy, and stainless steels (generally requires highly deoxidized wire)
8. Ar + 3% to 5% O_2	Oxidizing	Various steels using deoxidized wire
9. Ar + 20% to 30% O_2	Oxidizing	Various steels, chiefly with short-circuiting arc
10. Ar + 5% O_2 + 15% CO_2	Oxidizing	Various steels using deoxidized wire
11. CO_2	Oxidizing	Plain-carbon and low-alloy steels, deoxidized wire essential
12. CO_2 + 3% to 10% O_2	Oxidizing	Various steels using deoxidized wire
13. CO_2 + 20% O_2	Oxidizing	Steels

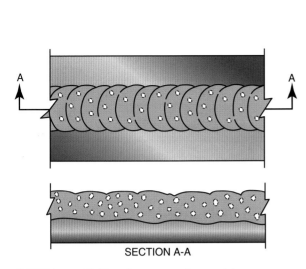

FIGURE 8.56 Uniformly scattered porosities

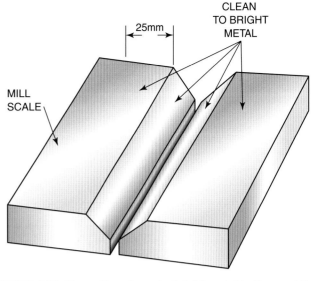

FIGURE 8.57 Clean all surfaces to bright metal before welding

PRACTICES

To make acceptable MIG welds consistently, the major skill required is the ability to set up the equipment and weldment correctly in response to variations in material thickness, position and type of joint. A correctly set-up MIG welding station can, in many cases, be operated with minimum skill. Ideally, only a few tests would be needed for the welder to make the necessary adjustments in setup and manipulation techniques to achieve a good weld. Previous welding experiments should have given a graphic set of comparisons to help the welder make the correct changes.

The practices in this chapter are grouped to keep the number of variables in the setup to a minimum. Often, the only change required before going on to the next weld is to adjust the power settings. Figures that are given in some of the practices will give general operating conditions, such as voltage, amperage and shielding gas or gas mixture. These are general values, and the welder will have to make some fine adjustments. Differences in the type of machine being used and the material surface condition will also affect the settings.

SHORT-CIRCUITING OR DIP TRANSFER

Low currents allow the liquid metal at the electrode tip to be transferred by direct contact with the molten weld pool. This process requires close interaction between the wire feeder and the power supply.

The short-circuiting mode of transfer is the most common process used with MIG welding:

- on thin or properly prepared thick sections of material;
- on a combination of thick to thin materials;
- with a wide range of electrode diameters;
- with a wide range of shielding gases;
- in all positions.

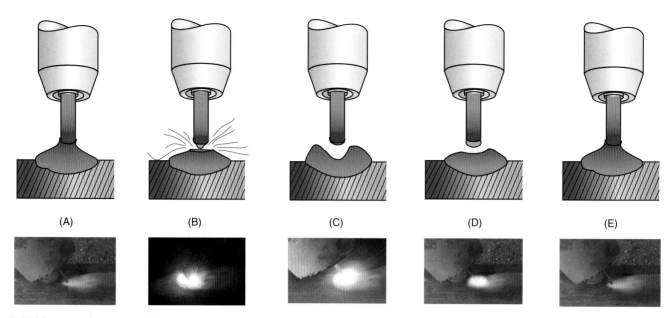

(A)   (B)   (C)   (D)   (E)

FIGURE 8.58 Schematic of short circuiting transfer

The 0.6mm–1.2mm wire electrodes are the most common diameters for the short-circuiting mode. The most popular shielding gases used on carbon steel are a combination of 80–95 per cent argon (Ar) and 5–20 per cent CO_2. The amperage range may be as low as 35A for materials of 0.8mm or as high as 225A for materials up to 3mm in thickness on square butt weld joints. Thicker base metals can be welded if the edges are bevelled to allow weld joint penetration.

The transfer mechanisms in this process are quite straightforward, as shown in Figure 8.58. To start, the wire is in direct contact with the molten weld pool, Figure 8.58A. Once the electrode touches the molten weld pool, the arc and its resistance are removed and the welding amperage quickly rises as it begins to flow freely through the tip of the wire into the molten weld pool. The resistance to current flow is highest at the point where the electrode touches the molten weld pool, where the temperature is very high. The higher the temperature, the higher the resistance to current flow. A combination of high current flow and high resistance causes a rapid rise in the temperature of the electrode tip.

As the current flow increases, the interface between the wire and molten weld pool is heated until it explodes into a vapour (Figure 8.58B), establishing an arc. This small explosion produces sufficient force to depress the molten weld pool. A gap between the electrode tip and the molten weld pool (Figure 8.58C) immediately opens. With the resistance of the arc re-established, the voltage increases as the current decreases. The low current flow is insufficient to continue melting the electrode tip off as fast as it is being fed into the arc. As a result, the arc length rapidly decreases (Figure 8.58D) until the electrode tip contacts the molten weld pool (Figure 8.58E). The liquid formed at the wire tip during the arc-on interval is transferred by surface tension to the molten weld pool, and the cycle begins again with another short circuit. If the system is properly tuned, the rate of short circuiting can be repeated from approximately 20 to 200 times per second, causing a characteristic buzzing sound. The spatter is low and the process easy to use. The low heat produced by dip transfer makes the system easy to use in all positions on sheet metal, low-carbon steel, low-alloy steel, and stainless steel ranging in thickness from 0.6mm to 3mm. The short-circuiting process does not produce enough heat to make quality welds in sections much thicker than 6mm unless it is used for the root pass on a butt weld or to fill gaps in joints. Although this technique is highly effective, lack-of-fusion defects can occur unless the process is perfectly tuned and the welder is highly skilled, especially on thicker metal.

Carbon dioxide works well with this process because it produces the forceful, high-energy arc needed during the arc-on interval to displace the weld pool. Helium can be used as well. Pure argon is not as effective because its arc tends to be sluggish and not very fluid. However, a mixture of 5–20 per cent carbon dioxide and 80–95 per cent argon produces a less harsh arc and a flatter, more fluid, and desirable weld profile. Although more costly, this gas mixture is preferred. A gas mixture of 98 per cent argon and 2 per cent oxygen may also be used on thinner carbon steels and sheet metal. This mixture produces lower-energy short-circuiting transfer and can be an advantage on thin gauge materials, producing minimal burn-through at voltages as low as 13 volts.

New technology in wire manufacturing has allowed smaller wire diameters to be produced. These have become the preferred size even though they are more expensive. The short-circuiting process works better with a short electrode stickout. The power supply is most critical. It must have a constant voltage, otherwise known as constant-potential output, and sufficient inductance to slow the rate of current increase during the short-circuit interval. Too little inductance causes spatter due to high-current surges, while too much causes the system to become sluggish.

DIP TRANSFER TECHNIQUE

WORKSHOP TASK 8.1

Stringer Bead Deposition

Using a properly set-up and adjusted MAGS welding machine, Table 8.9, appropriate PPE, 0.8 and 1.2mm diameter wire, and two or more pieces of mild steel sheet 150mm long and 1.6mm–3.0mm thick, you will make a stringer bead weld in the flat position, Figure 8.59.

- Starting at one end of the plate and using either a pushing or dragging technique, make a weld bead along the entire 150mm length of the metal.

- After the weld is complete, check its appearance.

- Make any needed changes in voltage, wire feed speed, or electrode extension to correct the weld.

- Repeat the weld and make additional adjustments. After the machine is set, start to work on improving the straightness and uniformity of the weld.

Keeping the bead straight and uniform can be hard because of the limited visibility due to the small amount of light and the size of the molten weld pool. The welder's view is further restricted by the shielding gas nozzle, Figure 8.60. Even with limited visibility, it is possible to make a satisfactory weld by watching the edge of the molten weld pool, the sparks and the weld bead produced. Watching the leading edge of the molten weld pool (forehand welding, pushing technique) will show you the molten weld pool fusion and width. Watching the trailing edge of the molten weld pool (backhand welding, dragging technique) will show you the amount of build up and the relative heat input, Figure 8.61. The quantity and size of sparks produced can indicate the relative location of the filler wire in the molten weld pool. The number of sparks will increase as the wire strikes the solid metal ahead of the molten weld pool. The gun itself may begin to vibrate or bump as the wire momentarily pushes against the cooler, un-melted base metal before it melts. Changes in weld width, build up and proper joint tracking can be seen by watching the bead as it appears from behind the shielding gas nozzle.

- Repeat each type of bead until consistently good beads are obtained. Turn off the welding machine and shielding gas and reinstate your work area when you are finished welding.

Complete a copy of the 'Student Welding Report' listed in Appendix I or provided by your instructor.

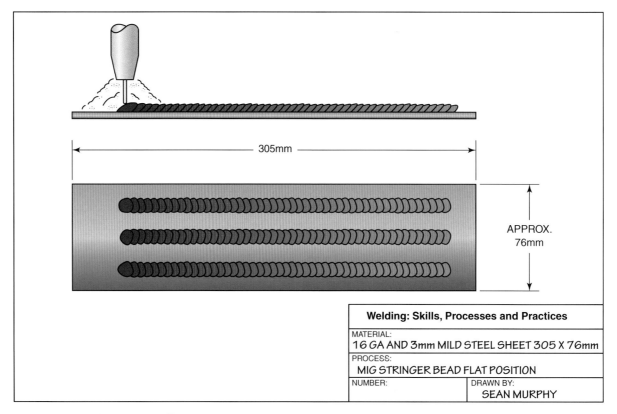

Welding: Skills, Processes and Practices	
MATERIAL: 16 GA AND 3mm MILD STEEL SHEET 305 X 76mm	
PROCESS: MIG STRINGER BEAD FLAT POSITION	
NUMBER:	DRAWN BY: SEAN MURPHY

FIGURE 8.59 Stringer beads in the flat position

TABLE 8.9 Typical welding current settings for mild steel

Process	Wire Diameter (mm)	Amperage Range (Optimum)	Voltage Range (Optimum)	Shielding Gas
Short-circuiting	0.76	60 (100) 140	14 (15) 16	100% CO_2
	0.88	90 (130) 150	16 (17) 20	75% Ar + 25% CO_2
				98% Ar + 2% O

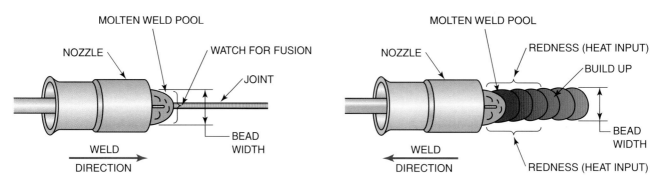

FIGURE 8.60 Watching forehand welding

FIGURE 8.61 Watching backhand welding

WORKSHOP TASK 8.2

Butt Joint in Flat (PA) Position

Using the same equipment, materials, and procedures listed previously make a welded butt joint in the flat position (PA), Figure 8.62A.

- Tack weld the sheets together and place them flat on the welding table, Figure 8.63.
- Starting at one end, with the torch held at 75–80° slope angle and 90° tilt angle run a bead along the joint.
- Watch the molten weld pool and bead for signs that a change in technique may be required.
- Make any needed changes as the weld progresses. By the time the weld is complete, you should be making the weld nearly perfectly.
- Using the same technique, make another weld. This time, the entire 150mm of weld should be flawless.
- Repeat with both thicknesses of metal until consistently good beads are obtained.
- Turn off the welding machine and shielding gas and reinstate when you are finished welding.

  Complete a copy of the 'Student Welding Report'.

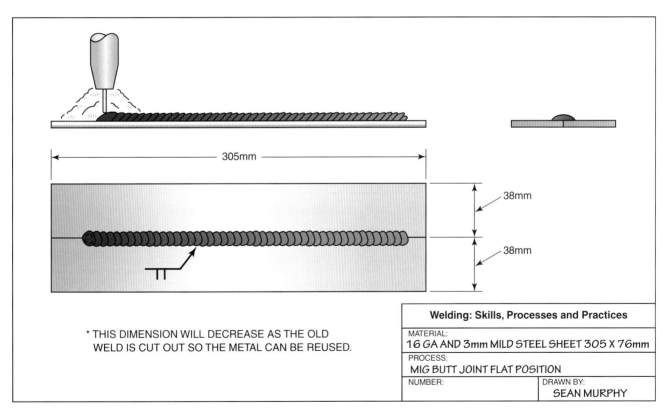

* THIS DIMENSION WILL DECREASE AS THE OLD
WELD IS CUT OUT SO THE METAL CAN BE REUSED.

Welding: Skills, Processes and Practices

MATERIAL:
16 GA AND 3mm MILD STEEL SHEET 305 X 76mm

PROCESS:
MIG BUTT JOINT FLAT POSITION

NUMBER:

DRAWN BY:
SEAN MURPHY

FIGURE 8.62A Butt joint in flat position

WORKSHOP TASK 8.3

Lap Joint in the Flat (PA) Position

Using the same equipment, materials, and procedures listed previously make a lap joint in the flat position (PA), Figure 8.62B.

● Tack weld the sheets together with a maximum overlap of 13mm and place them flat on the welding table.

● Starting at one end, with the torch held at 75–80° slope angle and 45° tilt angle run a bead along the joint.

● Watch the molten weld pool and bead for signs that a change in technique may be required.

● Make any needed changes as the weld progresses. By the time it is complete, you should be making the weld nearly perfectly.

● Using the same technique, make another weld. This time, the entire 150mm of weld should be flawless.

● Repeat with both thicknesses of metal until consistently good beads are obtained.

● Turn off the welding machine and shielding gas and reinstate when you are finished welding.

Complete a copy of the 'Student Welding Report'.

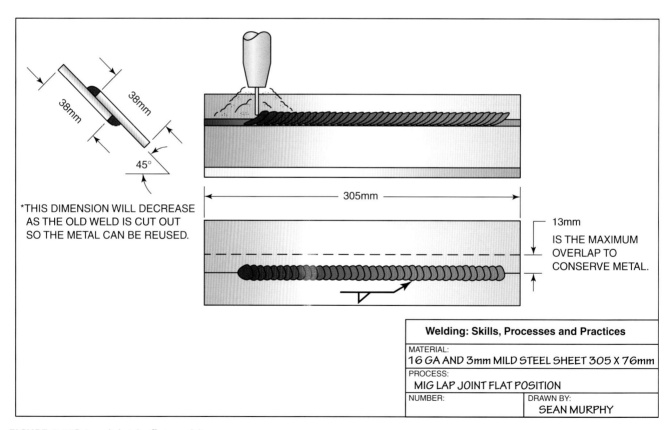

*THIS DIMENSION WILL DECREASE AS THE OLD WELD IS CUT OUT SO THE METAL CAN BE REUSED.

38mm 38mm 38mm 45°

305mm

13mm IS THE MAXIMUM OVERLAP TO CONSERVE METAL.

Welding: Skills, Processes and Practices

MATERIAL:
16 GA AND 3mm MILD STEEL SHEET 305 X 76mm

PROCESS:
MIG LAP JOINT FLAT POSITION

NUMBER:

DRAWN BY:
SEAN MURPHY

FIGURE 8.62B Lap joint in flat position

WORKSHOP TASK 8.4

T-joint in the Flat (PA) Position

Using the same equipment, materials, and procedures listed previously make a T-joint in the flat position (PA), Figure 8.62C.

- Tack weld the sheets together and place them flat on the welding table, Figure 8.63.

- Starting at one end, with the torch held at 75–80° slope angle and 45° tilt angle run a bead along the joint.

- Watch the molten weld pool and bead for signs that a change in technique may be required.

- Make any needed changes as the weld progresses. By the time it is complete, you should be making the weld nearly perfectly.

- Using the same technique, make another weld. This time, the entire 150mm of weld should be flawless.

- Repeat with both thicknesses of metal until consistently good beads are obtained.

- Turn off the welding machine and shielding gas and reinstate when you are finished welding.

  Complete a copy of the 'Student Welding Report'.

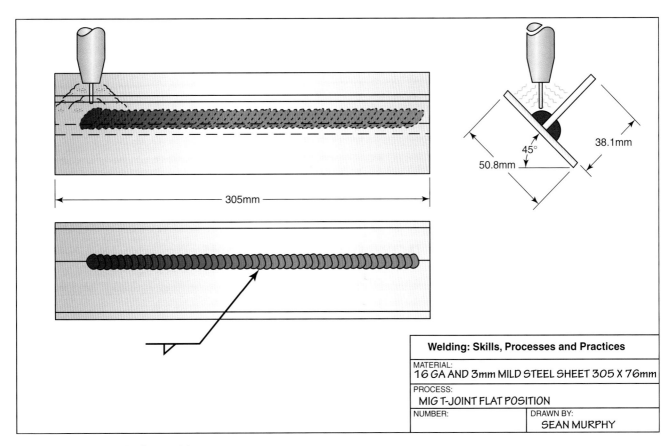

Welding: Skills, Processes and Practices	
MATERIAL: 16 GA AND 3mm MILD STEEL SHEET 305 X 76mm	
PROCESS: MIG T-JOINT FLAT POSITION	
NUMBER:	DRAWN BY: SEAN MURPHY

FIGURE 8.62C T-joint in flat position

Use enough tack welds to keep the joint in alignment during welding. Small tack welds are easier to weld over without adversely affecting the weld.

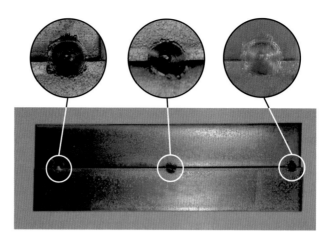

FIGURE 8.63 Tack welds

WORKSHOP TASK 8.5

Corner in the Flat (PA) Position

Using the same equipment, materials, and procedures listed previously make an outside corner joint in the flat position (PA).

- Tack weld the sheets together and place them flat on the welding table, Figure 8.63.
- Starting at one end, with the torch held at 75–80° slope angle and a tilt angle that bisects the joint, run a bead along the joint.
- Watch the molten weld pool and bead for signs that a change in technique may be required.
- Make any needed changes as the weld progresses. By the time it is complete, you should be making the weld nearly perfectly.
- Using the same technique, make another weld. This time, the entire 150mm of weld should be flawless.
- Repeat with both thicknesses of metal until consistently good beads are obtained.
- Turn off the welding machine and shielding gas and reinstate when you are finished welding.

Complete a copy of the 'Student Welding Report'.

VERTICAL DOWN (PG) POSITIONS

The vertical down technique can be useful when making some types of welds. The major advantages of this technique are:

- *Speed*. Very high rates of travel are possible.
- *Shallow penetration*. Thin sections or root openings can be welded with little burn-through.
- *Good bead appearance*. The weld has a nice width-to-height ratio and is uniform.

Vertical down welds are often used on thin sheet metals or in the root pass in butt joints, where the combination of controlled penetration and higher welding speeds are advantageous. Although it is easy to make welds with a good appearance, generally more skill is required to make sound welds with this technique than in the vertical up position. The most common problem is lack of fusion or cold lapping. To prevent this, the arc must be kept at or near the leading edge of the molten weld pool.

AXIAL SPRAY METAL TRANSFER

The freedom from spatter associated with the argon-shielded MAGS process results from a unique mode of metal transfer called axial spray metal transfer, Figure 8.64. This process is identified by the pointing of the wire tip from which droplets smaller than the diameter of the electrode wire are projected axially across the arc gap to the molten weld pool. Hundreds of drops per second cross from the wire to the base metal, propelled by arc forces at high velocity in the direction the wire is pointing. In the case of plain carbon steel and stainless steels, the molten weld pool may be too large and too hot to be controlled in vertical or overhead positions. Because the drops are very small and directed at the molten weld pool, the process is spatter free.

The spray transfer mode for carbon steels requires three conditions:

- argon-rich shielding gas mixtures;
- direct current electrode positive (DCEP) polarity (also called direct current reverse polarity, DCRP);
- a current level above a critical amount called the 'transition current'.

The shielding gas is usually a mixture of 95 per cent to 98 per cent argon and 5 per cent to 2 per cent oxygen, or 80 per cent to 90 per cent argon and 20 per cent to 10 per cent carbon dioxide. The added percentage of active gases allows greater weld penetration.

Figure 8.65 shows how the transfer rate of drops changes in relation to the welding current. At low currents, the drops are large and are transferred at rates below ten per second. They tend to bridge the gap between the electrode tip end and the molten weld pool. This produces a momentary short circuit that throws off spatter. However, the mode of transfer changes abruptly above the critical current, producing the desirable spray.

The transition current depends on the alloy being welded. It is also proportional to the wire diameter. Higher currents are needed with larger diameter wires, which imposes some restrictions

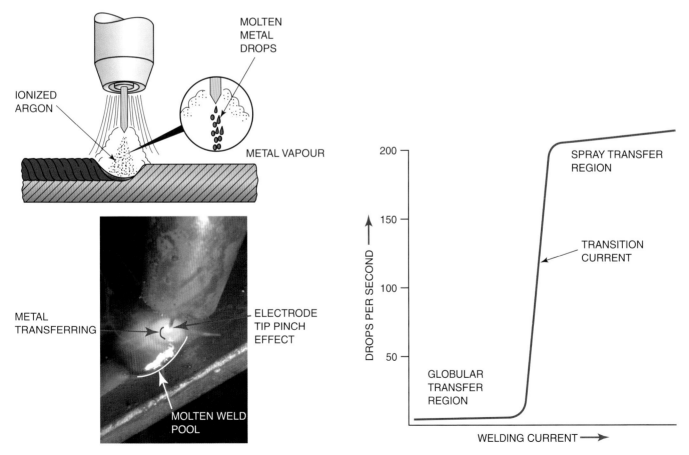

FIGURE 8.64 Axial sprayer metal

FIGURE 8.65 Desirable spray transfer

TABLE 8.10 MIG welding parameters for mild steel

| Mild Steel | Wire-feed Speed, mm/min | | Voltage, V | | | | |
Base-material Thickness, mm	0.89mm	1.14mm	CO_2	75 Ar-25 CO_2	Ar	98 Ar-2O_2	Current A
0.91	2667–2921	–	18	16	–	–	50–60
1.22	3556–4064	1178	19	17	–	–	70–80
1.52	4572–5588	2286–2794	20	17.7	–	–	90–110
1.90	6096–6604	3048–3302	20.7	18	20	–	120–130
3.16	7112–7620	3556–3810	21.5	18.5	20.5	–	140–150
4.75	8128–8636	4064–4445	22	19	21.5	23.5	160–170
6.35	9114–9652	4699–4953	22.7	19.5	22.5	24.5	180–190
7.94	10160–10668	5334–5588	23.5	20.5	23.5	25	200–210
9.52	10668–13208	5588–6858	25	22	25	26.5	220–250
12.70 and up	–	9525	28	26	29	31	300

on the process. The high current hinders welding of sheet metal because the high heat cuts through sheet metal. High current also limits its use to the flat and horizontal welding positions. Weld control in the vertical or overhead position is very difficult or impossible to achieve. Table 8.10 lists the welding parameters for a variety of gases, wire sizes, and metal thicknesses for MAGS welding of mild steel.

Globular Transfer

The *globular transfer process* is rarely used by itself because it transfers the molten metal across the arc in much larger droplets, but it is used in combination with pulsed-spray transfer.

Globular transfer can be used on thin materials and at a very low current range. It can be used with higher current but is not as effective as other welding modes of metal transfer.

Pulsed-Spray Transfer

CAUTION

The heat produced during pulsed-spray transfer welding using large-diameter wire or high current may be intense enough to cause the filter lens in a welding helmet to shatter. Be sure the helmet is equipped with a clear plastic lens on the inside of the filter lens. Avoid getting your face too close to the intense heat.

Because the change from spray arc to globular transfer occurs within a very narrow current range and the globular transfer occurs at the rate of only a few drops per second, a controlled spray transfer at significantly lower average currents is achievable. MIG welding with pulsed-arc metal transfer involves pulsing the current from levels below the transition current to those above it. The time interval below the transition current is short enough to prevent a drop from developing. About 0.1 seconds is needed to form a globule, so no globule can form at the electrode tip if the time interval at the low base current is about 0.01 second. Actually, the energy produced during this time is very low – just enough to keep the arc alive.

The arc's real work occurs during those intervals when the current pulses to levels above the transition current. The time of that pulse is controlled to allow a single drop of metal to transfer. In fact, with many power supplies, a few small drops could transfer during the pulse interval. As with conventional spray arc, the drops are propelled across the arc gap, allowing metal transfer in all positions. The average current can be reduced sufficiently to reduce penetration enough to weld sheet metal or reduce deposition rates enough to control the molten weld pool in all positions. This level control is achieved by changing the following variables, graphed in Figure 8.66:

- *Frequency.* The number of times the current is raised and lowered to form a single pulse, measured in pulses per second.

- *Amplitude.* The amperage or current level of the power at the peak or maximum, expressed in amperage.

- *Width of the pulses.* The amount of time the peak amperage is allowed to stay on.

Figure 8.67 shows a typical pulsed-arc welding system. Although developed in the mid-1960s, this technology did not receive much attention until solid state electronics were developed to handle the high power required of welding power supplies and provide a simpler, and more economical way to control the pulsing process. The newest generation of pulsed-arc systems interlocks the power supply and wire feeder so that the proper settings are obtained for any given job by adjusting a single knob. These are known as synergic systems. The greatest benefit to synergic pulsed arc is that the power supply reacts automatically to changes in the work conditions, such as inconsistent fit-ups or operator changes in stick out to maintain a more consistent weld.

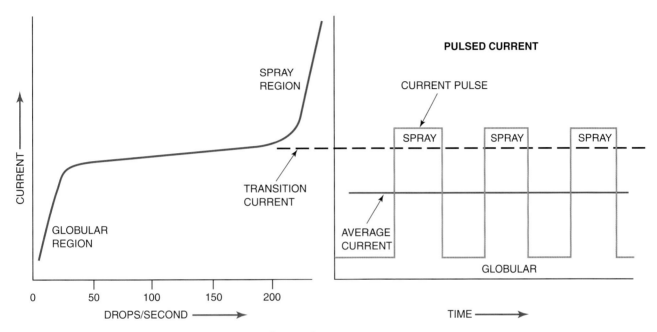

FIGURE 8.66 Mechanism of pulsed-arc spray transfer at a low average current

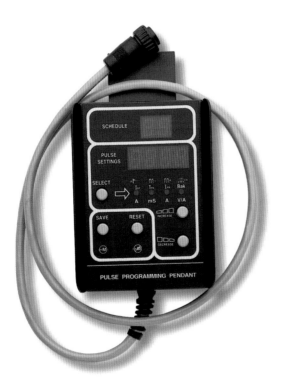

FIGURE 8.67 MIG pulsed-arc welding system controller

In some respects, synergic pulsed-arc systems are more complex to use because the correct relationships between the wire-feed speeds and power supply settings must be programmed into the equipment, and each wire composition, wire size, and shielding gas requires a special program. The manufacturer generally programs the most common combinations, allowing additional user input. Welding power supplies that produce synergic pulsation must be capable of both constant-current (CC) and constant-voltage (CV) power output.

Shielding Gases for Spray or Pulsed-arc Transfer

Axial spray transfer is impossible without shielding gases rich in argon. Pure argon is used with all metals except steels. As much as 80 per cent helium can be added to the argon to increase the heat in the arc without affecting the desirable qualities of the spray mode. With more helium, the transfer becomes progressively more globular, forcing the use of a different welding mode, to be described later. Since these gases are inert, they do not react chemically with any metals. This factor makes the MIG process the only productive method for welding metals sensitive to oxygen (essentially all metals except iron or nickel).

The cathodic cleaning action which helps to remove the thin layer oxides that form on metals is associated with argon at DCEP (and is also very important for fabricating metals such as aluminium, which quickly develop these undesirable surface oxides when exposed to air). This same cleaning action causes problems with steels. Iron oxide in and on the steel surface is a good emitter of electrons that attracts the arc. But these oxides are not uniformly distributed, resulting in very irregular arc movement and uneven weld deposits. This problem was solved by adding small amounts of an active gas such as oxygen or carbon dioxide to the argon, which produces a uniform film of iron oxide on the weld pool and provides a stable site for the arc. This discovery enabled uniform welds in ferrous alloys and expanded the use of MIG to welding those materials. The amount of oxygen needed to stabilize arcs in steel varies with the alloy. Generally, 2 per cent is sufficient for carbon and low-alloy steels. In the case of stainless steels, about 0.5 per cent should prevent a refractory scale of chromium oxide, which can be a starting point for corrosion. Carbon dioxide can substitute for oxygen. Between 8 per cent and 10 per cent is optimum for low-alloy steels. In many applications, carbon dioxide is the preferred addition because the weld bead has a better contour and the arc appears to be more stable. Gas mixes of 98 per cent argon – 2 per cent oxygen as well as 98 per cent argon – 2 per cent carbon dioxide are commonly used for spray transfer MAG welding of stainless steels.

WORKSHOP TASK 8.6

Stringer Bead Deposition

Using a properly set-up and adjusted MIG welding machine (see Table 8.11), correct safety protection, 1.0mm and/or 1.2mm-diameter wire, and two or more pieces of mild steel plate 150mm long × 6mm thick, you will make a welded stringer bead in the flat (PA) position.

- Start at one end of the plate and use either a push or drag technique to make a weld bead along the entire 150mm length of the metal using spray or pulsed-arc metal transfer.

- After the weld is complete, check its appearance and make any changes needed to correct the weld, Figure 8.63.

- Repeat the weld and make any additional adjustments required. After the machine is set, start working on improving the straightness and uniformity of the weld.

- Turn off the welding machine and shielding gas and reinstate your work area when you are finished welding.

Complete a copy of the 'Student Welding Report'.

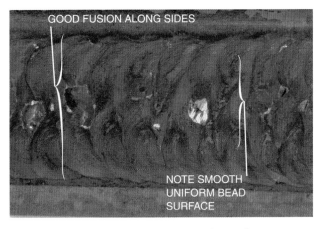

GOOD FUSION ALONG SIDES

NOTE SMOOTH UNIFORM BEAD SURFACE

FIGURE 8.68 Weld bead made with MIG axial spray metal transfer

TABLE 8.11 Typical welding current settings for axial spray metal transfer for mild steel

Process	Wire Diameter	Amperage Range (Optimum)	Voltage Range (Optimum)	Shielding Gas
Axial spray	0.8	115–200	15–27	98% Ar + 2% O_2
	1.2	165–300	18–32	

WORKSHOP TASK 8.7

Butt Joint in Flat (PA) Position

Using the same equipment, materials, and procedures listed previously make a welded butt joint in the flat position (PA), weld using spray transfer or pulsed-spray metal transfer, Figure 2.68.

- Tack weld the sheets together and place them flat on the welding table, Figure 8.63.
- Starting at one end, with the torch held at 75–80° slope angle and 90° tilt angle run a bead along the joint.
- Watch the molten weld pool and bead for signs that a change in technique may be required. Make any needed changes as the weld progresses.
- By the time the weld is complete, you should be making the weld nearly perfectly.
- Using the same technique, make another weld, this time, the entire 150mm of weld should be flawless.
- Repeat with both thicknesses of metal until consistently good beads are obtained.
- Turn off the welding machine and shielding gas and reinstate when you are finished welding.

Complete a copy of the 'Student Welding Report'.

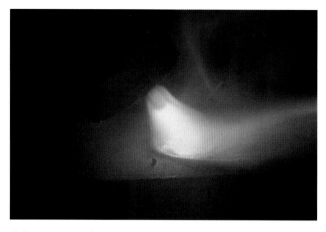

FIGURE 8.69 **MIG axial spray metal transfer**

WORKSHOP TASK 8.8

Lap Joint in the Flat (PA) Position

Using the same equipment, materials, and procedures listed previously make a lap joint in the flat position (PA), weld using spray transfer or pulsed-spray metal transfer.

- Tack weld the sheets together with a maximum overlap of 13mm and place them flat on the welding table, Figure 8.63.
- Starting at one end, with the torch held at 75–80° slope angle and 45° tilt angle run a bead along the joint.
- Watch the molten weld pool and bead for signs that a change in technique may be required. Make any needed changes as the weld progresses.
- By the time the weld is complete, you should be making the weld nearly perfectly.
- Using the same technique, make another weld, this time, the entire 150mm of weld should be flawless.
- Repeat with both thicknesses of metal until consistently good beads are obtained.
- Turn off the welding machine and shielding gas and reinstate when you are finished welding.

  Complete a copy of the 'Student Welding Report'.

WORKSHOP TASK 8.9

T-joint in the Flat (PA) Position

Using the same equipment, materials, and procedures listed previously make a T-joint in the flat position (PA), weld using spray transfer or pulsed-spray metal transfer, tack weld the sheets together and place them on the welding table.

- Starting at one end, with the torch held at 75–80° slope angle and 45° tilt angle run a bead along the joint.
- Watch the molten weld pool and bead for signs that a change in technique may be required, and watch the sides for undercutting.
- Make any needed changes as the weld progresses.
- By the time the weld is complete, you should be making the weld nearly perfectly.

- Using the same technique, make another weld. This time, the entire 150mm of weld should be flawless.

- Repeat with both thicknesses of metal until consistently good beads are obtained.

- Turn off the welding machine and shielding gas and reinstate when you are finished welding.

  Complete a copy of the 'Student Welding Report'.

WORKSHOP TASK 8.10

Outside Corner in the Flat (PA) Position

- Using the same equipment, materials, and procedures listed previously make an outside corner joint in the flat position (PA), weld using spray transfer or pulsed-spray metal transfer.

- Tack weld the sheets together and place them on the welding table.

- Starting at one end, with the torch held at 75–80° slope angle and a tilt angle that bisects the joint, run a bead along the joint.

- Watch the molten weld pool and bead for signs that a change in technique may be required.

- Make any needed changes as the weld progresses.

- By the time the weld is complete, you should be making the weld nearly perfectly.

- Using the same technique that was established in the last weld, make another weld. This time, the entire 150mm of weld should be flawless.

- Repeat with both thicknesses of metal until consistently good beads are obtained.

- Turn off the welding machine and shielding gas and reinstate when you are finished welding.

  Complete a copy of the 'Student Welding Report'.

FLUX CORED ARC WELDING

Flux cored arc welding (FCAW) is a fusion welding process in which weld heating is produced from an arc between the work and a continuously fed filler metal electrode. Atmospheric shielding is provided by the flux sealed within the tubular electrode, Figure 8.70. Extra shielding may or may not be supplied through a nozzle in the same way as in MIG. Improvements in the fluxes, smaller electrode diameters, increased reliability of the equipment, better electrode feed systems, and improved guns have all led to a rapid rise in the use of FCAW. Guns equipped with smoke extraction nozzles and electronic controls are the latest in a long line of improvements to this process, Figure 8.73. The flux inside the electrode protects the molten weld pool from the atmosphere, improves strength through chemical reactions and alloys, and improves the weld shape. Atmospheric contamination of molten weld metal, mainly from oxygen and nitrogen, occurs as it travels across the arc gap and within the pool before it solidifies. The addition of fluxing and gas-forming elements to the core electrode reduces or eliminates its effects. Small additions to the flux of alloying elements, deoxidizers, and gas-forming and slag agents all can improve the weld properties. Carbon, chromium and vanadium can be added to improve hardness, strength, creep resistance and corrosion resistance. Aluminium, silicon and titanium all help remove oxides and/or nitrides in the weld. Potassium, sodium and zirconium are added to the flux and form a slag.

The flux core additives that serve as deoxidizers, gas formers and slag formers either protect the molten weld pool or help to remove impurities from the base metal.

The slag covering of the weld is useful for several reasons. Slag is a non-metallic product resulting from the mutual dissolution of the flux and non-metallic impurities in the base metal. Deoxidizers may convert small amounts of surface oxides like mill scale back into pure metal. They work much like the elements used to refine iron ore into steel.

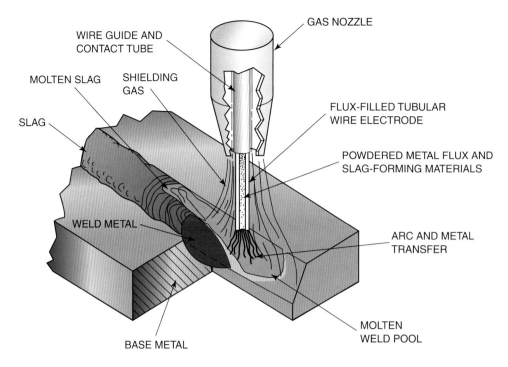

GAS NOZZLE

WIRE GUIDE AND
CONTACT TUBE

MOLTEN SLAG   SHIELDING
GAS

SLAG

FLUX-FILLED TUBULAR
WIRE ELECTRODE

POWDERED METAL FLUX AND
SLAG-FORMING MATERIALS

WELD METAL

ARC AND METAL
TRANSFER

MOLTEN
WELD POOL

BASE METAL

(A) GAS-SHIELDED FLUX CORED ARC WELDING (FCAW-G)

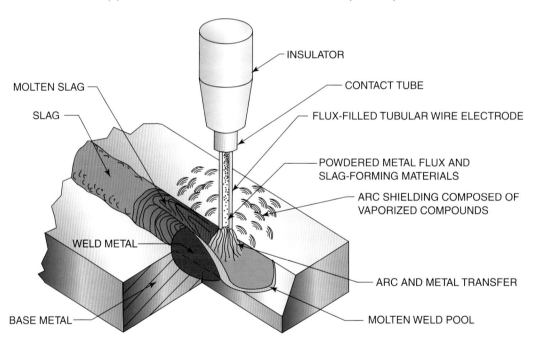

INSULATOR

MOLTEN SLAG

SLAG

CONTACT TUBE

FLUX-FILLED TUBULAR WIRE ELECTRODE

POWDERED METAL FLUX AND
SLAG-FORMING MATERIALS

ARC SHIELDING COMPOSED OF
VAPORIZED COMPOUNDS

WELD METAL

ARC AND METAL TRANSFER

BASE METAL

MOLTEN WELD POOL

(B) SELF-SHIELDED FLUX CORED ARC WELDING (FCAW-S)

FIGURE 8.70 Two types of flux cored arc welding

Gas formers rapidly expand and push the surrounding air away from the molten weld pool. If oxygen in the air were to come in contact with the molten weld metal, the metal would quickly oxidize. Sometimes this can be seen at the end of a weld when the molten weld metal erupts in a shower of tiny sparks.

Slag helps the weld by protecting the hot metal from the effects of the atmosphere, controlling the bead shape by serving as a dam or mould, and serving as a blanket to slow the weld's cooling rate, which improves its physical properties, Figure 8.72.

Smoke Extraction Nozzles

Because of the large quantity of smoke that can be generated during FCAW, systems for smoke extraction that fit on the gun have been designed, Figure 8.73. These use a vacuum to pull the smoke back into an extraction nozzle on the welding gun. The disadvantage of this slightly heavier gun is offset by the system's advantages:

* cleaner air for the welder to breathe because the smoke is removed before it rises to the welder's face;
* reduced heating and cooling cost because the smoke is concentrated, so less shop air must be removed with it.

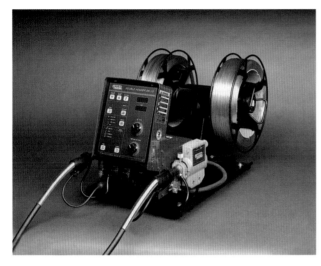

FIGURE 8.71 **Large capacity wire feed unit**

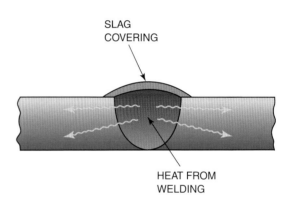

SLAG COVERING

HEAT FROM WELDING

FIGURE 8.72 **Slag blanketing the weld**

FIGURE 8.73 **Smoke extraction**

Electrode Feed

Electrode feed systems are similar to those used for MIG, and many feed systems are designed with dual feeders so that solid wire and flux core may be run in sequence. The major difference is that the more robust FCAW feeders are designed to use large-diameter wire and most often have two sets of feed rollers which help reduce the drive pressure on the electrode. Excessive pressure can distort the electrode wire diameter, which can allow some flux to be dropped inside the electrode guide tube.

Advantages of FCAW

FCAW offers the welding industry a number of important advantages.

- *High deposition rate.* High rates of depositing weld metal are possible. FCAW deposition rates of more than 11kg/hr of weld metal can be achieved. This compares to about 5kg/hr for manual metal arc (MMA) welding using a very large-diameter electrode of 6mm.

- *Minimum electrode waste.* The FCAW method makes efficient use of filler metal; from 75 per cent to 90 per cent of the weight of the FCAW electrode is metal, the remainder being flux. MMA electrodes have a maximum of 75 per cent filler metal; some MMA electrodes have much less. Also, a stub must be left at the end of each MMA welding electrode, average 50mm in length, resulting in a loss of 11 per cent or more of the MMA filler electrode. FCAW welding has no stub loss, so nearly 100 per cent of the FCAW electrode is used.

- *Reduced included angle.* Because of the deep penetration characteristic of FCAW, no edge-bevelling preparation is required on some joints in metal up to 13mm in thickness. When bevels are cut, the joint-included angle can be reduced to as small as 35°, Figure 8.74. This results in a smaller-sized weld and can save 50 per cent of filler metal with about the same savings in time and weld power used.

TIP

The narrower groove angle for FCAW compared to other welding processes saves on filler metal, welding time, and heat input into the page.

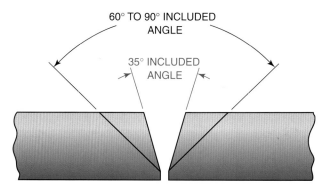

FIGURE 8.74 Narrow angle for FCAW

- *Minimum pre-cleaning.* The addition of *deoxidizers*, which combine with and remove harmful oxides on the base metal or its surface, and other fluxing agents permits high-quality welds to be made on plates with light surface oxides and mill scale. This eliminates most of the pre-cleaning required before MIG welding can be performed. It is often possible to make excellent welds on plates in the 'as cut' condition; no clean up is needed.

- *All-position welding.* Small-diameter electrode sizes in combination with special fluxes allow excellent welds in all positions. The slags produced assist in supporting the weld metal. When properly adjusted, this process is much easier to use than other all-position arc welding processes.

- *Flexibility.* Changes in power settings can permit welding to be done on thin-gauge sheet metals or thicker plates using the same electrode size. Multi-pass welds allow joining metals with no limit on thickness. This, too, is attainable with one size of electrode.

- *High quality.* Many codes permit welds to be made using FCAW. The addition of the flux gives the process the high level of reliability needed for welding on boilers, pressure vessels and structural steel.

- *Excellent control.* The molten weld pool is more easily controlled with FCAW than with MIG. The surface appearance is smooth and uniform even with less operator skill. Visibility is improved by removing the nozzle when using self-shielded electrodes.

Limitations of FCAW

- The main limitation of flux cored arc welding is that it is confined to ferrous metals and nickel-based alloys. Generally, all low and medium-carbon steels and some low-alloy steels, cast irons, and a limited number of stainless steels are presently weldable using FCAW.
- The equipment and electrodes used for the FCAW process are more expensive. However, the cost is quickly recoverable through higher productivity.
- The removal of post-weld slag requires another production step. The flux must be removed before the weldment is finished (painted) to prevent crevice corrosion.
- With the increased welding output comes an increase in smoke and fume generation, which may require improvements to the existing ventilation system in a workshop.

ELECTRODES

Methods of Manufacturing

The electrodes have flux tightly packed inside. One method used to make them is to first form a thin sheet of metal into a U-shape, Figure 8.75. A measured quantity of flux is poured into the U-shape before it is squeezed shut. The wire is passed through a series of dies to size it and further compact the flux. A second method of manufacture is to start with a seamless tube, usually about 25mm in diameter. One end of the tube is sealed, and the flux powder is poured into the open end. The tube is vibrated to ensure that it fills completely and the open end is sealed. The tube is now sized using a series of dies, Figure 8.76.

In both these methods of manufacturing, the sheet and tube are made up of the desired alloy and the flux is compacted inside the metal skin. This compacting helps make the electrode operate more smoothly and consistently. Electrodes are available in sizes from 0.8mm to 4mm in diameter. Smaller-diameter electrodes are much more expensive per kilogram than the same type in

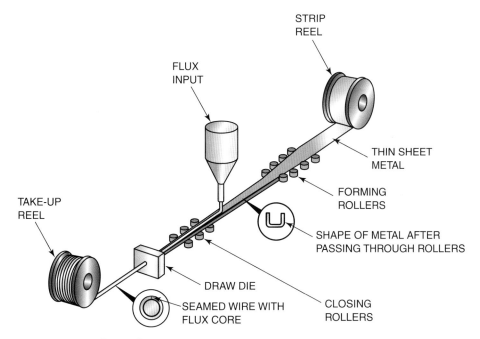

FIGURE 8.75 Putting flux in flux cored wire

a larger diameter. Larger-diameter electrodes produce such large welds they cannot be controlled in all positions. The most popular diameters range from 0.8mm to 2.4mm. The finished FCAW filler metal is packaged in a number of forms for purchase, Figure 8.77. The standard packing units for FCAW wires are spools, coils, reels and drums, Table 8.12.

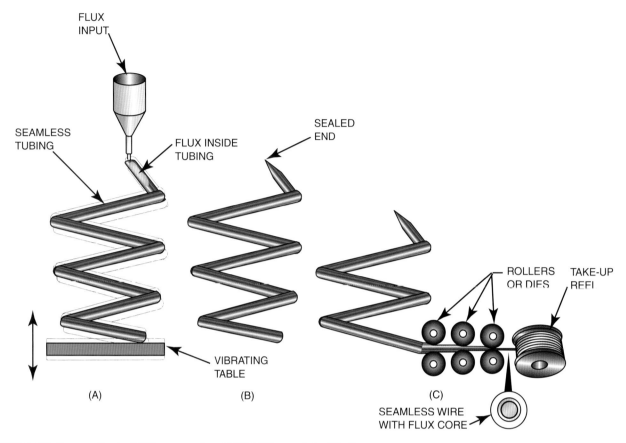

FIGURE 8.76 **One method of filling seamless FCAW filler metal with flux**

FIGURE 8.77 **A few packaged forms of FCAW filler metal**

TABLE 8.12 **Packaging size specification for commonly used FCAW filler wire**

Packaging	Outside Diameter	Width	Arbor (Hole) Diameter
Spools	(102mm)	(44.5mm)	16mm
	(203mm)	(57mm)	52.3mm
	(305mm)	(102mm)	52.3mm
	(356mm)	(102mm)	52.3mm
Reels	(559mm)	(318mm)	33.3mm
	(762mm)	(406mm)	33.3mm
Coils	(413mm)	(102mm)	305mm

	Outside Diameter	Inside Diameter	Height
Drums	584mm	406mm	864mm

Spools are made of plastic or wire-caged and are disposable. They are completely self-contained and are available in approximate weights from 0.5kg to 25kg. The smaller spools, 102mm and 203mm, weighing from 0.5 to 3kg, are most often used for smaller production runs or for DIY use; 305mm and 356mm spools are often used in colleges and welding fabrication shops.

Coils come wrapped or wire-tied together. They are un-mounted, so they must be supported on a frame on the wire feeder in order to be used. Coils are available in weights around 25kg. Because FCAW wires on coils do not have the expense of a disposable core, they cost a little less per kilogram, and are more desirable for higher-production shops. Reels are large wooden spools, and drums are shaped like barrels. Both are used for high-production jobs and can contain approximately 136kg to 454kg of FCAW wire. Because of their size, they are used primarily at fixed welding stations which are often associated with some form of automation, such as turntables or robotics.

Electrode Cast and Helix

To see the cast and helix of a wire, feed out a length of wire electrode and cut it off. Lay it on the floor and observe that it forms a circle. The diameter of the circle is known as the cast of the wire, Figure 8.78. Note that one end is slightly higher than the other. This height is the helix of the wire and this, coupled with the cast, causes the wire to rub on the inside of the contact tube, Figure 8.79. The slight bend in the electrode wire ensures a positive electrical contact between the contact tube and filler wire.

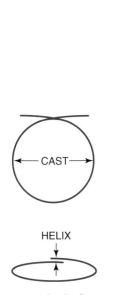

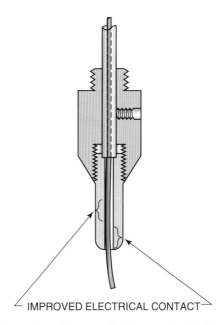

FIGURE 8.78 Method of measuring cast and helix of FCAW filler wire

FIGURE 8.79 Cast forces the wire to make better electrical contact with the tube

FLUX

The fluxes used are mainly based on lime or rutile (a mineral compound consisting of titanium dioxide, usually with a little iron). Their purpose is the same as in the manual metal arc welding (MMA) process. They can provide all or part of the following to the weld:

- *Deoxidizers.* Oxygen that is present in the welding zone has two forms. It can exist as free oxygen from the atmosphere surrounding the weld and also as part of a compound such as an iron oxide or carbon dioxide. In either case it can cause porosity in the weld if it is not removed or controlled. Chemicals are added that react to the presence of oxygen and combine to form a harmless compound, Table 8.13. The new compound can become part of the slag that solidifies on top of the weld, or some of it may stay in the weld as very small inclusions. Both methods result in a weld with better mechanical properties because of lower porosity.

- *Slag formers.* Slag can react with the molten weld metal chemically, and it can affect the weld bead physically. In the molten state it moves through the molten weld pool and acts as a magnet or sponge to chemically combine with impurities in the metal and remove them, Figure 8.80. Slags can be refractory, so that they become solid at a high temperature. As they solidify over the weld, they help it hold its shape and slow its cooling rate.

- *Fluxing agents.* Molten weld metal tends to have a high surface tension, which prevents it from flowing outward toward the edges of the weld. This causes undercutting along the junction of the weld and the base metal. Fluxing agents make the weld more fluid and allow it to flow outward, filling the undercut.

TABLE 8.13 Deoxidizing elements added to filler wire

Deoxidizing Element	Strength
Aluminum (Al)	Very strong
Manganese (Mn)	Weak
Silicon (Si)	Weak
Titanium (Ti)	Very strong
Zirconium (Zr)	Very strong

FIGURE 8.80 Impurities being floated to the surface by slag

- *Arc stabilizers.* Chemicals in the flux affect the arc resistance. As the resistance is lowered, the arc voltage drops and penetration is reduced. When the arc resistance is increased, the arc voltage increases and weld penetration is increased. Although the resistance within the ionized arc stream may change, the arc is more stable and easier to control, with reduced spatter.

- *Alloying elements.* Because of the difference in the mechanical properties of metal that is formed by rolling or forging and metal that is melted to form a weld bead, the metallurgical requirements of the two also differ. Some alloying elements change the weld's strength, ductility, hardness, brittleness, toughness and corrosion resistance. Others can be added in the form of powder metal to increase deposition rates.

- *Shielding gas.* As elements in the flux are heated by the arc, some of them vaporize and form gaseous clouds hundreds of times larger than their original volume. This rapidly expanding cloud forces the air around the weld zone away from the molten weld metal, Figure 8.81. Without the protection this process affords the molten metal, it would rapidly oxidize, severely affecting the weld's mechanical properties and rendering it unfit for service.

FCAW fluxes are divided into two groups based on the acid or basic chemical reactivity of the slag.

The rutile-based flux is acidic, and produces a smooth, stable arc and a refractory high-temperature slag for out-of-position welding. These electrodes produce a fine drop transfer, a relatively low fume, and an easily removed slag. The main limitation of the rutile fluxes is that their fluxing elements do not produce as high a quality deposit as do lime based fluxes.

The lime-based flux, is very good at removing certain impurities from the weld metal, but its low-melting-temperature slag is fluid, which makes it generally unsuitable for out-of-position welding. These electrodes produce a more globular transfer, more spatter, more fume and a more adherent slag than do the rutile systems. These characteristics are tolerated when it is necessary to deposit very tough weld metal and for welding materials with a low tolerance for hydrogen. Some rutile-based electrodes allow the addition of a shielding gas. With the weld partially protected by the shielding gas, more elements can be added to the flux, which produces welds with the best of both flux systems, high-quality welds in all positions. Some fluxes can be used on both single- and multiple-pass welds, and others are limited to single-pass welds only. Using a single-pass welding electrode for multi-pass welds may result in an excessive amount of manganese.

The manganese is necessary to retain strength when making large, single-pass welds. However, with the lower dilution associated with multi-pass techniques, it can strengthen the weld metal too much and reduce its ductility. In some cases, small welds that deeply penetrate the base metal can help control this problem. Table 8.14 lists the shielding and polarity for the flux classifications of mild steel FCAW electrodes. The letter G is used to indicate that the electrode has not been classified. Often the exact composition of fluxes is kept as a manufacturer's trade secret the only information supplied is current, type of shielding required, and some strength characteristics.

As a result of the relatively rapid cooling of the weld metal, the weld may become hard and brittle. This can be controlled by adding elements to the flux that affect the content of both the

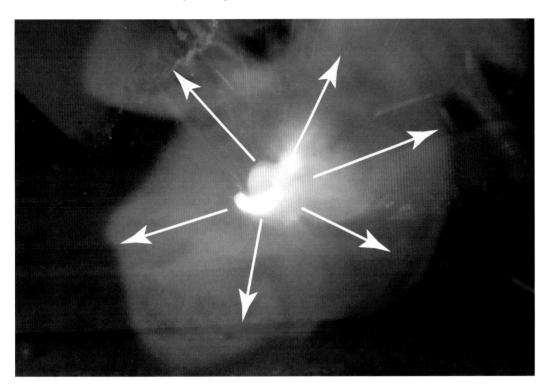

FIGURE 8.81 **Rapidly expanding gas cloud**

TABLE 8.14 **Welding characteristics of seven flux classifications**

Classification	Comments	Shielding Gas
T-1	Requires clean surfaces and produces little spatter. It can be used for single- and multiple-pass welds in all positions.	Carbon dioxide (CO_2) or argon/carbon dioxide mixes
T-2	Requires clean surfaces and produces little spatter. It can be used for single-pass welds in the flat (1G and 1F) and horizontal (2F) positions only.	Carbon dioxide (CO_2)
T-3	Used on thin-gauge steel for single-pass welds in the flat (1G and 1F) and horizontal (2F) positions only.	None
T-4	Low penetration and moderate tendency to crack for single- and multiple-pass welds in the flat (1G and 1F) and horizontal (2F) positions.	None
T-5	Low penetration and a thin, easily removed slag, used for single- and multiple-pass welds in the flat (1G and 1F) position only.	With or without carbon dioxide (CO_2)
T-6	Similar to T-5 without externally applied shielding gas.	None
T-G	The composition and classification of this electrode are not given in the preceding classes. It may be used for single- or multiple-pass welds.	With or without shielding

TABLE 8.15 Ferrite-forming elements used in FCAW fluxes

Element	Reaction in Weld
Silicon (Si)	Ferrite former and deoxidizer
Chromium (Cr)	Ferrite and carbide former
Molybdenum (Mo)	Ferrite and carbide former
Columbium (Cb)	Strong ferrite former
Aluminum (Al)	Ferrite former and deoxidizer

weld and the slag, Table 8.15. Ferrite is the softer, more ductile form of iron. The addition of ferrite-forming elements can control the hardness and brittleness of a weld. Refractory fluxes are sometimes called 'fast-freeze' because they solidify at a higher temperature than the weld metal. By becoming solid first, this slag can cradle the molten weld pool and control its shape. This property is very important for out-of-position welds.

The impurities in the weld pool can be metallic or nonmetallic compounds. Metallic elements that are added to the metal during the manufacturing process in small quantities may be concentrated in the weld. These elements improve the grain structure, strength, hardness, resistance to corrosion or other mechanical properties in the metal's as-rolled or formed state. But the deposited weld metal, or weld nugget, is like a small casting because the liquid weld metal freezes in a controlled shape, and some alloys adversely affect the properties of this casting. Non-metallic compounds are primarily slag inclusions left in the metal from the fluxes used during manufacturing. The welding fluxes form slags that are less dense than the weld metal so that they will float to the surface before the weld solidifies.

FLUX CORED STEEL ELECTRODE IDENTIFICATION

Mild Steel

The electrode number *E70T-10* is used as an example to explain the classification system for mild steel FCAW electrodes (Figure 8.82):

- *E.* Electrode.
- *7.* Tensile strength in units.

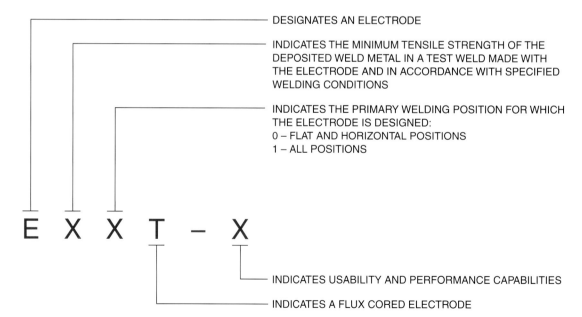

FIGURE 8.82 Identification system

- *0. 0* is used for flat and horizontal fillets only, and *1* is used for all-position electrodes.
- *T*. Tubular (flux cored) electrode.
- *10*. The number in this position can range from *1* to *14* and is used to indicate the electrode's shielding gas if any, number of passes that may be applied one on top of the other, and other welding characteristics of the electrode. The letter *G* is used to indicate that the shielding gas, polarity, and impact properties are not specified. If the letter *G* is followed by the letter *S* this indicates an electrode suitable only for single-pass welding.

The electrode classification *E70T–10* can have some optional identifiers added to the end of the number, as in *E70T–10MJH8*. These additions are used to add qualifiers to the general classification so that specific codes or standards can be met:

- *M* – Mixed gas of 75 per cent to 80 per cent Ar and CO_2 for the balance. If there is no *M*, either the shielding gas is CO_2 or the electrode is self-shielded.
- *J* – Describes the Charpy V-notch impact test value.
- *H8* – Describes the residual hydrogen levels in the weld: *H4* equals less than 4ml/100 g; *H8*, less than 8ml/100g; *H16*, less than 16 ml/100g.

Stainless Steel Electrodes

The classification for stainless steel for FCAW electrodes starts with the letter *E* as its prefix, followed by the American Iron and Steel Institute's (AISI) three-digit stainless steel number. This number indicates the type of stainless steel in the filler metal.

The AISI number is followed by a dash and a suffix number. 1 indicates an all-position filler metal, and 3 indicates an electrode to be used in the flat and horizontal positions only.

Metal Cored Steel Electrode Identification

The addition of metal powders to the flux core of FCAW electrodes has produced a new classification of filler metals. Some of the earlier flux cored filler metals that already had powder metals in their core had their numbers changed to reflect the new designation. The letter *T* for *tubular* was changed to the letter *C* for *core*. For example, E70T–1 became E70C–3C. The complete explanation of the cored electrode *E70C-3C* follows:

- *E*. Electrode.
- *7*. Tensile strength in units.
- *0. 0* is used for flat and horizontal fillets only, and *1* is used for all-position electrodes.
- *C*. Metal-cored (tubular) electrode.
- *3*. 3 is used for a Charpy impact value.
- *C*. The second letter *C* indicates CO_2. The letter *M* in this position indicates a mixed gas, 75 per cent to 80 per cent argon, with the balance being CO_2. If there is no *M* or *C*, then the shielding gas is CO_2. The letter *G* is used to indicate that the shielding gas, polarity and impact properties are not specified. If the letter *G* is followed by the letter *S* this indicates an electrode suitable only for single-pass welding.

Care of Flux Core Electrodes

Wire electrodes may be wrapped in sealed plastic bags or wrapped in special paper to protect them from the elements. Some are shipped in cans or cardboard boxes.

A small paper bag of a moisture-absorbing material, crystal desiccant, is sometimes placed in the shipping containers to protect wire electrodes from moisture. Some wire electrodes require

NOTE

The powdered metal added to the core flux can provide additional filler metal and/or alloys. This is one way the micro-alloys can be added in very small and controlled amounts, as low as 0.0005 per cent to 0.005 per cent. These are very powerful alloys that dramatically improve the metal's mechanical properties.

storage in an electric rod oven to prevent contamination from excessive moisture. Read the manufacturer's recommendations located in or on the electrode shipping container for information on use and storage.

Weather conditions affect your ability to make high-quality welds. Humidity increases the chance of moisture entering the weld zone. Water (H_2O), which consists of two parts hydrogen and one part oxygen, separates in the weld pool. When only one part of hydrogen is expelled, hydrogen entrapment occurs, which can cause weld beads to crack or become brittle. The evaporating moisture will also cause porosity. To prevent hydrogen entrapment, porosity and atmospheric contamination, it may be necessary to pre-heat the base metal to drive out moisture. Storing the wire electrode in a dry location is recommended. The electrode may develop restrictions due to the tangling of the wire or become oxidized with excessive rusting if the wire electrode package is mishandled, thrown, dropped or stored in a damp location.

> **CAUTION** ⓘ
>
> Always keep the wire electrode dry and handle it as you would any important tool or piece of equipment.

SHIELDING GAS

FCAW wire can be manufactured so that all of the required shielding of the molten weld pool is provided by the vaporization of some of the flux within the tubular electrode. When the electrode provides all of the shielding, it is called 'self-shielding' and the welding process is abbreviated FCAW–S (S for self-shielding). Other FCAW wire must use an externally supplied shielding gas to protect the molten weld pool. When a shielding gas is added, the combined shielding is called 'dual shield' and the process is abbreviated FCAW–G (G for gas). Care must be taken to use the cored electrodes with the recommended gases, and not to use gas at all with the self-shielded electrodes which may produce a defective weld. The shielding gas will prevent the proper disintegration of much of the deoxidizers. This results in the transfer of these materials across the arc to the weld. In high concentrations, the deoxidizers can produce slags that become trapped in the welds, causing undesirable defects. Lower concentrations may cause brittleness only. In either case, the chance of weld failure is increased. If these electrodes are used correctly, there is no problem.

The selection of a shielding gas will affect the arc and weld properties. The weld bead width, build up, penetration, spatter, chemical composition and mechanical properties are all affected by the shielding gas selection. Gases used for FCAW include CO_2 and mixtures of argon and CO_2. Argon gas is easily ionized by the arc, resulting in a highly concentrated path from the electrode to the weld. This produces a smaller droplet size that is associated with the axial spray mode of metal transfer, Figure 8.83. A smooth, stable arc results and there is a minimum of spatter. This transfer mode continues as CO_2 is added to the argon until the mixture contains more than 20 per cent per cent of CO_2. As the percentage of CO_2 increases further, weld penetration increases but arc stability decreases, causing an increase in spatter. A mixture of 80 per cent per cent argon and 20 per cent per cent CO_2 works best for jobs requiring a mixed gas.

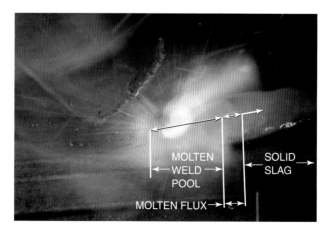

FIGURE 8.83 Axial spray transfer mode

Straight CO_2 is used for some welding, but the CO_2 gas molecule is easily broken down in the welding arc to form carbon monoxide (CO) and free oxygen (O). Both gases are reactive to some alloys in the electrode. As these alloys travel from the electrode to the molten weld pool, some of them form oxides. Silicon and manganese are the primary alloys that become oxidized and lost from the weld metal. Most FCAW electrodes are specifically designed to be used with or without shielding gas and for a specific shielding gas or percentage mixture. For example, an electrode designed specifically for use with 100 per cent CO_2 will have higher levels of silicon and manganese to compensate for the losses to oxidization. But if 100 per cent argon or a mixture of argon and CO_2 is used, the weld will have an excessive amount of silicon and manganese which will affect its mechanical or metallurgical properties. Although the weld may look satisfactory, it will probably fail prematurely.

> **CAUTION**
>
> Never use a FCAW electrode with a shielding gas it is not designated to be used with. The weld it produces may be rejected.

WELDING TECHNIQUES

A welder can control weld beads made by FCAW by making the following changes in the techniques used.

Gun Angle

Gun angle, work angle and *travel angle* are terms used to refer to the relation of the gun to the work surface, Figure 8.84. The gun angle can be used to control the weld pool. The electric arc produces an electrical force known as the arc force, which can be used to counteract the gravitational pull that tends to make the liquid weld pool sag or run ahead of the arc. By manipulating the electrode travel angle for the flat and horizontal position of welding to a 20° to 45° angle from the vertical, the weld pool can be controlled. A 40° to 50° tilt angle from the vertical plate is recommended for **fillet welds.**

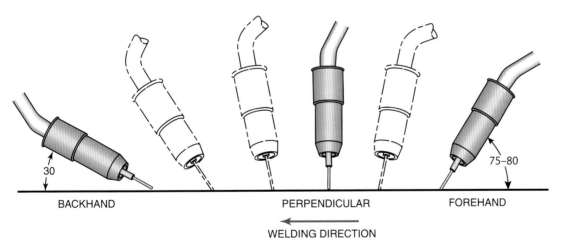

BACKHAND   PERPENDICULAR   FOREHAND

← WELDING DIRECTION

FIGURE 8.84 Welding gun angles

Changes in this angle will affect the weld bead shape and penetration. Shallower angles are needed when welding thinner materials to prevent burn-through. Steeper, perpendicular angles are used for thicker materials. FCAW electrodes have a flux that is mineral based, often called low-hydrogen. These fluxes are refractory and become solid at a high temperature. If too steep a forehand, or pushing, angle is used, slag from the electrode can be pushed ahead of the weld bead and solidify quickly on the cooler plate, Figure 8.85. It can be trapped under the edges of the weld by the molten weld metal. To avoid this problem, most flat and horizontal welds should be performed with a backhand angle. Vertical up welds require a forehand gun angle to direct the arc deep into the groove or joint for better control of the weld pool and deeper penetration, Figure 8.86. Slag entrapment associated with most forehand welding is not a problem for vertical welds.

A gun angle around 90° to the metal surface either slightly forehand or backhand works best for overhead welds, Figure 8.87. The slight angle aids with visibility of the weld, and helps control spatter build up in the gas nozzle.

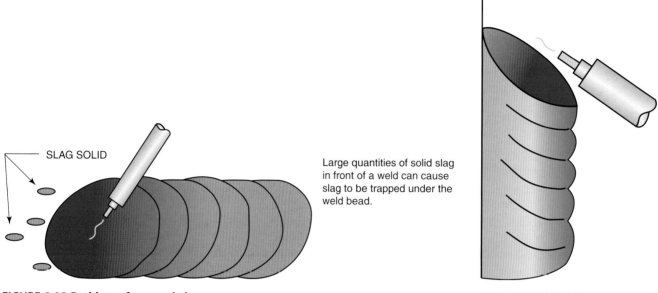

SLAG SOLID

Large quantities of solid slag in front of a weld can cause slag to be trapped under the weld bead.

FIGURE 8.85 Problem of trapped slag

FIGURE 8.86 Vertical up gun angle

Forehand/Perpendicular/Backhand Techniques

Forehand, *perpendicular* and *backhand* are the terms most often used to describe the gun angle as it relates to the work and the direction of travel. The forehand technique is sometimes referred to as *pushing* the weld bead, and backhand may be referred to as *pulling* or *dragging* the weld bead. The term *perpendicular* is used when the gun angle is at approximately 90° to the work surface, Figure 8.87.

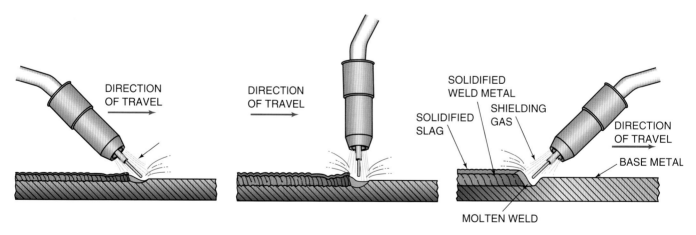

DIRECTION OF TRAVEL

DIRECTION OF TRAVEL

SOLIDIFIED WELD METAL

SHIELDING GAS

SOLIDIFIED SLAG

DIRECTION OF TRAVEL

BASE METAL

MOLTEN WELD

FIGURE 8.87 Changing the welding gun angle between forehand, perpendicular and backhand angles will change the shape of the weld bead produced

The forehand technique has the following advantages:

- *Joint visibility*. You can easily see the joint where the bead will be deposited, Figure 8.88.
- *Electrode extension*. The contact tube tip is more visible, making it easier to maintain a constant extension length.

- *Less weld penetration.* It is easier to weld on thin sheet metal without melting through.
- *Out-of-position welds.* This technique works well on vertical up and overhead joints for better control of the weld pool.

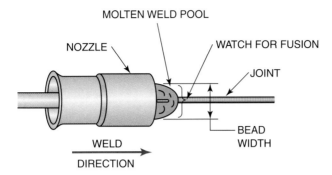

FIGURE 8.88 Welder's view with a forehand angle

The forehand technique also has these disadvantages:

- *Weld thickness.* Thinner welds may occur because less weld reinforcement is applied to the weld joint.
- *Welding speed.* Because less weld metal is being applied, the rate of travel along the joint can be faster, which may make it harder to create a uniform weld.
- *Slag inclusions.* Some spattered slag can be thrown in front of the weld bead and be trapped or included in the weld, resulting in a weld defect.
- *Spatter.* Depending on the electrode, the amount of spatter may be slightly increased.

The perpendicular technique has the following advantages:

- *Machine and robotic welding.* This gun angle is used on automated welding because there is no need to change the gun angle when the weld changes direction.
- *Uniform bead shape.* The weld's penetration and reinforcement are balanced between those of forehand and backhand techniques.

These are the disadvantages of the perpendicular technique:

- *Limited visibility.* Because the welding gun is directly over the weld, there is limited visibility of the weld unless you lean your head way over to the side.
- *Weld spatter.* Because the weld nozzle is directly under the weld in the overhead position, more weld spatter can collect in the nozzle, causing gas flow problems or even shorting the tip to the nozzle.

The backhand technique has these advantages:

- *Weld bead visibility.* It is easy to see the back of the molten weld pool, which makes it easier to control the bead shape, Figure 8.89.
- *Travel speed.* Because of the larger amount of weld metal being applied, the rate of travel may be slower, making it easier to create a uniform weld.
- *Depth of fusion.* The arc force and the greater heat from the slower travel rate both increase the depth of weld joint penetration.

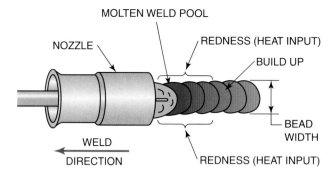

FIGURE 8.89 Welder's view with a backhand angle

The backhand technique also has the following disadvantages:

- *Weld build up.* The weld bead may have a convex (raised or rounded) weld face.
- *Post-weld finishing.* Because of the weld bead shape, more work may be required if the product has to be finished by grinding smooth.
- *Joint following.* It is harder to follow the joint because your hand and the FCAW gun are positioned over the joint, and you may wander from the seam.
- *Loss of penetration.* An inexperienced welder sometimes directs the wire too far back into the weld pool causing the wire to build up in the face of the weld pool reducing joint penetration.

Travel Speed

Travel speed has been defined as the linear rate at which the arc is moved along the weld joint. Fast travel speeds deposit less filler metal. If the rate of travel increases, the filler metal cannot be deposited fast enough to adequately fill the path melted by the arc. This causes the weld bead to have an unfilled groove melted into the base metal next to the weld. This condition is known as undercut and occurs along the edges or toes of the weld bead. Slower travel speeds will, at first, increase penetration and increase the filler weld metal deposited. As the filler metal increases, the weld bead will build up in the weld pool. Because of the deep penetration of flux cored wire, the angle at which you hold the gun is very important for a successful weld. If all welding conditions are correct and remain constant, the preferred rate of travel for maximum weld penetration is one that allows you to stay within the selected welding variables and still control the fluidity of the weld pool. Another way to work out correct travel speed is to consult the manufacturer's recommendations chart for the metres per minute (m/min) burn-off rate for the selected electrode.

Mode of Metal Transfer

The mode of metal transfer describes how the molten weld metal is transferred across the arc to the base metal. The mode that is selected, the shape of the completed weld bead, and the depth of weld penetration depend on the welding power source, wire electrode size, type and thickness of material, type of shielding gas used, and best welding position for the task.

Spray Transfer with FCAW-G

The spray transfer mode is the most common process used with gas-shielded FCAW (FCAW–G), Figure 8.82. As the gun trigger is depressed, the shielding gas automatically flows and the electrode bridges the distance from the contact tube to the base metal, making contact with the base metal to complete a circuit. The electrode shorts and becomes so hot that the base metal melts and forms a weld pool. The electrode melts into the weld pool and burns back toward the contact tube. A combination of high amperage and the shielding gas along with the electrode size produces a pinching effect on the molten electrode wire, causing the end of the electrode wire to spray across the arc. The characteristic of spray-type transfer is a smooth arc, through which hundreds of small droplets per second are transferred through the arc from the electrode to the weld pool. Spray transfer can produce a high quantity of metal droplets, up to approximately 250 per second. The current required for a spray transfer to take place is dependent on the electrode size, composition of the electrode, and shielding gas. Below the transition current (critical current), globular transfer takes place.

In order to achieve a spray transfer, high current and larger-diameter electrode wire are needed. A shielding gas of carbon dioxide a mixture of carbon dioxide and argon or an argon and oxygen mixture is required. FCAW–G is a welding process that, with the correct variables, can be used

- on thin or properly prepared thick sections of material;
- on a combination of thick to thin materials;
- with small or large electrode diameters;
- with a combination of shielding gases.

Globular Transfer with FCAW-G

Globular transfer occurs when the welding current is below the transition current, Figure 8.90. The electrode forms a molten ball at its end that grows in size to approximately two to three times the original electrode diameter. These large molten balls are transferred across the arc at the rate of several drops per second. The arc becomes unstable because of the weight of these large drops. When argon gas is introduced it causes the ball to spin as it transfers across the arc to the base metal. This unstable globular transfer can produce excessive spatter.

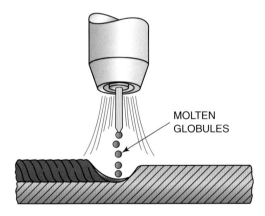

MOLTEN GLOBULES

FIGURE 8.90 Globular transfer method

Both FCAW–S and FCAW–G use direct current electrode negative (DCEN) when welding on thin-gauge materials to keep the heat in the base metal and the small-diameter electrode at a controllable burn-off rate. The electrode can then be stabilized, and it is easier to manipulate and control the weld pool in all weld positions. Larger-diameter electrodes are welded with direct current electrode positive (DCEP) because the larger diameters can keep up with the burn-off rates.

The recommended weld position means the position in which the work piece is placed for welding. All welding positions use either spray or globular transfer, but we will concentrate here on the flat and horizontal welding positions. In the flat welding position the work piece is placed flat on the work surface. In the horizontal welding position the work piece is positioned perpendicular to the workbench surface. The amperage range may be from 30 to 400 amperes or more for welding materials from gauge thickness up to 37mm. On square butt weld joints, thicker base metals can be welded with little or no edge preparation. This is one of the great advantages of

FCAW. If edges are prepared and cut at an angle (bevelled) to accept a complete joint weld penetration, the depth of penetration will be greatly increased. FCAW is commonly used for general repairs to mild steel in the horizontal, vertical, and overhead welding positions, sometimes referred to as all positional welding.

Electrode Extension

The electrode extension is measured from the end of the electrode contact tube to the point the arc begins at the end of the electrode, Figure 8.91. Compared to MIG welding, the electrode extension required for FCAW is much greater. The electrical resistance of the longer extension causes the wire to heat up, which can drive out moisture from the flux, resulting in a smoother arc with less spatter.

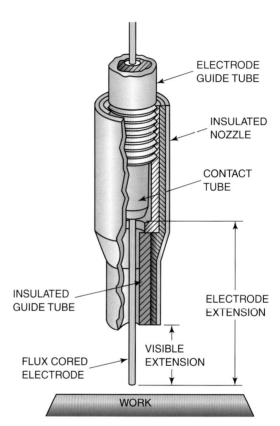

ELECTRODE
GUIDE TUBE

INSULATED
NOZZLE

CONTACT
TUBE

ELECTRODE
EXTENSION

INSULATED
GUIDE TUBE

VISIBLE
EXTENSION

FLUX CORED
ELECTRODE

WORK

FIGURE 8.91 **Self-shielded electrode nozzle**

Porosity

FCAW can produce high-quality welds in all positions, although porosity in the weld can be a persistent problem. Porosity can be caused by moisture in the flux, improper gun manipulation or surface contamination.

The flux used in the FCAW electrodes can pick up moisture from the surrounding atmosphere, so the electrodes must be stored in a dry area. Once the flux becomes contaminated with moisture, it is very difficult to remove. Water (H_2O) breaks down into free hydrogen and oxygen in the presence of an arc, Figure 8.92. The hydrogen can be absorbed into the molten weld metal, where it can cause post-weld cracking. The oxygen is also absorbed into the weld metal, where it forms oxides. If a shielding gas is used, the FCAW gun gas nozzle must be close enough to the weld to provide adequate shielding gas coverage. If there is a wind or if the

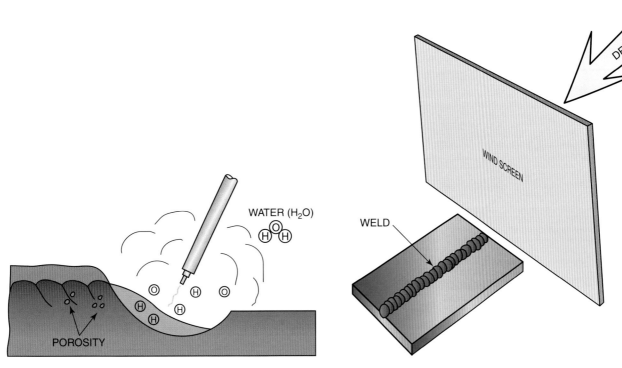

FIGURE 8.92 Water and porosity

FIGURE 8.93 Wind and draft protection

nozzle-to-work distance is excessive, the shielding will be inadequate. If welding is to be done out-side or in an area subject to draughts, the gas flow rate must be increased or a wind shield placed to protect the weld, Figure 8.93. A common misconception is that the flux within the electrode will either remove or control weld quality problems caused by surface contaminations. That is not true. The addition of flux makes FCAW more tolerant to surface conditions than MIG welding, but there are still adverse effects. New hot-rolled steel has a layer of dark grey or black iron oxide. Although this layer is very thin, it may provide a source of enough oxygen to cause porosity which is usually uniformly scattered through the weld, Figure 8.94. Unless it is severe, porosity is usually not visible in the finished weld but is trapped under the surface as the weld cools.

Because porosity is under the surface, nondestructive testing methods, including X-ray, magnetic particle and ultrasound, must be used to locate it in a weld. It can also be detected by mechanical testing such as guided bend, free bend and nick-break testing for establishing weld parameters. Often it is better to remove the mill scale before welding rather than risk the production of porosity. All welding surfaces within the weld preparation and the surrounding surfaces

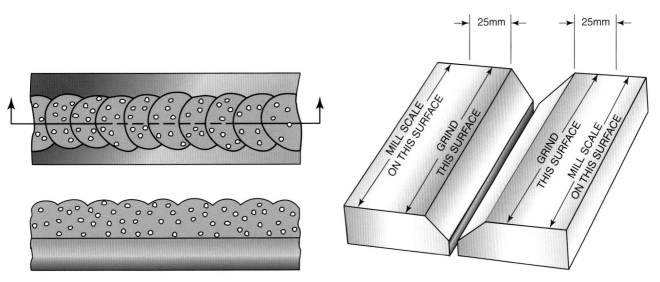

FIGURE 8.94 Uniformly scattered porosity

FIGURE 8.95 Grinding requirements

within 25mm must be cleaned to bright metal, Figure 8.95. Cleaning may be by grinding, filing, sanding or shot blasting. Any time FCAW are to be made on metals that are dirty, oily, rusty or wet or that have been painted, the surface must be pre-cleaned, either chemically or mechanically. One advantage of chemically cleaning oil and paint is that it is easier to clean larger areas. Both oil and paint smoke easily when heated, causing weld defects, and must be removed from the area of the weld. In the case of small parts the entire part may need to be cleaned.

PRACTICE WELDS

WORKSHOP TASK 8.11

Stringer Beads in the Flat (PA) Position

Using a properly set-up and adjusted FCAW machine, Table 8.16, proper safety protection, E70T-1 and/or E71T-11 electrodes of diameter 1.0mm and 1.2mm, and one or more pieces of mild steel plate, 300mm long and 6mm or thicker, you will make a striger bead weld in the flat position, Figure 8.96.

- Starting at one end of the plate and using a dragging technique, make a weld bead along the entire 300mm length of the metal.
- After the weld is complete, check its appearance. Make any needed changes to correct the weld.
- Repeat the weld and make additional adjustments.
- After the machine is set, start to work on improving the straightness and uniformity of the weld. Use weave patterns of different widths and straight stringers without weaving.
- Repeat with both classifications of electrodes until beads can be made straight, uniform, and free from any visual defects.
- Turn off the welding machine and shielding gas and clean up your work area when you are finished welding.

  Complete a copy of the 'Student Welding Report'.

TABLE 8.16 FCAW parameters for use if specific settings are unavailable from electrode manufacturer (base metal thickness 6-13mm)

Electrode			Welding Power			Shielding Gas		Base Metal	
Type	Size	Amps	Wire-feed Speed, ipm (cm/min)	Volts	Type	Flow	Type	Thickness
E70T-1 E71T-1	0.9mm	130 to 150	288 to 380 (732 to 975)	22 to 25	None	n/a	Low-carbon steel	6mm to 13mm
E70T-1 E71T-1	1.2mm	150 to 210	200 to 300 (508 to 762)	28 to 29	None	n/a	Low-carbon steel	6mm to 13mm
E70T-5 E71T-11	0.9mm	130 to 200	288 to 576 (732 to 1463)	20 to 28	75% argon 25% CO_2	30 cfh	Low-carbon steel	6mm to 13mm
E70T-5 E71T-11	1.2mm	150 to 250	200 to 400 (508 to 1016)	23 to 29	75% argon 25% CO_2	35 cfh	Low-carbon steel	6mm to 13mm

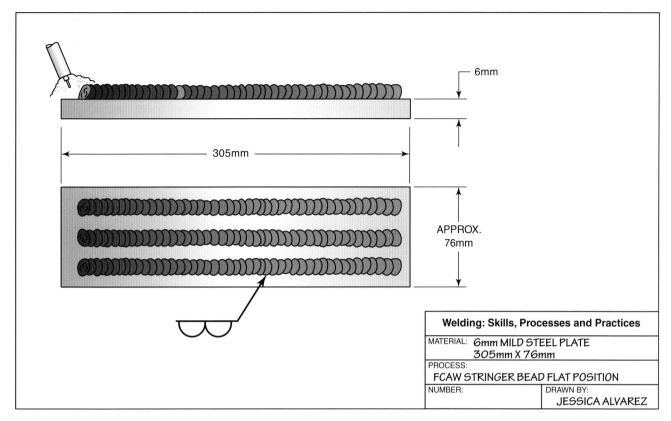

FIGURE 8.96 FCAW stringer bead, 6mm mild steel, flat position

Square Butt Welds

One advantage of FCAW is the ability to make 100 per cent-joint-penetrating welds without bevelling the edges of the plates. These full-joint-penetrating welds can be made in plates that are 6mm or less in thickness. Welding on thicker plates risks the possibility of a lack of fusion on both sides of the root face, Figure 8.97.

There are several disadvantages of having to bevel a plate before welding:

- It adds an operation to the fabrication process.
- More filler metal and welding time are required to fill a bevelled joint than are required to make a square jointed weld.
- Bevelled joints have more heat from the thermal bevelling and additional welding required to fill the groove. The lower heat input to the square joint means less distortion.

The major disadvantage of making square jointed welds is that as the plate thickness approaches 6mm or if the weld is out of position, a much higher level of skill is required. It is much easier to practise making this type of weld in metal 3mm thick and then move up in thickness as your skills improve.

WORKSHOP TASK 8.12

Square Butt Welds in Flat (PA) Position

Using a properly set-up and adjusted FCAW machine, proper safety protection, E70T-1 and/or E71T-11 electrodes of diameter 1.0mm and 1.2mm, and one or more pieces of mild steel plate, 300mm long and 6mm or less in thickness, you will make a butt weld in the flat position, Figure 8.98.

- Tack weld the plates together and place them in position to be welded.
- Starting at one end, run a bead along the joint. Watch the molten weld pool and bead for signs that a change in technique may be required.
- Make any needed changes as the weld progresses in order to produce a uniform weld.
- Repeat with both classifications of electrodes until defect-free welds can consistently be made in the 6mm-thick plate.
- Turn off the welding machine and shielding gas and reinstate your work area when you are finished welding.

    Complete a copy of the 'Student Welding Report'.

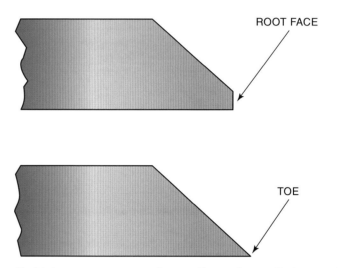

FIGURE 8.97 A bevelled joint may or may not have a flat surface, called a root face

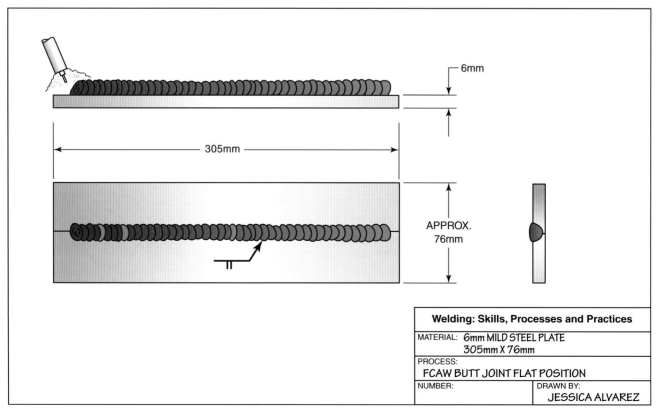

FIGURE 8.98 FCAW butt joint, 6mm mild steel, flat position

V-Butt Welds in the Flat (PA) Position

Although for speed and economy engineers try to avoid specifying welds that require bevelling the edges of plates, it is not always possible. Any time the metal being welded is thicker than 6mm and a 100 per cent joint penetration weld is required, the edges of the plate must be prepared with a bevel. Fortunately, FCAW allows a narrower bevel preparation to be made and still achieve a thorough thickness weld, due to the deeper penetration characteristics of the FCAW process, Figure 8.99.

All FCAW butt welds are made using three different types of weld runs, Figure 8.100.

- *Root run.* The first weld bead of a multiple-run weld. The root run fuses the two parts together and establishes the depth of weld metal penetration.
- *Filler run.* Made after the root pass is completed and used to fill the weld preparation with weld metal. More than one pass is often required.
- *Capping run.* The last weld pass on a multi-pass weld. The capping run may be made with one or more welds. It must be uniform in width, reinforcement and appearance.

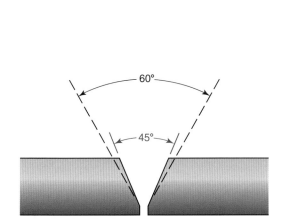

FIGURE 8.99 Reduced groove angle for FCAW

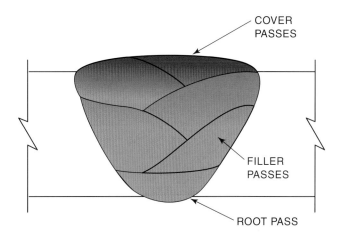

FIGURE 8.100 Three types of weld passes make up a weld

Root Run

A good root run is needed in order to obtain a sound weld. The root may be either open or closed using a backing strip, Figure 8.101.

The backing strips are usually made from a piece of metal 6mm thick and 25mm wide, and should be 50mm longer than the base plates. The strip is attached to the plate by tack welds made on the sides of the strip, Figure 8.102. Most production welds do not use backing strips, so they are made as open root welds. Because of the difficulty in controlling root weld face contours in FCAW, open-root joints are often avoided on critical welds. If an open-root weld is needed, the root run may be put in with an MMA electrode or the root face of the FCAW can be retouched by grinding and/or back welding. Care must be taken with any root run not to have the weld face too convex, Figure 8.103. Convex weld faces tend to trap slag along the toe of the weld, which can be extremely difficult to remove it, especially if there is any undercutting. To avoid this, adjust the welding power settings, speed and weave pattern so that a flat or slightly concave weld face is produced, Figure 8.104.

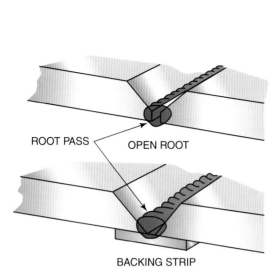

FIGURE 8.101 **Root run**

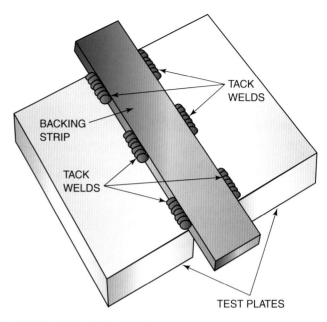

FIGURE 8.102 **Backing strip**

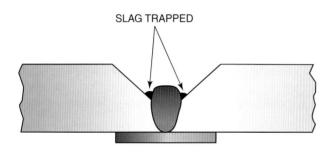

FIGURE 8.103 **Slag trapped beside weld bead is hard to remove**

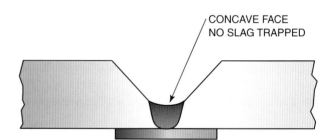

FIGURE 8.104 **Flat or concave weld faces are easier to clean off**

Filler Run

Filler runs are made with either stringer beads, which are made with a straight progression and very little gun manipulation, or weave beads, in which the operator oscillates the gun from side to side in order to widen the weld profile. Either bead type works well for flat or vertically positioned welds, but stringer beads work best for horizontal and overhead positioned welds. When multiple-run filler welds are required, each weld bead must overlap the others along the edges. Edges should overlap smoothly enough so that the finished bead is uniform, Figure 8.105. Stringer beads usually overlap about 25 per cent to 50 per cent, and weave beads overlap approximately 10 per cent to 25 per cent.

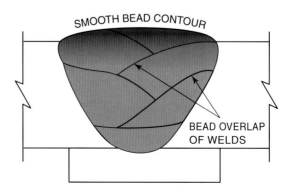

FIGURE 8.105 **The surface of a multipass weld should be as smooth as if it were made by one weld**

Each weld bead must be cleaned before the next bead is started. The filler run ends when the weld preparation has been filled to a level just below the plate surface.

Capping Run

The capping runs may or may not simply be a continuation of the weld beads used to make the filler run. The major difference between the filler run and the capping run is the importance of the weld face. Keeping the face and toe of the capping run uniform in width, reinforcement and appearance and free of defects is essential. Most welds are not tested beyond a visual inspection and the capping run must meet a strict visual inspection standard. The weld should be uniform in width and reinforcement. There should be no arc strikes or hammer marks from chipping or slag removal operations on the plate other than those on the weld itself. The weld must be free of incomplete fusion and cracks. It must be free of overlap, and undercut must not exceed either 10 per cent of the base metal or 1mm, whichever is less. Reinforcement must have a smooth transition with the base plate and be no higher than 3mm.

WORKSHOP TASK 8.13

V-butt Joint in the Flat (PA) Position

Use a properly set-up and adjusted FCAW machine, Table 8.18; appropriate PPE; E70T-1 and/or E71T-11 electrodes of diameter 1.0mm and or 1.2mm; one or more pieces of mild steel plate, bevelled, 305mm long and 10mm thick; and a backing strip 350mm long, 25mm wide and 6mm thick. You will make a V-butt weld in the flat position, Figure 8.107.

- Tack weld the backing strip to the plates. There should be a root gap of approximately 3mm between the plates. The bevelled surface can be made with or without a root face, Figure 8.109.
- Place the test plates in position at a comfortable height and location. Be sure that you have complete and free movement along the full length of the weld joint. It is often a good idea to make a practice run along the joint with the welding gun without power to make sure nothing will interfere with your making the weld. Be sure the welding cable is free and will not get caught on anything during the weld.
- Start the weld outside the weld preparation on the backing strip tab, Figure 8.109. This is done so that the arc is smooth and the molten weld pool size is established at the beginning of the V-butt preparation.
- Continue the weld out onto the tab at the outer end of the weld joint. This process ensures that the end of the V-butt is completely filled with weld.
- Repeat with both classifications of electrodes until consistently defect-free welds can be made.
- Turn off the welding machine and shielding gas and reinstate your work area when you are finished welding.

Complete a copy of the 'Student Welding Report'.

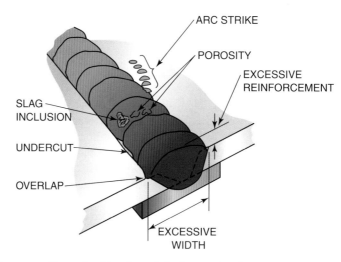

FIGURE 8.106 Common discontinuities found during a visual examination

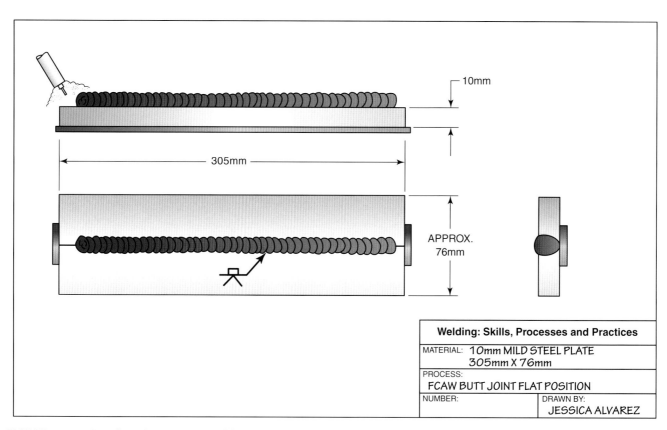

FIGURE 8.107 FCAW butt joint, 10mm mild steel, flat position

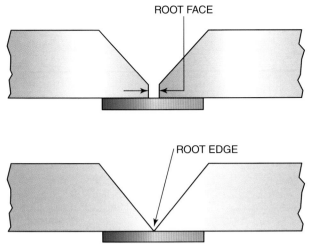

ROOT FACE

ROOT EDGE

FIGURE 8.108 Groove layout with and without a root face

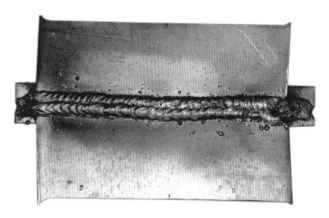

FIGURE 8.109 Using the ends of a backing strip

FILLET WELDS

A fillet weld is made on a lap joint and a T-joint. It should be built up to the thickness of the plate, Figure 8.110. On thick plates the fillet must be made up of several passes as with a groove weld. The difference with a fillet weld is that a smooth transition from the plate surface to the weld is required. If this transition is abrupt, it can cause stresses that will weaken the joint. The lap joint is made by overlapping the edges of the plates. They should be held together tightly before tack welding them together. A small tack weld may be added in the centre to prevent distortion during welding, Figure 8.111. Chip and wire brush the tacks before you start to weld. The T-joint is made by tack welding one piece of metal on another piece of metal at a right angle, Figure 8.112. After the joint is tack welded together, the slag is chipped from the tack welds so that it is not included in the final weld.

Holding thick plates tightly together on T-joints may cause under-bead cracking, or lamellar tearing, Figure 8.113. On thick plates the weld shrinkage can be great enough to pull the metal apart well below the bead or its heat-affected zone. In production welds, cracking can be controlled by not assembling the plates tightly together. The space between the two plates can be set by placing a small wire spacer between them, Figure 8.114.

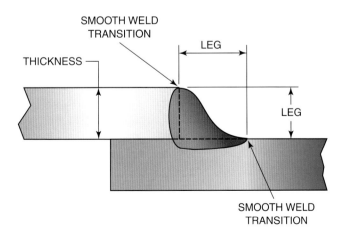

SMOOTH WELD
TRANSITION

LEG

THICKNESS

LEG

SMOOTH WELD
TRANSITION

FIGURE 8.110 The legs of a fillet weld should generally be equal to the thickness of the base metal

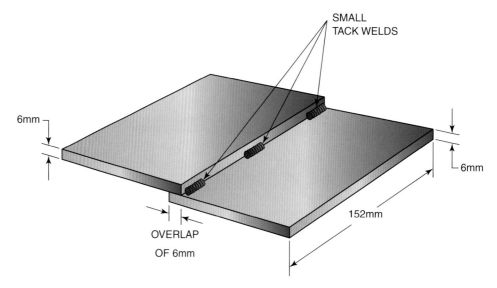

FIGURE 8.111 Tack welding the plates of a lap joint together

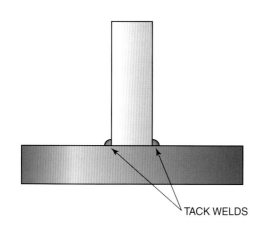

FIGURE 8.112 Tack welding both sides of a T-joint

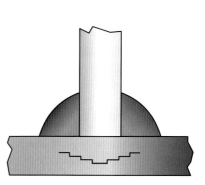

FIGURE 8.113 Underbead cracking, or lamellar tearing, of the base plate

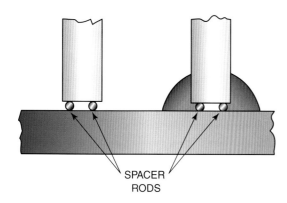

FIGURE 8.114 Spacers in T-joints

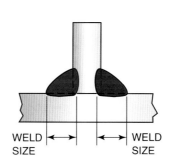
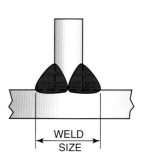

FIGURE 8.115 Fillet weld size

A fillet welded lap or T-joint can be strong if it is welded on both sides, even without having deep penetration, Figure 8.115. Some T-joints may be prepared for welding by cutting either a bevel or a J-groove in the vertical plate. This cut is not required for strength but may be necessary because of design limitations. Unless otherwise specified, most fillet welds will be equal in size to the plates welded. A fillet weld will be as strong as the base plate if the size of the two welds equals the total thickness of the base plate. The weld bead should have a flat or slightly concave appearance to ensure the greatest strength and efficiency, Figure 8.116. The root of fillet welds must be melted to ensure a completely fused joint. A notch along the root of the weld pool is an indication that the root is not being fused together, Figure 8.117. To achieve complete root fusion, move the arc to a point as close as possible to the leading edge of the weld pool, Figure 8.118. If the arc strikes the un-melted plate ahead of the molten weld pool, it may become erratic, increasing weld spatter.

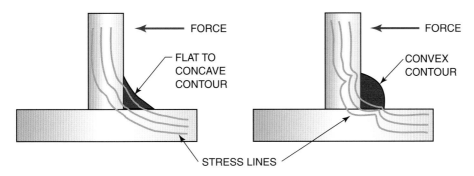

FIGURE 8.116 Fillet weld shape

FIGURE 8.117 Watch the root of the weld bead to be sure there is complete fusion

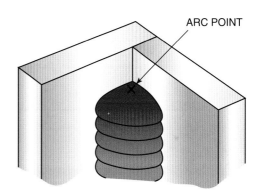

FIGURE 8.118 Moving the arc as close as possible to the leading edge of the weld will provide good root fusion

WORKSHOP TASK 8.14

T-fillet Weld in the Flat (PA) Position

Use a properly set-up and adjusted FCAW machine, appropriate PPE; E70T-1 and/or E71T-11 electrodes of diameter 1.0mm and or 1.2mm, and one or more pieces of mild steel plate, bevelled, 300mm long and 10mm thick. You will make a fillet weld in the flat position.

- Tack weld the pieces of metal together and brace them in position. When making the lap or T-joints in the flat position, the plates must be at a 45° angle so that the surface of the weld will be flat, Figure 8.119A and Figure 8.119B.
- Starting at one end, make a weld along the entire length of the joint.
- Repeat each type of joint with both classifications of electrodes until consistently defect-free welds can be made.
- Turn off the welding machine and shielding gas and reinstate your work area when you are finished welding.

Complete a copy of the 'Student Welding Report'.

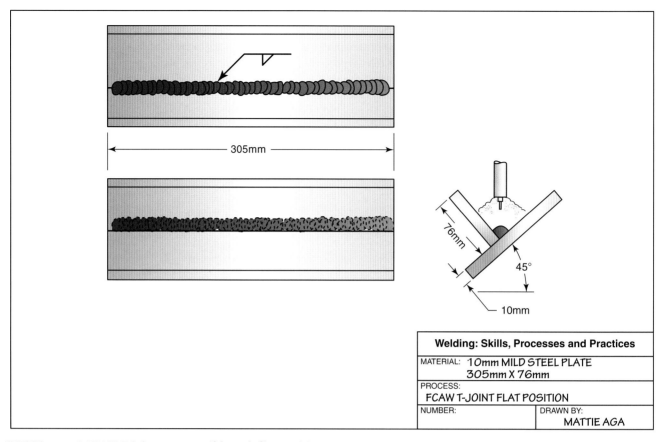

FIGURE 8.119A **FCAW T-joint, 10mm mild steel, flat position**

WORKSHOP TASK 8.15

Lap Fillet in the Flat (PA) Position

Use a properly set-up and adjusted FCAW machine, Table 8.17; appropriate PPE; E70T-1 and/or E71T-11 electrodes of diameter 1.0mm and or 1.6mm; one or more pieces of mild steel plate, bevelled, 150mm long and 20mm thick or thicker. You will make a fillet weld in the flat position.

● Follow the same instructions for the assembly and welding procedure outlined previously.

● Repeat each type of joint with both classifications of electrodes until consistently defect-free welds can be made.

● Turn off the welding machine and shielding gas and reinstate your work area when you are finished welding.

Complete a copy of the 'Student Welding Report'.

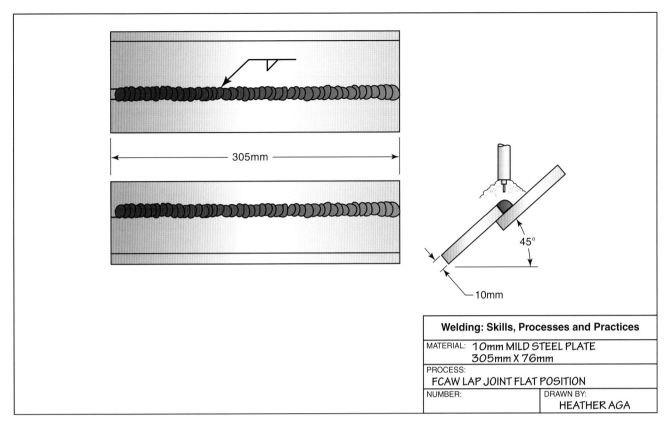

FIGURE 8.119B FCAW lap joint, 10mm mild steel, flat position

TABLE 8.17 FCAW parameters for use if specific settings are unavailable from electrode manufacturer (base metal thickness 12 – 20mm)

Electrode		Welding Power			Shielding Gas		Base Metal	
Type	Size	Amps	Wire-feed Speed, (cm/min)	Volts	Type	Flow	Type	Thickness
E70T-1 E71T-1	0.9mm	130 to 150	(732 to 975)	22 to 25	None	n/a	Low-carbon steel	13mm to 19mm
E70T-1 E71T-1	1.2mm	150 to 210	(508 to 762)	28 to 29	None	n/a	Low-carbon steel	13mm to 19mm
E70T-1 E71T-1	1.4mm	150 to 300	(381 to 889)	25 to 33	None	n/a	Low-carbon steel	13mm to 19mm
E70T-1 E71T-1	1.6mm	200 to 400	(381 to 762)	27 to 33	None	n/a	Low-carbon steel	13mm to 19mm
E70T-5 E71T-11	0.9mm	130 to 200	(732 to 1463)	20 to 28	75% argon 25% CO_2	30 cfh	Low-carbon steel	13mm to 19mm
E70T-5 E71T-11	1.2mm	150 to 250	(508 to 1016)	23 to 29	75% argon 25% CO_2	35 cfh	Low-carbon steel	13mm to 19mm
E70T-5 E71T-11	1.4mm	150 to 300	(381 to 889)	21 to 32	75% argon 25% CO_2	35 cfh	Low-carbon steel	13mm to 19mm
E70T-5 E71T-11	1.6mm	180 to 400	(368 to 889)	21 to 34	75% argon 25% CO_2	40 cfh	Low-carbon steel	13mm to 19mm

SUMMARY

Slight changes in welding gun angle, electrode extension and welding position can make significant differences in the quality of the weld produced. As a new welder you might find it difficult to see the effect of these changes if they are slight. It is a good idea to make more radical changes so it is easier for you to see their effects on the weld. Later as you develop your skills you can use these slight changes to control the weld's quality and appearance as it progresses along the joint. Small adjustments in your welding technique are required to compensate for slight changes that occur along a welding joint, such as joint gap and the increasing temperature of the base metal and the ability to see the weld joint. Variations in conditions can significantly affect welding setup for the MIG process. Before starting an actual weld in the field you should practise on scrap metal of a similar thickness and type of metal to be welded to test your setup. This can significantly increase the chances that your weld will meet standards or specifications. Making these sample or test welds is more important when you are welding on site, since welds outside the workshop are more difficult to control.

You will find it helpful when you are initially setting up your welder to have someone assist you, so that he or she can make changes in the welding machine's settings as you are welding. This teamwork can significantly increase your set-up accuracy and reduce set-up time. Later on when you have developed a keen eye for watching the weld, you can then make these adjustments for yourself more rapidly and accurately. Working with another student will also give you a better understanding of how other individuals' set-up preferences affect their welds. Welding is an art, and each welder may have slight differences in preference for voltage, amperage, gas flow and other set-up variables. Flux cored arc welding produces a large quantity of fumes. Make sure that you are welding so that your face is well out of this rising plume of welding fumes. On site, welders sometimes use fans to gently blow the fumes away from them. However, if the fan is too close to the welding zone, excessive air velocity will blow the shielding away from the weld, which may result in weld porosity. Take precautions to protect yourself from any potential health hazards.

ACTIVITY AND REVIEW

1   List the items that make up a basic semi-automatic welding system.

2   Identify what must be done to the shielding gas cylinder before the valve protection collar is removed.

3   Explain why the shielding gas valve is 'cracked' before the flow meter regulator is attached.

4   List the most likely cause of a bird-nest.

5   Indicate why all fittings and connections must be tight.

6   State which parts are activated by depressing the gun switch.

7   Identify what benefit a welding wire's cast provides.

8   State how you can determine the location of a problem that stops the wire from being successfully fed through the conduit.

9   State the advantages of using a feed roller pressure that is as light as possible.

10   Explain why the feed roller drag should prevent the spool from coasting to a stop when the feed stops.

11   State why you must always wind the wire tightly into a ball or cut it into short lengths before discarding it in the proper waste container.

12   Explain why the flowmeter ball would float at different heights with different shielding gases if the shielding gases are flowing at the same rate.

13   State how the amperage adjusted on a MIG welder.

14   Explain what happens to the weld as the electrode extension is lengthened.

15   Indicate what the effect on the weld would be by changing the welding angle from a dragging to a pushing angle.

16   State the advantages of adding oxygen or CO_2 to argon for welds on steel.

17   Describe what mill scale is.

18   State what type of porosity is most often caused by mill scale.

19   Explain what the welder should watch if the view of the weld is obstructed by the shielding gas nozzle.

20   State the advantages of making vertical down welds.

21   List in what form metal is transferred across the arc in the dip and spray transfer processes.

22   Identify the type of liner to be used with aluminium wires.

23   Describe the spot welding process.

24   Explain what the flux provides in a weld made with FCAW.

25   State how the electrode extension is measured in FCAW.

26   Identify what can cause porosity in a FCAW joint.

27   Identify the major safety concerns that a FCAW operator should be aware of.

28   List the problems that high pressure on the feed roller can cause.

Tungsten Inert Gas Shielded Welding

LEARNING OBJECTIVES

After completing this chapter, you should be able to:

- describe the tungsten inert gas shielded welding process
- list precautions taken to limit tungsten erosion
- contrast the various types of tungsten electrodes and how they are used
- shape the end of the tungsten electrode and clean it
- remove a contaminated tungsten end
- choose an appropriate nozzle for the job
- accurately read a flow meter
- compare and contrast the three types of welding current used for TIG welding
- list three problems that can occur as a result of an incorrect gas flow rate
- set up a TIG workstation for carbon steel, stainless steel and aluminum operations
- strike a TIG welding arc on carbon steel, stainless steel and aluminum
- explain why the filler rod end must be kept inside the protective zone of the shielding gas and how to accomplish this
- determine the correct machine settings for the minimum and maximum welding current for the machine used, the types and sizes of tungsten, and the metal types and thicknesses

- compare the characteristics of low-carbon and mild steels, stainless steel, and aluminum with respect to TIG welding
- prepare carbon steel, stainless steel, and aluminum for TIG welding.

UNIT REFERENCES

NVQ units:

Complying with Statutory Regulations and Organizational Safety Requirements.
Joining Materials by the Manual TIG and Plasma-arc Welding Processes.
Producing Fillet Welded Joints using a Manual Welding Process.

VRQ units:

Engineering Environment Awareness.
Manual Welding Techniques.

KEY TERMS

etching the process of using acids, bases or other chemicals to dissolve and reveal any unwanted matter on the surface of the material.

venturi effect which acts like a slipstream to draw other gases into the weld.

INTRODUCTION

The aircraft industry developed the TIG process for welding magnesium during the late 1930s and the early 1940s. Over the years it has been referred to as heli-arc (helium shielding), argon arc (argon shielding) and in America is referred to as gas shielded tungsten arc welding (GTAW). In this process, an arc is established between a non-consumable tungsten electrode and the work piece. Under the correct welding conditions, the electrode does not melt. The work melts at the spot where the arc impacts on its surface, producing a molten weld pool, into which the filler metal is fed directly. Since hot tungsten is sensitive to oxygen contamination, a good inert shielding gas is required to keep the air away from the hot tungsten and molten weld metal. The gas provides the needed arc characteristics and protects the molten weld pool from atmospheric contamination. Because fluxes are not used, the welds produced are sound, free of slag, and as corrosion-resistant as the parent metal.

Before development of the TIG process, welding aluminum and magnesium was difficult and the welds produced were porous and prone to corrosion. When argon became plentiful and DCEN was recognized as more suitable than DCEP, the TIG process became more common. Until the development of metal inert gas shielded welding (MIG) in the late 1940s, TIG was the only acceptable process for welding such reactive materials as aluminum, magnesium, titanium, and some grades of stainless steel, regardless of thickness. Reactive metals are ones that are easily contaminated when heated to their molten welding temperatures.

Contamination can come from the air or can be picked up from surfaces containing oxides, oils, paints or similar materials. Although economical for welding sheet metal, the process proved tedious and expensive for joining sections much thicker than 6mm. The eye-hand co-ordination required to make TIG welds is very similar to the co-ordination required for oxy-fuel gas welding. TIG welding is often easier to learn when a person can oxy-fuel gas weld. With the development of combination gases (active), a whole new range of materials have been successfully welded by this variation and are referred to as tungsten active gas shielded welding process TAGS. Both processes use the same equipment.

TIG WELDING EQUIPMENT

Figure 9.1 and Figure 9.2 show two industrial applications of tungsten inert gas shielded welding.

FIGURE 9.1 Semi-automatic operation allows a stainless steel part to be TIG welded as it is turned past the torch

FIGURE 9.2 An operator welds cap ring

TIG Power Sources

All three types of welding current, or polarities, can be used for TIG welding. Each has features that make it more desirable for specific conditions or with certain types of metals. The major differences are in the heat distributions and the presence or degree of arc cleaning. Figure 9.3 shows the heat distribution for each of the three types of currents.

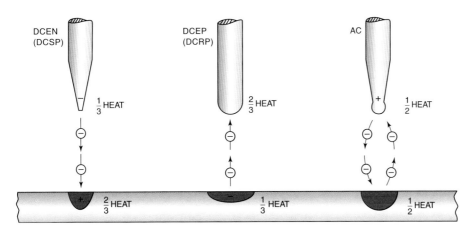

FIGURE 9.3 Heat distribution between tungsten electrode and work with each type of welding current

Direct current electrode negative (DCEN), which used to be called direct current straight polarity (DCSP), concentrates about two-thirds of its welding heat on the work and the remaining one-third on the tungsten. The higher heat input to the weld results in deep penetration. The low heat input into the tungsten means that a small sized tungsten can be used without erosion problems.

Direct current electrode positive (DCEP), which used to be called direct current reverse polarity (DCRP), concentrates only one-third of the arc heat on the plate and two-thirds of the heat on the electrode. This produces wide welds with shallow penetration, but has a strong cleaning action upon the base metal. The high heat input to the tungsten means that a large size tungsten is required, preferably prepared with a balled tip for zirconiated electrodes. This is a good current choice for thin, heavily oxidized metals and magnesium. The metal being welded will not emit electrons as freely as tungsten, so the arc may wander or be more erratic than DCEN. Because of the near molten state of the electrode tip, DCEP operations are almost always done in the flat position.

There are many theories as to why DCEP has a cleaning action. The most probable explanation is that the electrons accelerated from the cathode surface lift the oxides that interfere with their movement. The positive ions accelerated to the metal's surface provide additional energy and, together, the electrons and ions cause surface erosion. Although this theory is disputed, it is important to note that cleaning does occur, that it requires argon-rich shielding gases and DCEP polarity, and that it can be used to advantage, Figure 9.4.

Alternating current (AC) concentrates about half of its heat on the work and the other half on the tungsten. Alternating current is DCEN half of the time and DCEP the other half of the time. The frequency at which the current cycles, is the rate at which it makes a full change in direction, Figure 9.5. In the United Kingdom, it cycles at the rate of 100 times per second, or 50 Hertz (50 Hz). Referring again to Figure 9.5, make a note of where the current is at its maximum peak at points A and B. The rate gradually decreases until it stops at points C and D. The arc is extinguished at these points and, as the current reversal begins, must be re-established. This event requires the emission of electrons from the cathode to ionize the shielding gas. When the hot, emissive electrode becomes the cathode, re-establishing the arc is easy. However, it is often quite difficult to when the colder and less emissive work piece becomes the cathode. Because voltage from the power supply is designed to support a relatively low voltage arc, it may be insufficient to initiate electron flow. When the arc does not reignite consistently, it becomes destabilized and

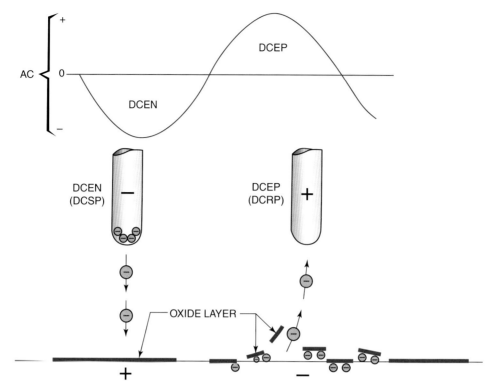

FIGURE 9.4 Electrons collect under the oxide layer

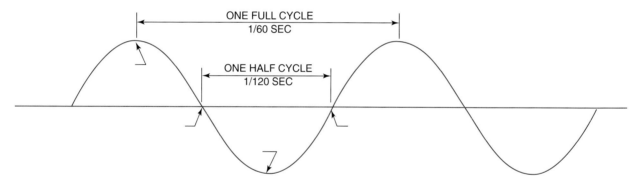

FIGURE 9.5 Sine wave of alternating current

can cause poor welding performance. This phenomenon is called rectification. A voltage assist from another source (auxiliary) is needed. A high-voltage but low-current spark gap oscillator is commonly used at a relatively low cost. The high frequency ensures that a voltage peak will occur reasonably close to the current reversal in the welding arc, creating a low-resistance ionized path for the welding current to follow, Figure 9.6A and Figure 9.6B. This same device is often used to initiate direct current arcs, a particularly useful technique for mechanized welding.

The high-frequency current is established by capacitors discharging across a gap set on points inside the machine. Reducing the point gap setting will raise the frequency of the current. The voltage is stepped up with a transformer from the primary voltage supplied to the machine. The available amperage to the high-frequency circuit is very low. Thus, when the circuit is complete, the voltage quickly drops to a safe level. The high frequency is induced on the primary welding current in a coil, which may be set so that it automatically cuts off after the arc is established, when welding with DC. It is kept on continuously with AC and transformer rectifier power supplies. When used in this manner, it is referred to as alternating current, high-frequency stabilized, or ACHF. Most of the newer inverter power supplies have advanced circuitry that switches between DCEN and DCEP so quickly that rectification cannot occur and high frequency is not used.

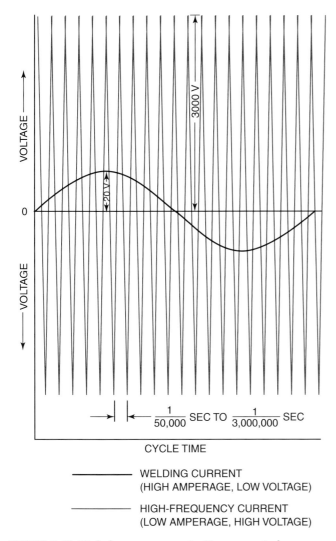

VOLTAGE ————▶

3000 V

0

20 V

◀———— VOLTAGE

$\frac{1}{50,000}$ SEC TO $\frac{1}{3,000,000}$ SEC

CYCLE TIME

———————— WELDING CURRENT
(HIGH AMPERAGE, LOW VOLTAGE)

———————— HIGH-FREQUENCY CURRENT
(LOW AMPERAGE, HIGH VOLTAGE)

FIGURE 9.6A High frequency arc starting current shown over the low-frequency welding current

FIGURE 9.6B High frequency first appears as a blue glow around the tungsten before the welding current state its arc

AC Balance Control

While older transformer-rectifier welding power supplies can only produce AC power in the form of a sine wave (Figure 9.5), newer transformer rectifiers and inverter power supplies can provide enhanced AC output by controlling the dwell time spent on each side of the AC cycle.

The two additional AC waveforms are called square wave and advanced square wave. Figure 9.7 shows a comparison of the three waveforms. The waveforms represent how the current is measured in amplitude and time as the current moves through a cycle.

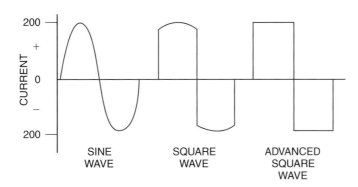

CURRENT

200
+
0
−
200

SINE
WAVE

SQUARE
WAVE

ADVANCED
SQUARE
WAVE

FIGURE 9.7 Comparison of the three different AC waveforms

CAUTION

If you wish to change from DC to AC using a transformer-rectifier, it is recommended that you switch off the machine before engaging the silicon bridge rectifier to obtain AC power supply. This is because when engaging the rectifier unit, additional heat is produced, and by turning off you prevent a sudden surge going on the rectifying unit and allowing the heat sink to prevent burning out diodes.

Welding machines with square wave capabilities have a knob or digital readout that controls the amount of time the current spends on the electrode negative and electrode positive half of the AC cycle. More dwell time on the electrode negative side results in deeper penetration and reduced cleaning action. The dial usually says 'max penetration', Figure 9.8. An additional benefit of an AC arc balanced to favour electrode negative is that the extra current delivered to the work will reduce the heat at the tip of the tungsten, which may allow a smaller diameter electrode, or even a pointed end, to be used.

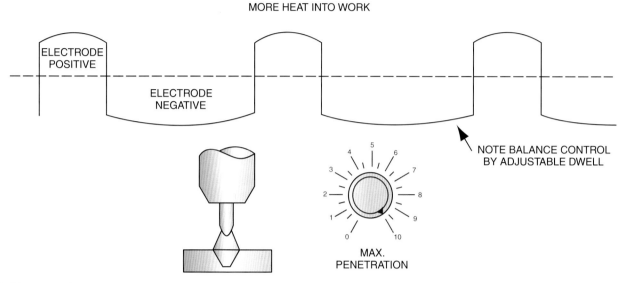

MORE HEAT INTO WORK

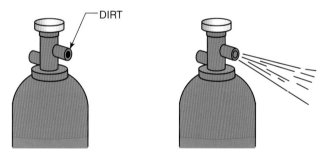

FIGURE 9.8 Maximum penetration balance control setting

FIGURE 9.9 During transportation or storage, dirt may collect in the gas outlet

If the machine is adjusted to have more dwell time spent on the electrode positive side of the half cycle, the result will be a wider, shallower bead profile with greater cleaning action. This machine setting (called 'max cleaning') is useful when it is impractical to completely remove small amounts of surface contaminants, or when welding on aluminum castings. The extra electrode positive will help to etch away more or thicker oxides. When balance controls are set toward 'max cleaning', a balled end will be required on the electrode. If the ball becomes larger than 1–1.5 times the electrode diameter, the next larger size should be selected.

When the control knob is set inbetween max cleaning and max penetration, it is called a balanced waveform. This is a good place to start for general AC applications, Figure 9.10. Balance control can be increased toward maximum electrode negative if greater penetration or a pointed electrode is required, although operators must be careful not to go too far, or the oxides on top of the aluminum will not be completely removed and poor fusion or porosity may result. It is also important to note that transformer-rectifiers, when set at max penetration, have 68 per cent dwell time on electrode negative, while advanced square wave inverter power supplies may produce as much as 99 per cent dwell time on electrode negative. This means that when a transformer

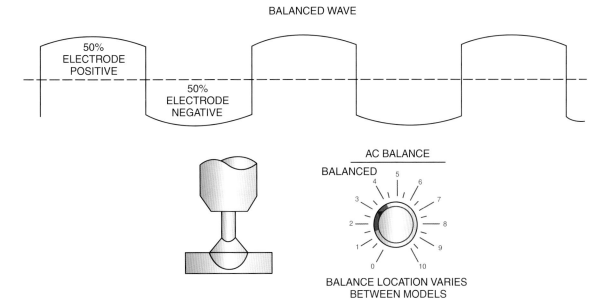

FIGURE 9.10 Balanced control setting

square wave is set at 10, it delivers about the same amount of electrode negative as when an advanced square wave inverter is set to 7.

If you have a square wave power supply, try adjusting the balance control toward max cleaning and max penetration when working on aluminium test pieces, and note the results in etching at the weld toes and the overall bead profile.

Contactor

This device allows the arc to be extinguished without removing the electrode and gas shield from the weld zone. It functions via a trigger switch on the torch or the foot pedal. The contactor allows the weld metal to cool in a controlled environment, and at the same time it protects the welder by switching off the open circuit voltage (OCV) when the torch is not being used.

Slope-in Control

This is a feature of many modern machines which allows the welder to gradually build up the current at the start of a weld and prevent burn-through. It is especially suited to thin section material and also prevents contamination of the electrode on arc initiation. The control is normally graduated from 0–10, refering to the amount of seconds set to reach full welding current. It is often referred to as a 'soft start' and can be switched off when not required.

Slope-Out Control

Sometimes referred to as a 'crater filling device', this reduces or 'decays' the welding current in gradual steps to allow the welder to fill up the crater that forms at the end of a weld. The normal control is graduated from 0–20, timed in seconds to the point where the arc is extinguished.

Hot Start

The hot start allows a controlled surge of welding current as the arc is started to quickly establish a molten weld pool, which can otherwise be difficult on metals with a high thermal conductivity. Adjustments can be made in the length of time and the percentage above the normal current, Figure 9.11.

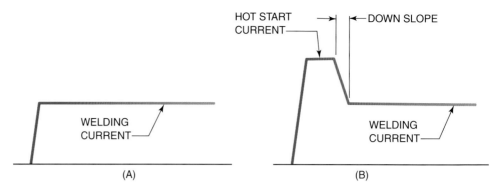

FIGURE 9.11 Standard method of starting welding current (A); hot start method of starting welding current (B)

Pre-Flow Control

Pre-flow time is the time during which gas flows to clear out any air in the nozzle or surrounding the weld zone before the welding current is started. The control allows the operator to set this time, Figure 9.12. Because some machines do not have pre-flow, many welders find it hard to hold a position while waiting for the current to start. One solution is to use the post-flow for pre-flow. Switch on the current to engage the post-flow. Now, with the current off, the gas is flowing, and the TIG torch can be moved to the welding position. Lower your helmet and restart the current before the post-flow stops.

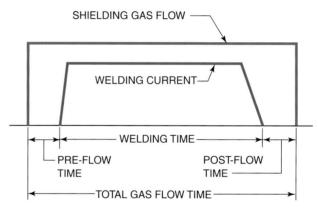

FIGURE 9.12 Welding time compared to shielding gas flow time

Post-Flow Control

The post-flow time is the time during which the gas continues flowing after the welding current has stopped. This period serves to protect the molten weld pool, the filler rod, and the tungsten electrode as they cool to a temperature at which they will not oxidize rapidly. The time of the flow is determined by the welding current and the tungsten size, Table 9.1.

TABLE 9.1 Post-welding gas flow times

Electrode Diameter (mm)	Post-welding Gas Flow Time*
1.6	5 sec
2.4	8 sec
3.2	15 sec

*The time may be longer if either the base metal or the tungsten electrode does not cool below the rapid oxidation temperatures within the postflow times shown.

Gas-Flow Control

The shielding gas flow rate is measured in cubic feet per hour (cfh) or in metric measure as litres per minute (L/min). The rate of flow should be as low as possible while still giving adequate coverage. High gas flow rates waste shielding gases and may lead to contamination from turbulence in the gas. Air is drawn into the gas envelope by a *venturi effect* (slipstream) around the edge of the nozzle. Also, the air can be drawn in under the nozzle if the torch is held at too low an angle to the metal, Figure 9.13.

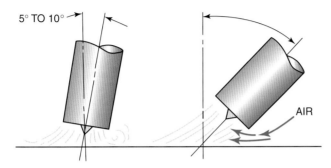

FIGURE 9.13 Too low an angle between the torch and work may draw in air

The larger the nozzle size, the higher the flow rate permissible without causing turbulence. Table 9.2 shows the average and maximum flow rates for most nozzle sizes. A gas lens can be used in combination with the nozzle to stabilize the gas flow, thus eliminating some turbulence. A gas lens will add to the turbulence problem if there is any spatter or contamination on its surface.

TABLE 9.2 Suggested argon gas flow rate for given cup sizes

Nozzle Inside Diameter	Gas Flow*	
(mm)	*cfh*	*(L/min)*
6	10–14	4.7–6.6
8	11–15	5.2–7.0
10	12–16	5.6–7.5
11	13–17	6.1–8.0
13	17–20	8.0–9.4
16	17–20	8.0–9.4

*The flow rates may need to be increased or decreased depending upon the conditions under which the weld is to be performed.

Remote Controls

A remote control can be used to start and stop the weld, and to vary the current. It can be either foot-operated or hand-operated. The foot control works adequately if the welder can be seated. Welds performed away from a welding station may use a hand or thumb control, or may not have any remote controls.

Most remote controls have an on-off switch that is activated at the first or last part of the control movement. A variable resistor increases the current as the control is pressed more, similar to the accelerator pedal on a car, Figure 9.14. The operating amperage range is determined by the value that has been set on the main controls of the machine.

FIGURE 9.14 Foot operated device can be used to increase current

Welding Return Lead

Ensure that the welding return lead matches the maximum amperage of the machine and that the insulation is free from any burns, cuts or any physical damage. It should have a good positive connection into the machine and the clamp connection should show no signs of fraying as this will reduce the current output and could cause the machine to overheat. The contact point should be flat and cleaned of any corrosion, paint or surface protection that could offer resistance to the flow of current and prevent the arc striking up.

Shielding Gases

The shielding gases that are used for the TIG welding process are argon (Ar), helium (He), hydrogen (H), nitrogen (N), or a mixture of two or more of these gases. The shielding gas protects the molten weld pool and the tungsten electrode from the harmful effects of air. It also affects the amount of heat produced by the arc and the resulting weld bead appearance.

Argon and helium are inert gases. They will not combine chemically with any other material and will not affect the molten weld pool in any way. Argon and helium may be found in mixtures but never as compounds.

Argon

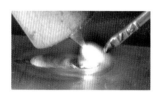

FIGURE 9.15 Highly concentrated ionized argon gas column

This is a by-product in air separation plants, where air is cooled and fractionally distilled. The primary products are oxygen and nitrogen. Before these gases were produced on a large scale, argon was a rare gas. Now it is distributed in cylinders as gas or in bulk as a liquid. Because argon is denser than air, it effectively shields welds in deep weld preparations in the flat position. However, this higher density can be a hindrance when welding overhead because higher flow rates are necessary. Argon is relatively easy to ionize and thus suitable for alternating current applications and easier starts. This property also permits fairly long arcs at lower voltages, making it virtually insensitive to changes in arc length. Argon is also the only commercial gas that produces the cleaning effect discussed earlier. These characteristics are most useful for manual welding, especially with filler metals added, as shown in Figure 9.15.

Helium

This is a by-product of the natural gas industry. It offers the advantage of deeper penetration. The arc force is sufficient to displace the molten weld pool with very short arcs. In some mechanized applications the tip of the tungsten electrode is positioned below the work piece surface to obtain very deep and narrow penetration. It is also very effective at high welding speeds, as for tube mills. However, helium is less forgiving for manual welding. Penetration and bead profile are sensitive to arc length, and the long arcs that are needed for feeding filler wires are more difficult to control. Helium has been mixed with argon to gain the combined benefits of cathode cleaning and deeper penetration, particularly for manual welding. The most common of these mixtures is 75 per cent helium and 25 per cent argon.

Although the TIG process was developed with helium as the shielding gas, argon is now used whenever possible because it is much cheaper. Helium also has some disadvantages. Because it is lighter than air, helium does not allow for good shielding. Its flow rates must be about twice as high as argon's for acceptable stiffness in the gas stream, and proper protection is difficult in drafts unless high flow rates are used. It is difficult to ionize, necessitating higher voltages to support the arc and making the arc more difficult to ignite. Alternating current arcs are very unstable. However, helium is not used with alternating current because the cleaning action does not occur.

Hydrogen

This is not an inert gas and is not used as a primary shielding gas. However, it can be added to argon in small amounts (1–3 per cent) when deep penetration and high welding speeds are needed. It also improves the weld surface cleanliness and bead profile on some grades of stainless steel that are very sensitive to oxygen. Hydrogen additions are restricted to stainless steel because hydrogen is the primary cause of porosity in aluminum welds. It can cause porosity in carbon steel and, in highly restrained welds, underbead cracking in carbon and low alloy steel.

Nitrogen

This is not an inert gas. Like hydrogen, nitrogen has been used as an additive to argon. It cannot be used with some materials, such as ferritic steel, because it produces porosity. In other cases, such as with austenitic stainless steel, nitrogen is useful as an austenite stabilizer in the alloy. It is used to increase penetration when welding copper. Because of its low relative cost, nitrogen is also sometimes used as a backing gas on some austenitic stainless steel pipe and tubing. When using nitrogen to protect the back side of a weld, care must be taken to be sure that all fit ups are tight and that no nitrogen mixes with the shield gas at the top of the weld. Because of the general success with inert gas mixtures and because of potential metallurgical problems, nitrogen has not received much attention as an additive for TIG welding.

Flow Meter

The flow meter may be merely a flow regulator used on a manifold system or it may be a combination flow and pressure regulator used on an individual cylinder, Figure 9.16 and Figure 9.17.

The flow is metered or controlled by opening a small valve at the base of the flow meter. The rate of flow is then read in units of cfh (cubic feet per hour), or L/min (litres per minute). The reading is taken from a fixed scale that is compared to a small ball floating on the stream of gas. Meters from various manufacturers may read from the top, centre or bottom of the ball, Figure 9.18. The ball floats on top of the stream of gas inside a tube that gradually increases in diameter in the upward direction. The increased size allows more room for the gas flow to pass by the ball. If the tube is not vertical, the reading is *not* accurate. Changes in pressure will also affect the accuracy of the flow meter reading. It is important to be sure the gas being used is read on the proper flow scale. Less dense gases, such as helium and hydrogen, will not support the ball on as high a column with the same flow rate as a denser gas, such as argon.

> **CAUTION**
>
> Because the gases used with TIG are predominately inert they pose a health risk with regards to asphyxiation. Localized exhaust ventilation (LEV) is recommended, especially in confined spaces. Care must be taken to ensure that the extraction rate is not set too high as to remove the shielding gas from the weld zone.

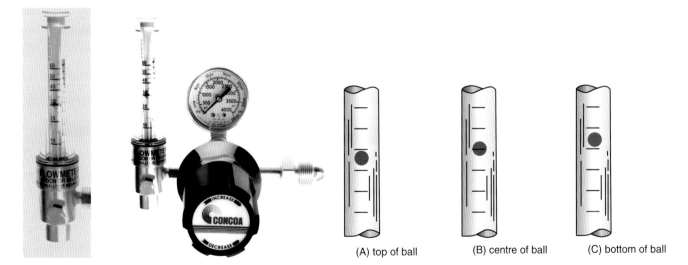

FIGURE 9.16 Flow meter

FIGURE 9.17 Flow meter regulator

FIGURE 9.18 Three methods of reading a flow meter

(A) top of ball   (B) centre of ball   (C) bottom of ball

Torches

TIG welding torches are available as water-cooled or air-cooled (gas cooled). The heat transfer efficiency for TIG welding may be as low as 20 per cent. This means that 80 per cent of the heat generated does not enter the weld. Much of this heat stays in the torch and must be removed by some type of cooling method.

FIGURE 9.19 Power cable safety fuse

NOTE

With the introduction of multi-purpose electrode types, such as cerium and lanthanum, it is often no longer necessary to produce a balled end on an electrode when welding aluminum with alternating current. However, when making surfacing welds or build ups on aluminum castings, a hemispherical end preparation on an electrode will produce a wider arc and weld pool, which is desirable in these types of operations.

The water-cooled TIG welding torch is more efficient at removing waste heat than an air-cooled torch and operates at a lower temperature, resulting in a lower tungsten temperature and less erosion.

The air-cooled torch is more portable than the water-cooled torch because it has fewer hoses, and is easier to manipulate. Also, the water-cooled torch requires a water reservoir or other system, which needs to contain some type of safety device (in-line fuse), Figure 9.19, to make it possible to shut off the power if the water flow is interrupted. The power cable is surrounded by the return water to keep it cool so that a smaller-size cable can be used. Without the cooling water, the cable quickly overheats and melts through the hose.

The water can be stopped or restricted for a number of reasons, such as a kink in the hose, a heavy object sitting on the hose or failure to turn on the system. When an open system is used, a pressure regulator must be installed to prevent pressures that are too high from damaging the hoses.

TIG welding torch heads are available in a variety of amperage ranges and designs, Figure 9.20. The amperage listed on a torch is the maximum rating and cannot be exceeded without possible damage to the torch. The various head angles allow better access in tight places and some can be swivelled easily to new angles. The back cap that both protects and tightens the tungsten can be short or long. Short caps are beneficial for getting the torch into restricted areas. Long caps can accommodate a full-length electrode, which will carry more heat away from the tip and allow slightly more current to be used before the electrode overheats, Figure 9.21 and Figure 9.22.

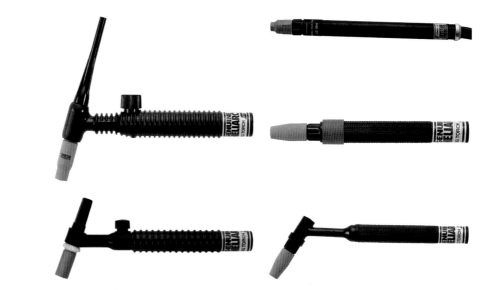

FIGURE 9.20 TIG welding torches

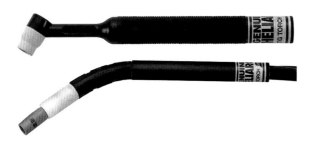

FIGURE 9.21 Short back caps

FIGURE 9.22 Long back caps

Gas Lens

Gas turbulence can cause poor shielding of the weld area especially in deep V-preparations. This can be improved by using a gas lens which attaches to the torch in front of the nozzle. The lens helps to produce a laminar flow in the shielding gas which is more directional and produces a 'stiffer' gas flow parallel to the electrode. This is done by using layers of gauze to regularize the gas flow which normally is 'bell' like when it emerges from the nozzle. This not only improves gas coverage but also allows the electrode stick-out to be increased, improving visibility and access in difficult situations.

Hoses

A water-cooled torch has three hoses connecting it to the welding machine. The hoses are for shielding gas to the torch, cooling water to the torch and cooling water return and housing the power cables to the torch, Figure 9.23. Air-cooled torches may have one hose for shielding gas attached to the power cable, Figure 9.24.

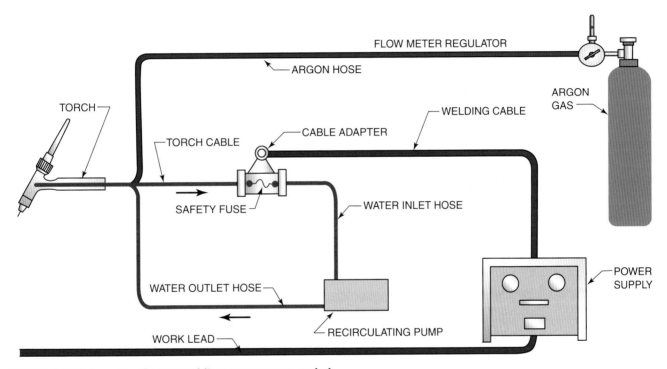

FIGURE 9.23 **Schematic of a TIG welding setup water cooled**

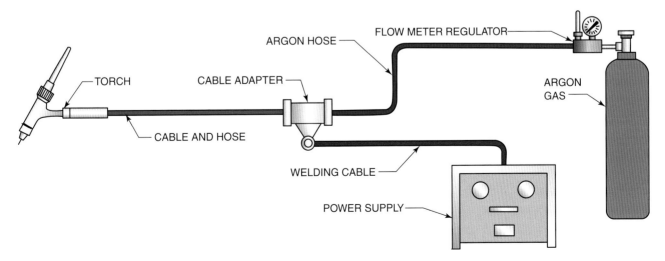

FIGURE 9.24 **Schematic of TIG welding setup air cooled**

The shielding gas hose must be plastic. Rubber hoses contain oils that can be picked up by the gas, resulting in weld contamination. The water-in hose may be made of any sturdy material. Water hose fittings have left-hand threads (denoted by nicks on the nut), and gas hose fittings have right-hand threads.

The return water hose also contains the welding power cable. This permits a much smaller-size cable to be used because the water keeps it cool. The water must be supplied to the torch head and return around the cable. This allows the head to receive the maximum cooling from the water before the power cable warms it. Running the water through the torch first has another advantage: when the water solenoid is closed, there is no water pressure in the hoses, which is particularly important. This feature also prevents condensation in the torch. If a water leak should occur during welding, the welding power is stopped, closing the water solenoid and thus stopping the leak and potential shock hazard.

A protective covering can be used to prevent the hoses from becoming damaged by hot metal, Figure 9.25. Even with this protection, the hoses should be supported, Figure 9.26, so that they are not underfoot on the floor and the chance of their being damaged by hot sparks is reduced.

FIGURE 9.25 Zip-on protective covering

FIGURE 9.26 A bracket holds the leads off the floor

Nozzles

The nozzle, or cup, directs the shielding gas directly on the welding zone. The nozzle size is determined by the diameter of the opening and its length, Table 9.3., and is often the welder's personal preference. Occasionally, a specific choice must be made based upon joint design or location. Small nozzle diameters allow the welder a better view of the molten weld pool and can be operated with lower gas flow rates. Larger nozzle diameters can give better gas coverage, even in draughty places.

Nozzles may be made from ceramic, such as alumina or silicon nitride (opaque), or from fused quartz (clear), especially useful in situations where the nozzle covers the work area. Ceramic nozzles are heat resistant and have a relatively long life, although this is affected by the current level and proximity to the work. Silicon nitride nozzles will withstand much more heat, resulting in a longer useful life.

CAUTION

Uncoil all sheathed leads to prevent electromotive forces (EMF) building up resistance to the flow of electricity, especially when HF is used, as this can affect the insulation on the cables.

TABLE 9.3 Recommended cup sizes

Tungsten Electrode Diameter (mm)	Nozzle Orifice Diameter (mm)
1.6	6 to 10
2.4	10 to 11
3.2	11 to 13

The fused quartz (glass) used in a nozzle is a special type that can withstand the welding heat. These nozzles are no more easily broken than ceramic ones. The added visibility they offer is often worth their added expense.

The longer a nozzle, the longer the tungsten must be extended from the collet. This can cause higher temperatures, resulting in greater tungsten erosion. When using long nozzles, it is better to use low amperages or a large-sized tungsten.

Back Caps

The function of the back cap is to encase the tungsten electrode and at the same time exert a pressure to squeeze the front of the collet tightly against the body of the torch. This ensures that a positive electrical path is established to guarantee emission of electrons at the tip of the electrode. Back caps are available in a variety of sizes, including a screw-in blank for use in tight locations and with short length tungstens. Back caps come with an 'O' ring designed to ensure a gas tight connection. As the majority of the back caps are of moulded plastic design, care must be taken to ensure the torch is secured at all times when not in use to prevent damage.

Tungsten Electrodes

Tungsten, (green tip) atomic symbol W (named after Wolfram, the scientist who discovered), has the following properties:

- high tensile strength;
- high melting temperature: (3410°C);
- good electrical conductivity.

These electrodes are available in sizes varying from 0.25mm to 6mm in diameter. The high melting temperature and good electrical conductivity make tungsten the best choice for a non-consumable electrode. Its arc temperature, around 6000°C, is much higher than its melting temperature. As the tungsten electrode becomes hot, the arc between the electrode and the work will stabilize.

The thermal conductivity of tungsten and the heat input are prime factors in the use of tungsten as an electrode. Because of the intense heat of the arc, some erosion of the electrode will occur. This eroded metal is transferred across the arc, Figure 9.27. Slow erosion of the electrode results in limited tungsten inclusions in the weld, which are hard spots that cause stresses to concentrate, possibly resulting in weld failure. Although tungsten erosion cannot be completely eliminated, it can be limited by:

- having good mechanical and electrical contact between the electrode and the collet;
- using as low a current as possible;
- using a water-cooled torch;

CAUTION

Do not overtighten the back cap as the thread pattern is extremely fine and it is very easily stripped. If the collet is not securing the tungsten, check the collet is the correct type for that particular torch. Also check the shoulder of the collet for damage, which could affect its efficiency.

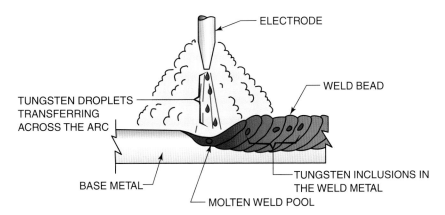

FIGURE 9.27 Some tungsten will erode from the electode

- using as large a size of tungsten electrode as possible;
- using DCEN current;
- using as short an electrode extension from the collet as possible;
- using the proper electrode end shape;
- using an alloyed tungsten electrode.

The torch end of the electrode is tightly clamped in a collet, which fits inside the torch and is cooled by air or water. The collet is a cone-shaped copper alloy sleeve that holds the electrode in the torch and provides the electrical path from the power source. Heat from both the arc and the tungsten electrode's resistance to the flow of current must be absorbed by the collet and torch. To ensure that the electrode is being cooled properly, make sure the collet connection is clean and tight. For water-cooled torches, make sure water flow is adequate.

Collet-tungsten connection efficiency is shown in Figure 9.28 and Figure 9.29.

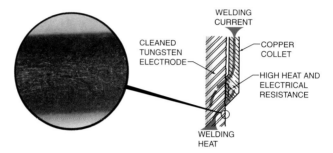

FIGURE 9.28 Irregular surface of a cleaned tungsten electrode (poor heat transfer to collet)

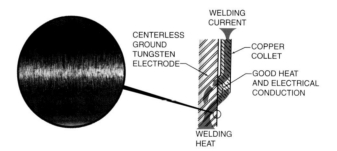

FIGURE 9.29 Smooth surface of a centreless ground tungsten electrode (good heat transfer to collet)

Large-diameter electrodes conduct more current because the resistance heating effects are reduced. However, excessively large sizes may result in too low a temperature for a stable arc. In general, the current-carrying capacity at DCEN is about ten times greater than that at DCEP. The electrode tip shape impacts the temperature and erosion of the tungsten. With DCEN, a pointed tip concentrates the arc as much as possible and improves arc, starting with either a high-voltage electrical discharge or a touch start. Because DCEN does not put much heat on the tip, it is relatively cool, the point is stable, and it can survive extensive use without damage, Figure 9.30A.

With alternating current, the tip is subjected to more heat than with DCEN. To allow a larger mass at the tip to withstand the higher heat, the tip is rounded. The melted end must be small to ensure the best arc stability, Figure 9.30B.

DCEP has the highest heat concentration on the electrode tip. For this reason, the electrode is ground at a 45° angle for 50% of its diameter so that it will form a ball when the arc is struck, Figure 9.30C.

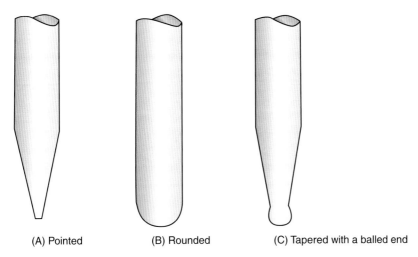

(A) Pointed   (B) Rounded   (C) Tapered with a balled end

FIGURE 9.30 **Basic tungsten electrode shapes**

Types of Tungsten Electrodes

Tungsten electrodes have a number of properties that make it an excellent non-consumable electrode for the TIG welding process. These properties can be improved by adding cerium, lanthanum, thorium, or zirconium to the tungsten.

For TIG welding, tungsten electrodes are classified as the following:

- pure tungsten, EWP;
- 1 per cent thorium tungsten, EWTh-1;
- 2 per cent thorium tungsten, EWTh-2;
- 0.8 per cent – 1 per cent zirconium tungsten, EWZr;
- 2 per cent cerium tungsten, EWCe-2;
- 1 per cent lanthanum tungsten, EWLa-1;
- alloy not specified, EWG.

See Table 9.4 for more information

- *Pure Tungsten Electrodes, EWP (green tip)* has the poorest heat resistance and electron emission characteristic of all the tungsten electrodes. It has a limited use with AC welding of metals such as aluminum and magnesium, and is not recommended for use with inverter type welding power supplies. New alloying elements with higher current carrying capacity; better emission

TABLE 9.4 **Tungsten electrode types and identification**

AWS Classification	Tungsten Composition	Tip Colour
EWP	Pure tungsten	Green
EWTh-1	1% thorium added	Yellow
EWTh-2	2% thorium added	Red
EWZr	1/4% to 1/2% zirconium added	Brown
EWCe-2	2% cerium added	Orange
EWLa-1	1% lanthanum added	Black
EWLa-1.5	1.5% lanthanum added	Gold
EWLa-2	2% lanthanum added	Blue
EWG	Alloy not specified	Grey

characteristics and greater durability have replaced pure tungsten electrodes for AC welding in many workshops.

● *Thoriated Tungsten Electrodes, EWTh-1(yellow tip) and EWTh-2(red tip)* thorium oxide (ThO_2), when added to tungsten in percentages of up to 0.6 per cent, improves tungsten's current-carrying capacity. The addition of 1 per cent to 2 per cent of thorium oxide does not further improve current-carrying capacity but does help with electron emission. This can be observed by a reduction in the electron force (voltage) required to maintain an arc of a specific length. Thorium also increases the serviceable life of the tungsten. The improved electron emission allows it to carry approximately 20 per cent more current, resulting in a corresponding reduction in electrode tip temperature, leading to less tungsten erosion and weld contamination. Thoriated tungstens also provide a much easier arc starting characteristic than pure or zirconiated electrodes. Thoriated tungsten works well with direct current electrode negative (DCEN). It can maintain a sharpened point and is very well suited for making welds on steel, steel alloys (including stainless), nickel alloys, and most other metals other than aluminum or magnesium. Thoriated electrodes do not work well with alternating current (AC). It is difficult to maintain a balled end, which is still commonly used for AC welding. A thorium spike, Figure 9.31, may also develop on the end of a thorium electrode when it is overheated, disrupting a smooth arc. Furthermore, whether balled or sharpened to a point, when a thoriated tungsten electrode is overheated by exceeding its maximum amperage rating, small vaporized droplets of tungsten will transfer across the arc and into the work, affecting the structural integrity of the weld, see Figure 9.27.
A visible thorium spike on an electrode tip, indicates that the electrode has overheated and the weld may be contaminated by vaporized droplets.

● *Zirconium Tungsten Electrodes (white tip and brown tip)* EWZr zirconium dioxide (ZrO_2) also helps tungsten emit electrons freely. The addition of zirconium to the tungsten has the same effect on the electrode characteristic as thorium, but to a lesser degree. Because zirconium tungsten is more easily melted than thorium tungsten, ZrO_2 electrodes are recommended to be used with AC current. Because of the ease in forming the desired balled end, zirconium tungsten is normally the electrode chosen for AC welding of aluminum and magnesium alloys. Zirconiated tungsten is more resistant to weld pool contamination than pure tungsten, thus providing excellent weld qualities with minimal contamination. Unlike thoriated tungsten, zirconiated tungsten is not radioactive.

● *Cerium Tungsten Electrodes (pink tip and grey tip),* EWCe-2 were developed to replace thoriated tungsten. Cerium dioxide (CeO_2), which is not radioactive, is added to tungsten. These electrodes have a current-carrying capacity similar to that of thoriated tungsten; however, they have an improved arc starting and arc stability characteristic, similar to that of thoriated tungsten. They also have a longer life than most other electrodes, including thorium. These electrodes have strong welder appeal for DC welding at low current settings; however, they are also used more and more with AC processes. In fact, ceriated electrodes, along with lanthanum oxide electrodes, are being recommended as a multi-purpose electrode by the manufacturers of inverter based welding power supplies. With excellent arc starts at low amperages, ceriated tungsten has become more and more common in orbital tube and pipe manufacturing operations as well as with thin sheet and delicate work. Cerium electrodes contain approximately 2 per cent cerium dioxide.

● *Lanthanum Tungsten Electrodes (gold, blue, black tip)* EWLa-1, EWLa-1.5, EWLa lanthanated tungsten is available as 1 per cent, 1.5 per cent, and 2 per cent. Lanthanum trioxide has the lowest work function of any of the materials, thus it usually starts easiest and has the lowest temperature at the tip, which resists grain growth and promotes longer service life. Lanthanum electrodes last longer than thorium electrodes when they are not overheated, and they are also resistant to thermal shock, which makes them a good choice for pulsed TIG operations. Lanthanum electrodes require about 15 per cent less power to initiate and maintain low current arcs. Lanthanum tungsten is a 'rare earth' material and is not radioactive. Whereas lanthanated electrodes have been used successfully in Europe and Japan since 1993, they are relatively new in the United Kingdom. This electrode has been primarily used for DC welding, but it is now being recommended, along with cerium electrodes, for AC welding, given its high current-carrying capacity and resistance to spitting during AC operations. These qualities allow lanthanum and cerium electrodes to be used with a pointed tip prep on AC operations

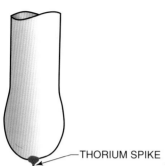

—THORIUM SPIKE

FIGURE 9.31 Thorium spike on balled end tungsten electrode

CAUTION

Thorium is a very low-level radioactive oxide, but the level of radioactive contamination from a thorium electrode has not been found to be a health hazard during welding. It is, however, recommended that grinding dust be contained. Because of concern in other countries regarding radioactive contamination to the welder and welding environment, thoriated tungsten has been replaced with other alloys.

with aluminum alloys. The pointed tip allows for a more controllable arc and the ability to make smaller weld profiles in fillet joints. While slightly more expensive than thoriated electrodes, lanthanum can be used for most workshop operations, thus eliminating confusion about electrode selection when multiple alloys are welded.

● *Alloy Not Specified, EWG* the EWG classification is for electrodes whose alloys have been modified by manufacturers. These alloys have been developed and tested by manufacturers to meet specific welding criteria. The blend of alloying oxides are proprietary, and specific alloy compositions are not normally revealed by manufacturers, although they do describe the welding characteristics of these electrodes.

Shaping the Tungsten Electrode

The desired end shape of tungsten can be obtained by grinding, breaking, re-melting the end or using chemical compounds. Tungsten is brittle and easily broken. Welders must be sure to make a smooth, square break where they want it to be located.

A grinder or belt sander is often used to clean a contaminated tungsten electrode or to put a point at the end of the tungsten. The grinder or sander used to sharpen tungsten should have a fine, hard stone or a fine grit media for the belt (80–120 grit). It should only be used for grinding tungsten, to avoid contamination. Because of the hardness of the tungsten and its brittleness, the grinding stone chips off small particles of the electrode. A coarse grinding stone will result in more tungsten breakage and a poorer finish.

Using an electric grinder with a fine grinding stone or a belt sander with 80–120 grit abrasive media, one piece of tungsten 150mm long, and safety glasses, you will grind a point on tungsten electrodes. The tungsten will become hot and the heat will be transmitted quickly to your fingers. To prevent overheating, only light pressure should be applied against the grinding wheel or belt. Hold the electrode between the open fingers to prevent it kicking back in to the palm of your hand, Figure 9.34 This will also reduce the possibility of accidentally breaking the tungsten. When grinding a point for DC applications, the length of the taper should not exceed three x diameter of the electrode. This is done so that there is sufficient cross-sectional area behind the taper to dissipate the heat from the tip.

Grind the tungsten so that the grinding marks run lengthwise, Figure 9.32 and Figure 9.33. to focus the electron emission at the tip and reduce the amount of small particles of tungsten contaminating the weld. Move the tungsten up and down as it is twisted during grinding. This will prevent the tungsten from becoming hollow-ground.

Tungsten is hard but brittle, resulting in low impact strength. If tungsten is struck sharply, it will break without bending. When it is held against a sharp corner and hit, a fairly square break will result. Figure 9.35 and Figure 9.36 show ways to break the tungsten correctly on a sharp corner using a hammer, and with two pairs of clamps.

NOTE

If an electrode has a large aluminum deposit at the end from contact with the work or filler metal, it should be broken off before grinding on a stone wheel. Non-ferrous materials like aluminum, brass, and copper can load the wheel, causing it to become dangerously out of balance.

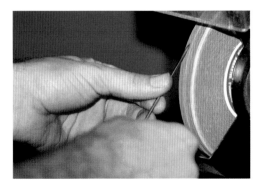

FIGURE 9.32 Correct method of grinding a tungsten electrode

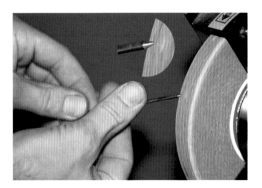

FIGURE 9.33 Incorrect method of grinding a tungsten electrode

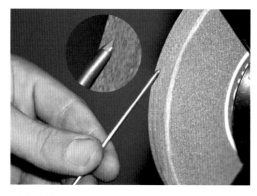

FIGURE 9.34 Correct way of holding a tungsten when grinding

FIGURE 9.35 Breaking the contaminated end from a tungsten by striking it with a hammer

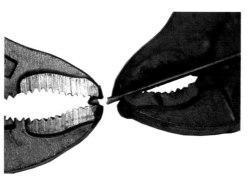

FIGURE 9.36 Correctly breaking the tungsten using pliers

Once the tungsten has been broken squarely, the end may be melted back so that it becomes somewhat rounded. This technique is appropriate for pure tungsten and zirconiated tungsten electrodes due to their lower current-carrying capacity and inability to hold a pointed end. Cerium and lanthanum electrodes will maintain a pointed end much longer and this technique is rarely used with them unless the wider bead profile produced by a hemispherical tip prep is desired. This is accomplished by switching the welding current to DCEP and striking an arc under argon shielding on a piece of copper, or another piece of clean metal. Do not use carbon, as it will contaminate the tungsten.

The tungsten can be cleaned and pointed using one of several compounds. The tungsten is heated by shorting it against the work and dipping it in the compound, a strong alkaline, where it rapidly dissolves. The chemical reaction occurs so quickly that enough additional heat is produced to keep the tungsten hot, Figure 9.37. When the tungsten is removed, cooled, and cleaned, the end will be tapered to a fine point. If the electrode is contaminated, the chemical compound will dissolve the tungsten, allowing the contamination to fall free.

TIP

Always wear eye protection as the end of the electrode can fly a good distance when struck.

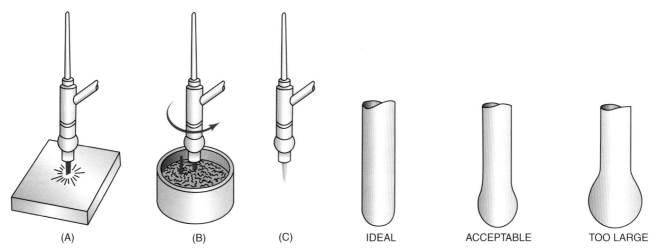

FIGURE 9.37 Chemically cleaning and pointing tungsten: (A) shortening the tungsten against the work to heat it to red hot, (B) inserting the tungsten into the compound and moving it around, and (C) cleaned and pointed tungsten ready for use

FIGURE 9.38 Melting the tungsten end shape

IDEAL   ACCEPTABLE   TOO LARGE

SETTING UP A TIG WELDER

Using a TIG welding machine; remote control welding torch; gas flow meter, gas source (cylinder or manifold), tungsten, nozzle, collet, collet body, cap and any other hoses, special tools, appropriate PPE and required equipment, you will set up the machine for TIG welding, Figure 9.39.

1   Start with the power switch off, Figure 9.40. Use a suitable spanner to attach the torch hose to the machine. The water hoses should have left-hand threads to prevent incorrectly connecting them. Tighten the fittings only as tightly as needed to prevent leaks, Figure 9.41. Attach the cooling water 'in' to the machine solenoid and the water 'out' to the power block.

FIGURE 9.39 TIG welding unit that can be added to a standard supply

FIGURE 9.40 Always be sure to turn off machine when making machine connections

2   The flow meter or flow meter regulator should be attached next. If a gas cylinder is used, secure it in place with a safety chain. Then remove the valve protection cap and 'crack' the valve to blow out any dirt. Attach the flow meter so that the tube is vertical.

FIGURE 9.41 Tighten each fitting as it is connected

FIGURE 9.42 Setting the current

3 Connect the gas hose from the meter to the gas 'in' connection on the machine.

4 With both the machine and main power switched off, turn on the water and gas so that the connection to the machine can be checked for leaks. Tighten any leaking fittings to stop the leakages.

5 Turn on both the machine and main power switches and watch for leaks in the torch hoses and fittings.

6 With the power off, switch the machine to the TIG welding mode.

7 Select the desired type of current and amperage range, Figure 9.42 and Figure 9.43.

FIGURE 9.43 Setting the amperage range

FIGURE 9.44 High frequency switch should be placed in the appropriate position

TABLE 9.5 Amperage range of tungsten electrodes

Electrode Diameter (mm)	DCEN	DCEP	AC
1.6	70–100	10–20	50–90
2.4	90–200	15–30	80–130
3.2	150–350	25–40	100–200
4	300–450	40–55	160–300

> **TIP** ❯
>
> Turn off all power before attempting to stop any leaks in the water system. The TIG welding system is now ready to be used.

8 Set the fine current adjustment to the proper range, depending upon the size of tungsten used, Table 9.5.

9 Place the high-frequency switch in the appropriate position, auto (HF start) for DC or continuous for AC, Figure 9.44.

10 The remote control can be plugged in and the selector switch set, Figure 9.45.

11 The collet and collet body should be installed on the torch first, Figure 9.46.

FIGURE 9.45 Setting the remote control switch   **FIGURE 9.46 Inserting collet and collet body**

12 The tungsten can be installed and the end cap tightened to hold the tungsten in place. Select and install the desired nozzle size. Adjust the tungsten length so that it does not stick out more than the diameter of the nozzle, Figure 9.47.

13 Check that all connections and settings are correct.

14 Turn on the power, depress the remote control and again check for leaks.

15 While the post-flow is still engaged, set the gas flow by adjusting the valve on the flow meter.

FIGURE 9.47 Installing the nozzle cup to the torch body

Striking Up an Arc

Using a properly set-up TIG welding machine, wearing the appropriate PPE and clean scrap metal, you will strike a TIG welding arc.

1 Position yourself so that you are comfortable and can see the torch, tungsten and plate while the tungsten tip is held about 6mm above the metal. Try to hold the torch at a vertical angle ranging from 0° to 15° from the perpendicular. Too steep an angle will not give adequate gas coverage, Figure 9.48.

2 Lower your arc welding helmet and depress the remote control. A high-pitched, erratic arc should be immediately jumping across the gap between the tungsten and the plate. If the high-frequency arc is not established, lower the torch until it appears, Figure 9.49. Or alternatively you can touch the nozzle on the work piece at an angle to the work, initiate the HF and lift the torch up to an angle of 75–80 degrees to the work surface.

3 Slowly increase the current until the main welding arc appears, Figure 9.50.

4 Observe the colour change of the tungsten as the arc appears.

5 Move the tungsten around in a small circle until a molten weld pool appears on the metal.

6 Slowly decrease the current and observe the change in the molten weld pool.

7 Reduce the current until the arc is extinguished.

8 Hold the torch in place over the weld until the post-flow stops.

9 Raise your helmet and inspect the weld. Repeat this procedure until you can easily start the arc and establish a molten weld pool using both AC and DCEN currents.

10 Turn off the welding machine, water, and shielding gas when you are finished.

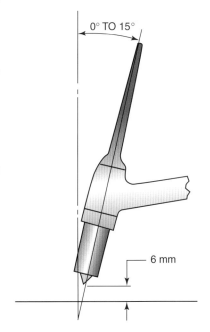

FIGURE 9.48 TIG torch position

A scratch start is an alternative method of initiating the arc by gently scratching the electrode on to the work surface. This technique can cause some contamination so is not recommended for critical applications.

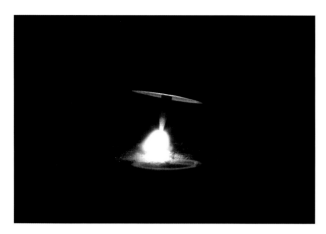

FIGURE 9.49 High frequency starting before arc starts

FIGURE 9.50 Stable gas tungsten arc

With a lift start, no HF is used, the electrode touches the work piece, the torch or foot pedal is engaged, but no current flows in the circuit at this point. When the electrode is lifted up from the work piece, this creates the arc gap at which point current flows and the arc is established. When the switch is released the current gently reduces until the arc is extinguished. This technique enables the welder to put the torch where they want to initiate the arc and eliminate stray arcing.

Torch Angle

The torch should be held as close to perpendicular as possible in relation to the plate surface. It may be angled from 0° to 15° from perpendicular for better visibility and still have the proper shielding gas coverage. As the gas flows out it must form a protective zone around the weld. Tilting the torch changes the shape of this protective zone, Figure 9.51. Too much tilting will cause the protective shielding gas zone to become so distorted that the weld may not be protected from contamination from the air. The closer the torch is held to perpendicular, the better the weld is shielded.

The velocity of the shielding gas also affects the protective zone as the torch angle changes. As the velocity increases, a low-pressure area develops behind the cup which can pull air into the shielding gas. The sharper the angle and the higher the flow rate, the greater the possibility contamination will occur from the onset of turbulence in the gas stream. Turbulence caused by the shielding gas striking the work will also cause air to mix with the shielding gas at high velocities.

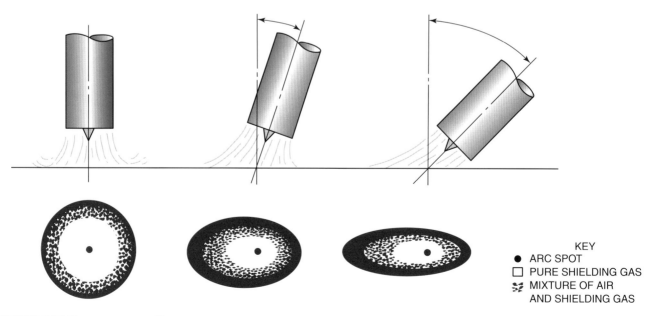

KEY
● ARC SPOT
□ PURE SHIELDING GAS
⠶ MIXTURE OF AIR
  AND SHIELDING GAS

FIGURE 9.51 Gas coverage patterns

FILLER ROD MANIPULATION

The filler rod end must be kept inside the protective zone of the shielding gas, Figure 9.52. The end of the filler rod is hot, and if it is removed from the gas protection, it will oxidize rapidly. The oxide will then be added to the molten weld pool. When a weld is stopped so that the welder can change position, the shielding gas must be kept flowing around the rod end to protect it until it is cool. If the end of the rod becomes oxidized, it should be cut off before restarting. The following method can be used both to protect the rod end and reduce the possibility of crater cracking – that is, breaking the arc but keeping the torch over the crater while, at the same time, sticking the rod in the molten weld pool before it cools, Figure 9.54. When the weld is restarted, the rod is simply melted loose again, Figure 9.55.

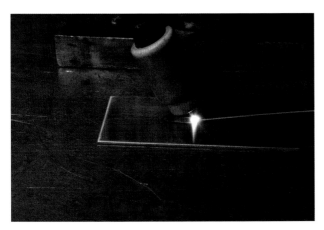

FIGURE 9.52 Hot filler end is well within the protective gas envelope

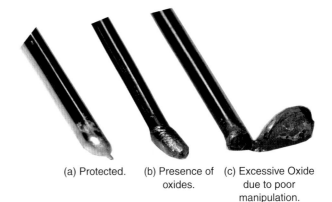

(a) Protected.   (b) Presence of oxides.   (c) Excessive Oxide due to poor manipulation.

FIGURE 9.53 Filler showing the effects of oxidation

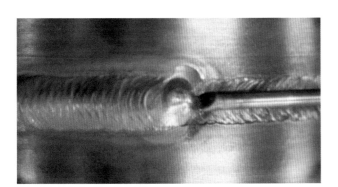

FIGURE 9.54 Filler being left in the molten weld pool

FIGURE 9.55 Filler being remelted as weld is continued

The rod should enter the shielding gas as close to the base metal as possible, Figure 9.56. A 15° angle or less to the plate surface prevents air from being pulled into the welding zone behind the rod, Figure 9.57.

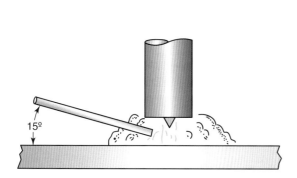

FIGURE 9.56 Keep the filler metal at approximately a 15 degee angle

FIGURE 9.57 Too much filler rod angle has caused oxides to be formed on the filler rod end

Tungsten Contamination

For new welding students, tungsten contamination is a frequently occurring and time-consuming problem. The tungsten becomes contaminated when it touches the molten weld pool or when it is touched by the filler metal. When this happens, especially with aluminum, surface tension pulls the contamination up onto the hot tungsten, Figure 9.58. The extreme heat causes some of the metal to vaporize and form a large, widely scattered oxide layer. On aluminum, this layer is black. On iron (steel and stainless steel), it is a reddish colour. This contamination forms a weak weld and, on a welding job, both the weld and the tungsten must be cleaned before any more welding can be done. The weld crater must be ground or chiseled to remove the tungsten contamination, and the tungsten end must be reshaped. Extremely tiny tungsten particles will show up if the weld is X-rayed. Failure to remove the contamination properly will result in the failure of the weld. When starting to weld, the new student may save weld practice time by burning off the contamination. On a scrap, usually copper plate, strike an arc using a higher than normal amperage setting. The arc will be erratic and discoloured at first, but, as the contamination vaporizes, the arc will stabilize. Contamination can also be knocked off by quickly flipping the torch head.

FIGURE 9.58 Contaminated tungsten

CURRENT SETTING

The amperage set on a machine and the actual welding current are often not the same. The amperage indicated on the machine's control is the same as that at the arc only for the following conditions:

- The power to the machine is exactly correct.
- The lead length is very short.
- All cable connections are perfect with zero resistance.
- The arc length is exactly the right length.
- The remote current control is in the full on position.

Any of these factors can change the actual welding amperage. There is also another, more significant difference between amperage and welding power. The welding power, in kilojoules is based on the formula:

$$\text{Arc energy (kj/mm)} = \frac{\text{arc voltage x welding current}}{\text{welding speed (mm/s) x 1000}}$$

Thus, the indicated power to a weld from two different types of welding machines set at 100 amperes will vary depending upon the voltage of the machine. The welding machine setting will vary within a range from low to high (cool to hot). The range for one machine may be different from that of another machine. The setting will also be different for various types and sizes of tungstens, polarities, types and thicknesses of metal, joint position or design, and shielding gas used. A chart, such as the one in Table 9.6, and a series of tests can be used to set the lower and upper limits for the amperage settings. As your welding skills improve with practice, you will become familiar with the machine settings. In industry, some welders mark a line on the dial to help in resetting the machine. If a welder is required to make a number of different machine setups, it is more professional to make up a chart and tape it to the machine.

TABLE 9.6 Sample chart used to record TIG welding machine settings

Current and Tungsten Electrode Size	Amperage/Machine Setting				
	Too Low	Low	Good	High	Too High

Setting the Amperage

Using a properly set-up TIG welding machine and torch, appropriate PPE, one of each available tungsten size and type, and 1.6mm–6mm thick, you will work with a group to develop a chart of the correct machine current setting for each type and size of tungsten.

1 Set the machine welding power switch for DCEN (DCSP) and the amperage control to its lowest setting, Figure 9.59.

2 Sharpen a point on each tungsten and install one of the smaller diameter tungstens in the TIG torch. Select a nozzle with a 13mm diameter hole and attach it to the torch head.

3 Set the pre-flow time to 0 and post-flow to 10 to 15 seconds. Connect the remote control if it is available.

4 Turn on the main power and hold the torch so that it cannot short out. Depress the remote controls to start the shielding gas so the flow rate can be set at 20cfh (9L/min). Switch the high frequency to start. All other functions, such as pulse, hot start, slope and so on, should be in the off position.

5 Place the piece of 1.6mm sheet metal flat on the welding table. Hold the torch vertically with the tungsten about 6mm above the metal. Lower your welding helmet and fully depress the remote control.

6 Watch the arc to see if it stabilizes and melts the metal. After a short period of time (15 to 30 seconds), stop, raise your helmet, and check the plate for a melted spot. If melting occurred, note the size of the spot and depth of penetration, Figure 9.60.

FIGURE 9.59 Lower the welding current to zero or as low as possible

FIGURE 9.60 Melting first occurring

7 Increase the amperage setting by 5 or 10A, note the setting on the chart, and repeat the process.

8 After each test, observe and record the results. The important settings to note are:

- when the tungsten first heats up and the arc stabilizes;
- when the metal first melts;
- when 100 per cent penetration of the metal first occurs;
- when burn-through first occurs;
- when the tungsten starts glowing white hot and/or melts.

9 The lowest (minimum) acceptable amperage setting is when the molten weld pool first appears on the base metal and the arc is stable. The highest (maximum) amperage setting is when the base metal burn-through or melting of the tungsten occurs. Any current setting between the high and low points is within the amperage range for that specific setup.

10 To establish the range for the next tungsten type or size, repeat the test. After each test, the metal should be cooled to prevent overheating.

11 After each type and size of tungsten has been tested and an operating range established, repeat the procedure using the next thicker metal.

12 Repeat this procedure until you have set up the operating ranges for all of the metals and tungstens you will be using.

13 Turn off the welding machine, shielding gas and cooling water, and reinstate your work area when you are finished welding.

Gas Flow

The gas pre-flow and post-flow times required to protect both the tungsten and the weld depend upon the following factors:

- wind or draft speed;
- nozzle size used;
- tungsten size used;
- amperage;
- joint design;
- welding position;
- type of metal welded.

The weld quality can be adversely affected by improper gas flow settings. The lowest possible gas flow rates and the shortest pre-flow or post-flow time can help reduce costs by saving the expensive shielding gas. The minimum flow rates and times must be increased when welding in draughty areas or for out-of-position welds, but can be somewhat lower for T-joints or welds made in tight areas. The maximum flow rates must never be exceeded because this causes weld contamination and increases the rejection rate.

Using a properly set-up TIG welding machine and torch, appropriate PPE, one of each available tungsten size, metal that is 1.6–6mm thick, and the welding current chart previously developed, you will work with a small group to make a chart of the minimum and maximum flow rates and times for each nozzle size, tungsten size and amperage setting. An assistant will also be needed to change and record the flow rate while you work.

Setting the Gas Flow

Set the machine welding power switch for DCEN (DCSP). Set the amperage to the lowest setting for the size of tungsten used. Set the pre-flow time to 0 and post-flow to 20 seconds, Figure 9.61. Turn on the main power. With the torch held so that it cannot short out, depress the remote control to start the shielding gas flow and set the flow at 20cfh (9L/min). Switch the high frequency to start. All other functions, such as pulse, hot start, slope and so on, should be in the off position.

FIGURE 9.61 **Setting the post-flow timer**

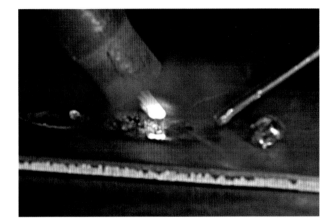

FIGURE 9.62 **Oxides forming due to inadequate gas shielding**

Starting with the smallest nozzle and tungsten size, strike an arc and establish a molten pool on a piece of metal in the flat position. Watch the molten weld pool and tungsten for signs of oxide formation as another person slowly lowers the gas flow rate. Have that person note this setting (where oxide formation begins), Figure 9.62, as the minimum flow rate on the chart next to the nozzle size and current setting. Now slowly increase the flow rate until the molten pool starts to be blown back or oxides start forming. This setting should be noted on the chart as the maximum flow rate for this current and nozzle size, Table 9.7. Lower the flow to a rate of 2cfh or 3cfh (1L/min or 2L/min) above the minimum value noted on the chart, and then stop the arc. Record the length of time from the point when the arc stops and the tungsten stops glowing as the post-flow time. Repeat this test at a medium and then high current setting for this nozzle and tungsten size. When using high current settings, it may be necessary to move the torch or use thicker plate to prevent burn-through.

TABLE 9.7 Sample chart for setting shielding gas flow rate and time

Electrode and Nozzle Size	Flow Rate					Post-flow Time		
	Too Low	Low	Good	High	Too High	Too Short	OK	Too Long

Repeat this test procedure with each available nozzle and tungsten size. Stainless steel or aluminum is preferred for this procedure because the oxides are more quickly noticeable than when mild steel is used. If aluminum is used, the welding current must be AC, and the high-frequency switch should be set on continuous.

PRACTICE WELDS

It is suggested that the stringer beads be done in each metal and position before the different joints are tried. Each metal has its own characteristics that may make one metal easier to work on than another. Mild steel is inexpensive and requires the least amount of cleaning. Slight changes in the metal have little effect on the welding skill required. Stainless steel is somewhat affected by cleanliness, requiring little pre-weld cleaning. However, the weld pool shows overheating or poor gas coverage. With aluminum, cleanliness is a critical factor. Oxides on aluminum may prevent the molten weld pool from flowing together. The surface tension helps hold the metal in place, giving excellent bead contour and appearance.

Low-Carbon and Mild Steels

These are the most common type of steels a new TIG welding student will experience. The TIG welding techniques required for welding steels begins with EWTh-1, EWTh-2, EWCe-2, or one of the EWLa pointed tungsten electrodes, Figure 9.63, with the welding machine set for DCEN (DCSP) welding current. Table 9.8 lists the types of filler metal used.

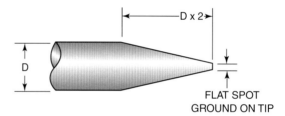

FIGURE 9.63 Tungsten tip shape for mild steel or stainless steel

TABLE 9.8 Filler metals for low-carbon and mild steels

SAE No.	Carbon %	AWS Filler Metal No.
	Low Carbon	
1006	0.08 max	RG60 or ER70S-3
1008	0.10 max	RG60 or ER70S-3
1010	0.08 to 0.15	RG60 or ER70S-3
	Mild Steel	
1015	0.11 to 0.16	RG60 or ER70S-3
1016	0.13 to 0.18	RG60 or ER70S-3
1018	0.15 to 0.20	RG60 or ER70S-3
1020	0.18 to 0.23	RG60 or ER70S-3
1025	0.22 to 0.29	RG60 or ER70S-3

During the manufacturing process, small pockets of primarily carbon dioxide gas sometimes become trapped inside low-carbon and mild steels. Only a few molecules of gas are trapped inside microscopic pockets within the steel, so they do not affect the steel strength. During most other types of welding, fluxes on the filler metal capture these gas pockets. TIG filler metals do not have fluxes, and these gas pockets expand during welding and can sometimes cause weld porosity. You are most likely to see this porosity when you are not using a filler metal. Most TIG filler metals have alloys, called deoxidizers that can help prevent porosity caused by gases trapped in the base metal.

Stainless Steel

The set-up and manipulation techniques required for stainless steel are nearly the same as those for low-carbon and mild steels, and skills transfer is easy. The major difference is that welds on stainless steels show the effects of contamination more easily and you must do a better job of pre-cleaning the base metal and filler metal. Also ensure you have adequate shielding gas coverage and do not overheat the weld.

The most common sign that there is a problem with a stainless steel weld is the bead colour after the weld. The greater the contamination, the darker the colour. The exposure of the weld bead to the atmosphere before it has cooled will also change the bead colour. It is impossible, however, to determine the extent of contamination of a weld with only visual inspection. Both light-coloured and dark-coloured welds may not be free from oxides. Thus, it is desirable to take the time and necessary precautions to make welds that are no darker than dark blue, Table 9.9. Welds with only slight oxide layers are better for multiple passes. Using a low arc current setting with faster travel speeds is important when welding stainless steel, because some stainless steels are subject to carbide precipitation, which is the combining of carbon with chromium. It occurs in some stainless steels when they are kept at a temperature between 625°C and 815°C for a long time and can be controlled by adding alloys to the stainless steel, lowering of the percentage of carbon or using special alloy filler metals during welding. The most important thing a welder can do is travel fast and use as little welding heat as possible. Black crusty spots may appear on weld beads, which are often caused by improper cleaning of the filler rod or failure to keep the end of the rod inside the shielding gas. Table 9.10 lists some common types of stainless steels and the recommended filler metals.

TABLE 9.9 Temperatures at which various coloured oxide layers form on steel

Surface Colour	Approximate Temperature at Which Colour is Formed (°C)
Light Straw	(200)
Tan	(230)
Brown	(275)
Purple	(300)
Dark Blue	(315)
Black	(425)

TABLE 9.10 Filler metals for stainless steels

AISI No.	AWS Filler No.	AISI No.	AWS Filler No.
303	ER308	310	ER310
304	ER308	316	ER316L
304L	ER308L	316L	ER316L
309	ER309	410	ER410

Aluminum

Aluminum is TIG welded using EWP, EWZr, EWCe-2 or EWLa rounded tip tungsten electrodes on transformer rectifier welding power supplies, Figure 9.64, with the welding machine set for ACHF welding current. EWZr, EWCe-2 or EWLa electrodes with a pointed tip may be used with some inverter welding power supplies due to the heat control afforded by their advanced

FIGURE 9.64 Tungsten tip shape for aluminium

circuitry. The alternating current provides good arc cleaning, and the continuous high frequency restarts the arc as the current changes direction.

The molten aluminum weld pool has high surface tension, which allows large weld beads to be controlled easily. Table 9.11 lists some basic types of filler metals used for aluminum welding. The high thermal conductivity of the metal may make starting a weld on thick sections difficult without first pre-heating the base metal. In most cases the preheat temperature is around 150° but will vary depending on metal thickness and alloy type. Specific pre-heat temperatures are available from the metal supplier. The processes of cleaning and keeping the metal clean can take a lot of time. Removal of the oxide layer is easy using a chemical or mechanical method, but ten minutes later it may require cleaning again. The oxide reduces the ability of the weld pool to flow together. Keep your hands and gloves clean and oil free so the base metal or filler rods do not become re-contaminated. Aluminum rapidly oxidizes at welding temperatures. If the filler rod is not kept inside the shielding gas, it will quickly oxidize, but, because of the low melting temperature of the filler rod, the end will melt before it is added to the weld pool if it is held too closely to the arc. Figure 9.65 and Figure 9.66.

TABLE 9.11 Filler metals for aluminum alloys

AISI No.	AWS Filler No.	AISI No.	AWS Filler No.
1100	ER1100	3004	ER4043
3003	ER4043	6061	ER4043

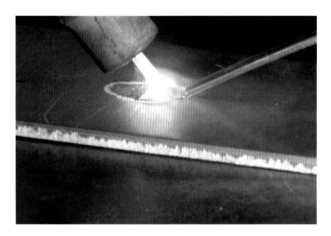

FIGURE 9.65 Aluminium filler being correctly added to the molten weld pool

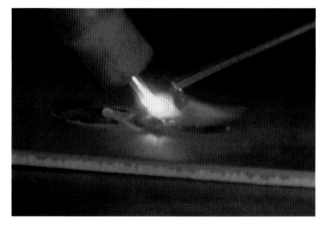

FIGURE 9.66 Filler rod being melted before it is added to the molten pool

Metal Preparation

Both the base metal and the filler metal used in the TIG process must be thoroughly cleaned before welding. Contamination left on the metal will be deposited in the weld because there is no flux to remove it. Oxides, oils, and dirt are the most common types of contaminants. They can be removed mechanically by grinding, wire brushing, scraping, machining or filing, or chemically with the use of acids, alkalis, solvents or detergents.

WORKSHOP TASK 9.1

Stringer Beads

Using a properly set-up and adjusted TIG welding machine on DCEN, appropriate PPE, and one or more pieces of mild steel, 150mm long and 1.6mm–3mm thick, using the leftward technique weld a stringer bead deposit in a straight line down the plate, Figure 9.68. Maintain uniform weld pool size and penetration.

FIGURE 9.67 Surfacing the weld

- Starting at one end of the piece of metal that is 3mm thick, hold the torch as close as possible to a 90° angle.

- Lower your welding helmet, strike an arc, and establish a weld pool. Tilt the torch to a 90° angle and slope the torch 0–10° to the perpendicular.

- Move the torch in a stepping or circular oscillation pattern down the plate toward the other end, adding filler wire to the front end of the molten pool as needed Figure 9.67.

- If the size of the weld pool changes, speed up or slow down the travel rate to keep the weld pool the same size for the entire length of the plate.

- The ability to maintain uniformity in width and keep a straight line increases as your skill improves and you are able to see more than just the weld pool.

- Repeat the process using both thicknesses of metal until you can consistently make the weld visually defect free.

- Turn off the welding machine, shielding gas and cooling water, and reinstate the work area when you are finished welding.

  Complete a copy of the 'Student Welding Report'.

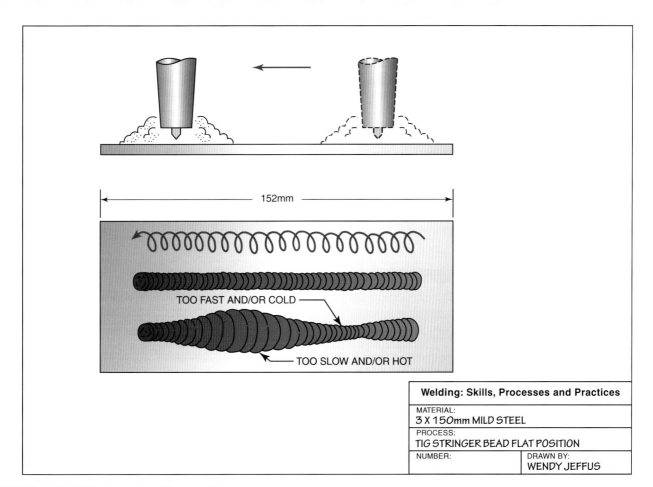

Welding: Skills, Processes and Practices		
MATERIAL: 3 X 150mm MILD STEEL		
PROCESS: TIG STRINGER BEAD FLAT POSITION		
NUMBER:	DRAWN BY: WENDY JEFFUS	

FIGURE 9.68 TIG Stringer bead flat position

WORKSHOP TASK 9.2

Outside Corner Joint

Using the same equipment and materials listed previously and wearing the appropriate PPE you will weld an outside corner joint in the flat position, Figure 9.69.

- Place one of the pieces of metal flat on the table and hold or brace the other piece of metal horizontally on it.
- Tack weld both ends of the plates together, Figure 9.70.
- Set the plates up and add two or three more tack welds on the joint as required, Figure 9.71.
- Starting at one end, tilt the torch to bisect the joint and slope the torch to 0–10° of the perpendicular, make a uniform weld, adding filler metal to the front edge of the molten pool as needed. In Figure 9.72, note the metal areas that are pre-cleaned before the weld is made.
- Repeat each weld as needed until all are mastered.
- Turn off the welding machine, shielding gas, and cooling water, and reinstate your work area when you are finished welding.

    Complete a copy of the 'Student Welding Report'.

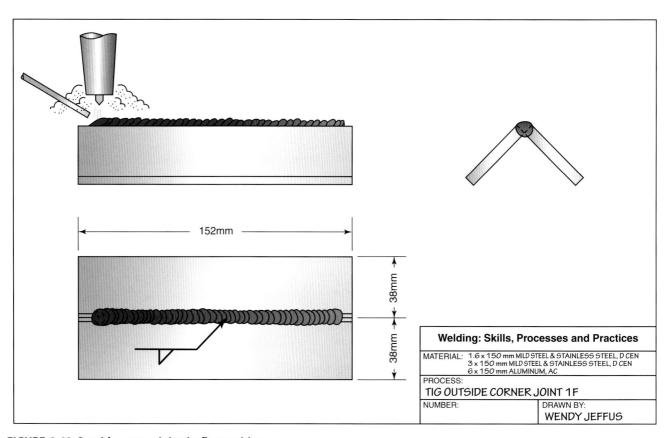

Welding: Skills, Processes and Practices

MATERIAL: 1.6 x 150 mm MILD STEEL & STAINLESS STEEL, D CEN
3 x 150 mm MILD STEEL & STAINLESS STEEL, D CEN
6 x 150 mm ALUMINUM, AC

PROCESS:
TIG OUTSIDE CORNER JOINT 1F

NUMBER:

DRAWN BY:
WENDY JEFFUS

FIGURE 9.69 Outside corner joint in flat position

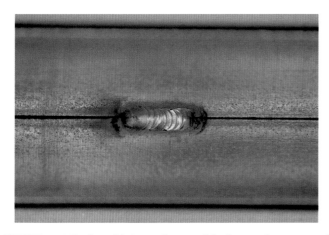

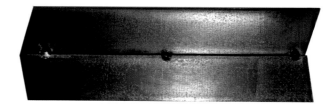

FIGURE 9.70 Tack weld. Note the good fusion at the start and crater fill at the end

FIGURE 9.71 Outside corner tack welded together

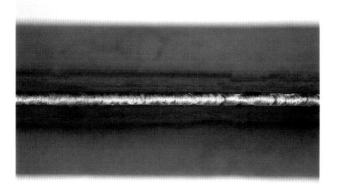

FIGURE 9.72 Outside corner. Note precleaning along weld

WORKSHOP TASK 9.3

Lap Joint

- Using the same equipment and materials as listed previously and wearing the appropriate PPE, you will weld a lap joint in the flat position, Figure 9.73.

- Place the two pieces of metal flat on the table with an overlap of 6–10mm. Hold the pieces of metal tightly together and tack weld them as shown in Figure 9.74 and Figure 9.75.

- Starting at one end, tilt the torch to a 45° angle and slope the torch to 0–10° of the perpendicular and make a uniform fillet weld along the joint, observing equal leg length.

- Ensure that the filler wire is placed to the top edge of the molten pool to offset any possible undercut of the top edge.

- Both sides of the joint can be welded.

- Repeat the process using all thicknesses of metal until you can consistently make the weld visually defect free.

- Turn off the welding machine, shielding gas and cooling water, and re-instate your work area when you are finished welding.

  Complete a copy of the 'Student Welding Report'.

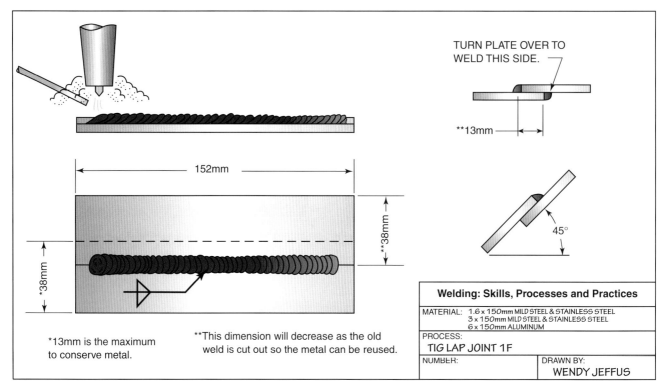

FIGURE 9.73 Lap joint in flat position

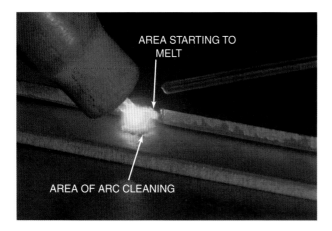

FIGURE 9.74 Be sure the top and bottom pieces are melted before adding filler metal

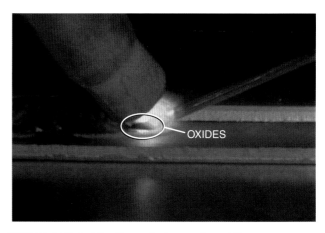

FIGURE 9.75 Oxides form during tack welding, ensure you clean tacks prior to welding

WORKSHOP TASK 9.4

T-fillet Joint

Using the same equipment and materials as listed previously and wearing the appropriate PPE, you will weld a T-joint in the flat position, Figure 9.76.

- Place one of the pieces of metal flat on the table and hold or brace the other piece of metal horizontally on it.

- Tack weld both ends of the plates together, Figure 9.77.

- Set up the plates in the flat position and add two or three more tack welds to the joint as required, Figure 9.78.

- On the metal that is 1.6mm thick, it may not be possible to weld both sides, but on thicker material a fillet weld can usually be made on both sides. The exception to this is if carbide precipitation occurs on the stainless steel during welding.

- Starting at one end, tilt the torch to 45° directing the heat into the bottom plate.

- Slope the torch at 0–10° off the perpendicular and add filler metal towards the top edge of the molten pool as needed to offset undercut and make a uniform weld observing equal leg length.

- Repeat the process using all thicknesses of metal until you can consistently make the weld visually defect free.

- Turn off the welding machine, shielding gas, and cooling water, and reinstate your work area when you are finished welding.

Complete a copy of the 'Student Welding Report'.

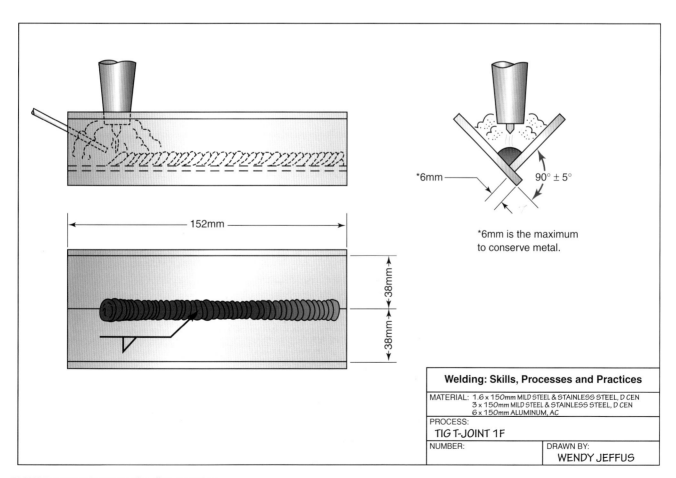

FIGURE 9.76 T-joint in the flat position

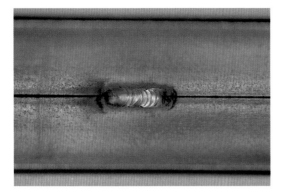

FIGURE 9.77 Tack weld on T-joint

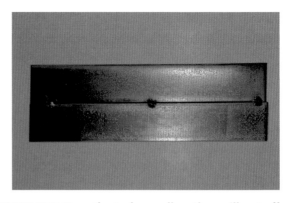

FIGURE 9.78 Keep the tacks small so they will not affect the joint

WORKSHOP TASK 9.5

Butt Joint

Using the same equipment and materials as previously used and wearing the appropriate PPE, you will weld a butt joint in the flat position, Figure 9.80.

- Place the metal flat on the table and tack weld both ends together, Figure 9.79.
- Two or three additional tack welds can be made along the joint as needed.
- Starting at one end, tilt the torch to a 90° angle.
- Slope the torch to 0–10° from the perpendicular, create a molten pool adding filler metal to the front of the molten pool and continue down the length of the joint.
- Repeat the process using all thicknesses of metal until you can consistently make the weld visually defect free.
- Turn off the welding machine, shielding gas, and cooling water and reinstate your work area when you are finished welding.

Complete a copy of the 'Student Welding Report'.

FIGURE 9.79 Tack weld on butt joint

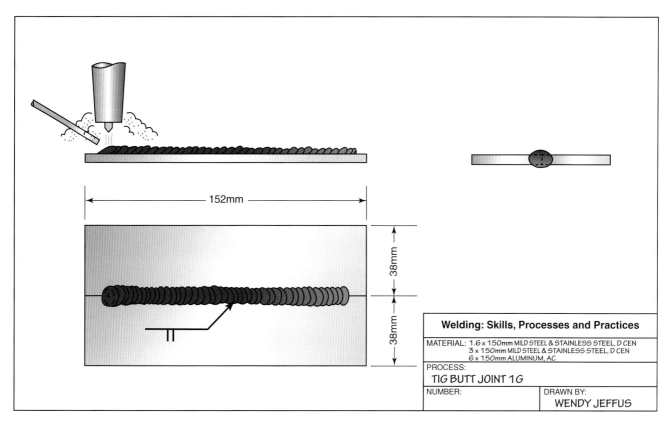

FIGURE 9.80 Square butt joint in flat position

BACK PURGING

When welding stainless steel, the penetrating weld metal will be oxidized and contaminated on butt joints and outside corner welds. Back purging can prevent this by ensuring that the weld metal enters an inert shielding. It is also used on pipe work to purge the internal volume of a pipe and to maintain a minimum gas flow during welding by using specially designed dams which temporarily block the internal bore of the pipe either side of the weld joint. A Y-piece is fitted to the flow meter with separate flow controls for shielding gas and purging requirements.

SUMMARY

Beginners find it difficult to position themselves so that they can control the electrode and filler metal and see the joint at the same time. Novice welders assume they must see the tungsten tip as they make the weld. Experienced welders, however, realize that they need to see only the leading edge of the molten weld pool to watch how the base metal is being melted, or depth of penetration, or whether the filler metal is being added at an appropriate rate. As you develop your skills, try to gradually reduce your need for seeing 100 per cent of the molten weld pool. This will be a significant advantage in the field because welding on site may have to be done out of position and in all environments.

Of all the welding processes, this is the most intolerant to any surface contamination, which will manifest itself by piping defects or contamination of the tungsten electrode. The process calls for a high degree of dexterity and hand to eye co-ordination. TIG welding is regarded as the gentleman's form of welding, in so much that it is regarded as a clean process devoid of slag and to a large percentage fumes.

ACTIVITY AND REVIEW

1 Identify which metals were only weldable by the TIG process before MIG was developed.

2 State what tungsten erosion is and how it occurs.

3 Explain what function, regarding tungsten heat, the collet and torch play.

4 Identify what problem an excessively large tungsten can cause.

5 Using the Table 9.4, answer the following:

a What colour identifies EWTh-2?
b What is the composition of EWCe-2?
c What colours identify EWLa?

6 Describe what adding thorium oxide does for the tungsten electrode.

7 Explain why a grinding stone that is used for sharpening tungstens should not be used for other metals.

8 State why the grinding marks should run lengthwise on the tungsten electrode.

9 Explain why we break off the contaminated end of a tungsten electrode.

10 Identify what advantage there is in having a water-cooled torch.

11 Describe why the shielding gas hoses must not be made of rubber.

12 Indicate why the water solenoid should be on the supply side of the water system.

13 State why the tube on a flow meter must be vertical.

14 Indicate what the heat distribution is with DCEN welding current.

15 State what the heat distribution is with AC welding current.

16 Explain why argon and helium are known as inert gases.

17 Identify the benefits of adding hydrogen to argon for welding.

18 State the purpose of a hot start.

19 Explain how air can be drawn into the shielding gas.

20 Identify the functions a remote control can provide the welder.

21 State what advantages can be gained by using a slope-in and slope-out control.

22 Describe what effect does torch angle has on the shielding gas protective zone.

23 State why the end of the filler rod must be kept in the shielding gas protective zone.

24 List four considerations to be taken into account when choosing the correct current setting.

25 Describe how the filler metal should be added to the molten weld pool.

Heat Input and Distortion Control Techniques

LEARNING OBJECTIVES

After completing this chapter, you should be able to:

- define expansion and contraction
- list the factors which affect distortion
- understand about thermal conductivity and co-efficient of linear expansion
- recognize the different forms of distortion
- use distortion control mechanisms
- understand how alternative joint design can be used to reduce distortion
- select corrective measures to deal with distortion
- understand the principles of pre-heating and post -heating.

UNIT REFERENCES

NVQ units:

Joining Materials by the Manual Metal Arc Welding Process.
Cutting and Shaping Materials using Gas Cutting Machines.
Heat Treating Materials for Fabrication Activities.

VRQ units:

Engineering Environment Awareness.
Fabrication and Engineering Principles.
Non-fusion Thermal Joining Methods.
Thermal Cutting Techniques.

INTRODUCTION

In order to complete a design successfully it is essential to understand the influence that the application of heat can have on a material, the structure, and the environment in which it will have to operate. Chapter 4 described the mechanical properties of a range of typical engineering materials and alloys. In this chapter we will look at the thermal conductivity of those materials, their thermal co-efficients of linear expansion and potential failures that can occur as a result of heating them. In terms of design we shall consider edge preparation and joint design. We will then look at how the fabricator or welder can help to eliminate the effects of distortion and avoid costly rework.

Every fabricator and welder should be aware of the effects of expansion and contraction in order to maintain dimensional accuracy and keep within given tolerances. This knowledge will allow you to plan for the effects of heat and use them to your advantage by including them in your weld sequence or preparatory work, thereby reducing time spent of correcting distortion, and wastage. If you understand these principles you can help reduce the carbon footprint by lowering operating costs, energy requirements and making the most effective use of materials – which are a finite resource.

EFFECTS OF HEAT INPUT

From the previous chapters you know that heat can be applied in a number of ways, including the combustion of gases, electric arcs, electrical resistance, the mixing of elements and frictional forces.

When metal is subjected to a source of heat it will increase in size due to 'expansion' taking place throughout its mass. This is measured by the co-efficient of linear expansion, which varies for different materials. If the heat is applied to a small 'localized' area only, the expansion will be local and uneven. The surrounding metal, which has remained comparatively cool, will exert a force to prevent expansion of the heated metal.

If the 'yield point' of the metal has been reached during the heating cycle, permanent deformation will take place. This means that when the metal cools down, it will not return to its original form but will remain distorted. The surrounding cold metal also offers resistance during cooling of the heated area and 'contractional stresses' will be set up during the plastic flow of the metal which will also contribute to distortion. The amount of distortion has a marked effect on the amount of structural strain that will remain in the metal after cooling. Also, if any restraint has been placed upon the metal to control distortion, then another form of stresses will be produced. 'Residual stresses' remain after the metal has cooled and place strains upon the structure itself.

We will now consider this series of effects in relation to welding. The amount of weld metal is relatively small in comparison to the mass of the parent metal, and therefore the greatest amount of heat is concentrated in the weld zone. The strength of the weld metal is greatly reduced at high temperatures, and because its mass is smaller than the structure, it will undertake most of the plastic flow during cooling. If the plastic flow exceeds the metal's ultimate tensile strength then 'fracture' may well occur.

Effects of Distortion

The relationship between welding and distortion can be shown in simple terms by using a flat plate which has a slot cut along its length as shown in the diagram below.

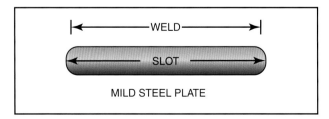

FIGURE 10.1 The relationship between welding and distortion

If a run of weld is now deposited above the slot, the heat input will be transferred by 'conduction' to the adjacent parent metal, which will then tend to expand in the direction shown in Figure 10.2.

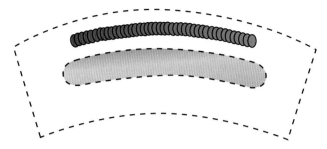

FIGURE 10.2 Direction of expansion

The bottom section of the plate, because it is comparatively cool, does not expand and starts to exert a force to resist any attempt to alter its shape. The top half will therefore absorb the expansion by an increase in thickness at the expense of an increase in length.

As both the weld and the parent metal have undergone expansion there will be a corresponding contraction on cooling which will be opposed by the bottom section of plate. As the cooling weld and material solidify their strength will increase until the opposition offered by the bottom plate is overcome and distortion takes place in the opposite direction as shown in Figure 10.3.

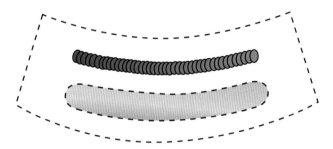

FIGURE 10.3 Opposite direction of expansion

An additional distortion will also take place on another plane. This results from unequal expansion between the weld metal and the surrounding plate, which tended to restrict movement upon heating, but cannot now resist the contraction of the weld area as it becomes stronger upon cooling. The result of this is a 'bowing' of plate towards the cooling deposit, Figure 10.4.

FIGURE 10.4 Bowing plate

TYPES OF DISTORTION

The fabricator or welder needs to be aware of three distinct types of distortion in welded structures.

Angular Distortion

When a root run is deposited in a single V-butt joint, the weld deposit will have undergone expansion during the welding operation and contracts as it cools down. This '**shrinkage**' will draw the edges of the welded plates together as shown in Figure 10.5.

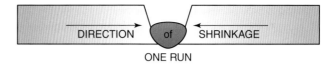

FIGURE 10.5 Angular distortion

If a second run is made, the contractional pulling force is opposed by the now solid first deposit, Figure 10.6. The force at the top of the V now tries to pull the plate edges together and the solidified weld metal at the bottom opposes this force, giving rise to 'angular' distortion along the joint.

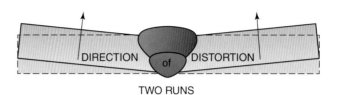

FIGURE 10.6 Contractional pulling force

In a T-fillet weld, angular distortion results in shrinkage across the 'face' of the weld with a reduction in the angle between the upright and base plate, Figure 10.7.

The golden rule governing angular distortion states that the more runs deposited, the greater the risk of distortion. Therefore the use of pre-setting is recommended. This is a method of compensating in advance for the distortion that will take place. It will be described later in this chapter.

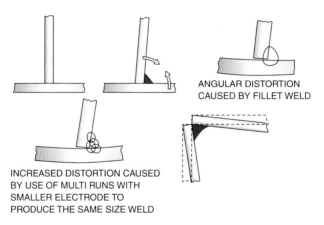

FIGURE 10.7 Increased distortion

Longitudinal Distortion

Longitudinal Distortion occurs along the length of the weld with a 'bowing' effect which means that the original length is reduced, Figure 10.8. For example in the case of a butt welded joint in material 4–7mm thick the amount of longitudinal shrinkage is 0.25mm per metre length of weld.

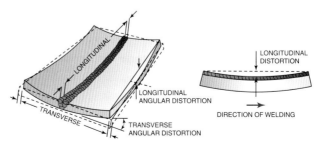

FIGURE 10.8 Longitudinal distortion

For material 8–10mm thick the shrinkage is 0.30mm per metre length of weld. If the length is less critical it is common practice to cut and weld oversize and machine back to the design specification.

Transverse Distortion

Transverse Distortion results in shrinkage across the face of the weld with a 'lifting' of the edges and a reduction in the width of the plate dimensions, Figure 10.9. The amount of distortion is proportional to the material thickness. Table 10.1 gives an indication of the amount of transverse 'shrinkage' for a single V-butt welded joint in low carbon steel plate.

Table 10.1 Transverse shrinkage for a single V-butt joint

SINGLE - V BUTT JOINT	
MATERIAL THICKNESS (mm)	TRANSVERSE SHRINKAGE (mm)
3	0.52
6	1.04
9	1.55

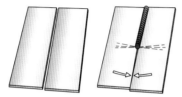

FIGURE 10.9 Transverse distortion

Warping and buckling is caused by weld shrinkage and may be reduced or avoided by careful location of the welds relative to the 'neutral plane' of the workpiece, see Figure 10.10. If the welds are located symmetrically to the neutral plane, the section being welded will warp only if the welding sequence has not been chosen correctly. With the proper sequence providing even distribution of the heat, the only visible result will be a 'shortening' caused by shrinkage. If the welds are located asymmetrically to the neutral plane, with unbalanced heat input, the workpiece will distort.

The extent of the distortion can be reduced by the use of the correct welding technique and by clamping the workpiece to a strongback.

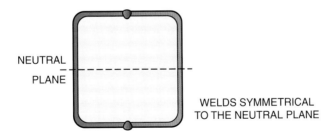

FIGURE 10.10 Warping and buckling

FACTORS AFFECTING DISTORTION

If it were possible to pre-heat all structures and components before welding, thereby obtaining equal expansion between the weld metal and parent metal, and an equal contraction by controlled cooling after welding, then no deformation would take place and there would be no distortion. However, this is not a practical solution because pre-heating is not always possible.

Other areas we need to consider are the 'co-efficient of linear expansion' which is a measure of the expansion a material undergoes when it is heated, and the 'thermal conductivity' which is a measure of the ease with which heat flows through a material.

If we take low carbon steel and stainless steels as an example; the thermal conductivity of stainless steel is approximately one-third that of low carbon steel and this would increase the shrinkage effect. The co-efficient of linear expansion of stainless steel is approximately 1.5 times that for low carbon steel, and this would increase shrinkage in the plate adjacent to the weld.

Size of weld is another critical factor which governs the expansion and contraction of the weld metal. The larger the angle of the V, the larger the weld deposit and the more strain in the cooling weld. Good selection of edge preparation and electrode sizes will reduce the amount of strain induced, and therefore the amount of distortion produced. In T-fillets this can be achieved by using fewer runs to deposit the required leg length, for example by using a larger sized electrodes.

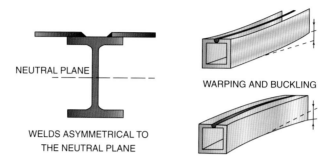

NEUTRAL PLANE

WELDS ASYMMETRICAL TO
THE NEUTRAL PLANE

WARPING AND BUCKLING

FIGURE 10.11 Clamping the workpiece

The 'root gap' and 'weld face' are used to ensure root penetration and control of the root run. Distortion here can be reduced by the use of 'weld sequences' and contol of the gap.

DISTORTION CONTROL TECHNIQUES

The fabricator or welder can use a wide range of techniques to reduce or even eliminate distortion.

Tack welding can be used with some structures and welding processes. Tack welding methods are often used in these circumstances.

Welding of auxiliary pieces to the main structure (constraints), which are removed after the completion of welding.

Tack welding of run-on and run-off plates to the ends of the joint to be welded to initiate and terminate the arc on and to ensure weld integrity, Figure 10.12.

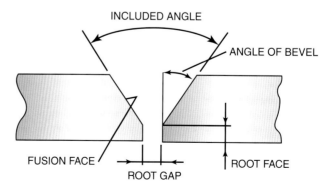

INCLUDED ANGLE

ANGLE OF BEVEL

FUSION FACE

ROOT GAP

ROOT FACE

FIGURE 10.12 Tack welding

The distance between tacks will vary with the type of material being welded and the rate of travel or speed, Figure 10.13. With materials that have a high co-efficient of linear expansion and when the rate of travel is slow, the tacks should be close together.

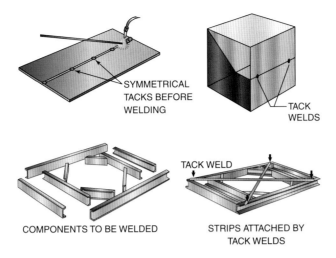

FIGURE 10.13 Varying distances between tacks

Taper spacing is used on thicker plate instead of tacking. The two edges to be joined are placed at an angle to each other and are pulled parallel by the welding heat, Figure 10.14. The edges can be held in position by clamps or special wedges which are moved along the gap as welding proceeds. The allowance varies with the type of material, plate thickness, speed of travel and welding process used.

FIGURE 10.14 Taper spacing

Pre-setting the plates in the opposite direction to the contraction will allow the plates to be pulled back in line as the contraction takes place, Figure 10.15. Pre-setting of fillet welded joints can be a problem, as it can result in a bad fit-up when the vertical is rocked away from the horizontal base plate, leaving a gap. In this case it may be necessary to use some form of restraining mechanism such as clamping, use of gussets, etc. which can be removed on completion of the weld. If pre-setting is to be used, then consideration must be given to the amount of angular distortion that is likely to occur. Factors which may have an impact on this include:

- speed of travel;
- number of runs deposited;
- type of edge preparation;
- heat input.

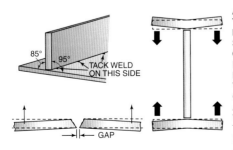

A 'SET' MAY BE PUT IN THE HORIZONTAL MEMBERS OF THE 'L' SECTION, SHOWN OPPOSITE PRIOR TO FILLET WELDING ON BOTH SIDES OF THE VERTICAL MEMBER ON CONTRACTION, THE ANGULAR DISTORTION WHICH RESULTS WILL TEND TO PULL THE MEMBERS SQUARE AFTER WELDING

FIGURE 10.15 Pre-setting

The amount of pre-setting which should be given to the plates before welding to compensate for angular distortion is a matter for experience. However, there are data sheets which give the average number of degrees of distortion which takes place under certain conditions.

Pre-bending or pre-springing (pre-cambering) the parts to be welded is a simple example of the use of opposing mechanical forces to counteract distortion due to welding, Figure 10.19. The top of the weld preparation, which will contain the bulk of the weld metal, is 'lengthened' when the plates are sprung, making the completed weld slightly longer than it would be if it had been made on the flat plate. When the clamps are released after welding, the plates return to the flat shape, allowing the weld to 'relieve' its longitudinal shrinkage stresses by 'shortening' to a straight line, and the welded plates assume the desired flatness

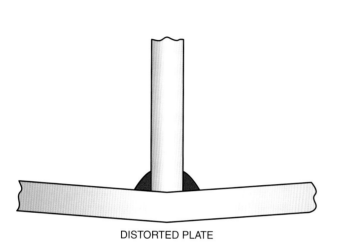

FIGURE 10.16 **Distorted plate**

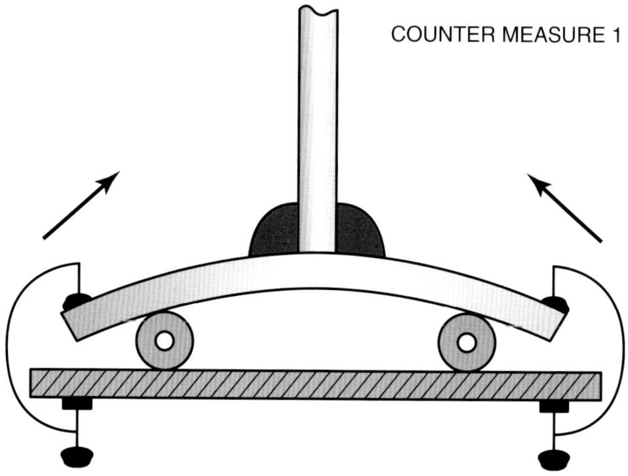

FIGURE 10.17 **Counter measure 1**

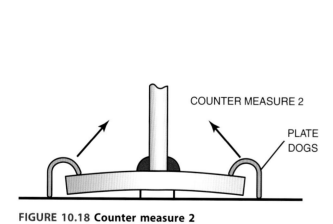

FIGURE 10.18 **Counter measure 2**

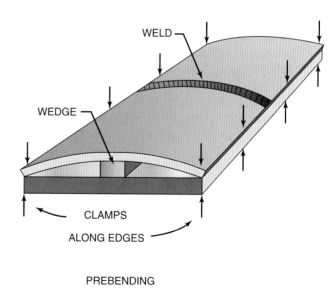

FIGURE 10.19 **Pre-bending**

Welding sequences can help control distortion by balancing the heat input to the workpiece, Figure 10.20. A well planned welding sequence involves placing weld deposits at different locations about the assembly so that, as the structure shrinks in one place, it counteracts the shrinkage forces of the welds already made.

'Back-step' welding is a technique in which a series of relatively short runs of weld is made, so that the subsequent run is laid down in the direction of the previous run and terminates where the previous run started. The progression is along the joint, but each portion of weld is deposited towards the previous run. This reduces the amount of heat input to any one portion of the weld. Because thermal stresses causing contraction in one section are 'neutralized' by stresses causing expansion in the neighbouring section, distortion is reduced or prevented altogether.

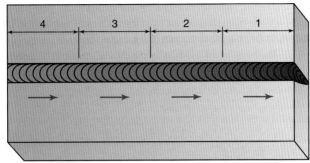

INDICATES DIRECTION OF WELDING 1.2.3.4. INDICATES RUN SEQUENCE

FIGURE 10.20 Welding sequences

Skip-welding or intermittent welding is another welding sequence used to minimize and control distortion, Figure 10.21. Sometimes known as 'planned wandering' this technique deposits weld runs at some distance from each other, so that the spread of heat is limited.

The diagram below shows one method where the runs of weld metal are placed alternatively on either side of the vertical plate. Another method is to deposit short runs of weld where they 'skip' from one section to another, once again the object of this procedure is to control the heat input into the welded joint.

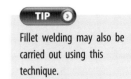

TIP

Fillet welding may also be carried out using this technique.

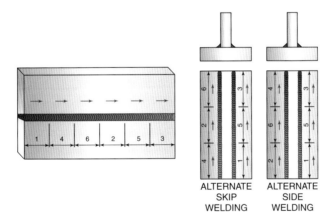

ALTERNATE SKIP WELDING   ALTERNATE SIDE WELDING

FIGURE 10.21 Skip-welding techniques

ALTERNATIVE JOINT DESIGN

If we look at a butt weld made with a 60° included angle preparation, it is apparent that the weld width at the top of the joint is appreciably greater than at the root. Working on the principle that shrinkage is proportional to the length of metal cooling, there is a greater contraction at the top of the weld. If the plates are free to move, as they mostly are in fabricating operations, they will rotate in relation to each other, causing angular distortion. A weld procedure is required to balance the amount of shrinkage about a 'neutral' axis.

In general, two approaches can be used:

● Weld both sides of the joint.

● Use an edge preparation which gives a more uniform width of weld through the thickness of plate.

Asymmetrical shrinkage shows up as longitudinal bowing in the direction of welding. This is a cumulative effect which builds up as the heating and cooling cycles progress along the joint. Some control can be achieved by the use of a double V-joint to balance the shrinkage so that more or less equal amounts of contraction occur on each side of the neutral axis. This gives less distortion than a single V-preparation.

It is very difficult to get a completely flat joint with a symmetrical double V-preparation as the first run *always* produces more angular distortion than subsequent runs. For this reason an asymmetrical (2/3–1/3) preparation is used so that the larger amount of weld metal on the second side pulls back the distortion which occurred when the first side was welded.

An alternative solution is to change the joint design to a 'U' or 'J' preparation with nearly parallel sides. This gives a uniform weld width through the section, achieved by welding short lengths on a planned or randomly distributed basis.

When welding double 'V', 'J' or 'U' butt joints, welds are deposited alternately each side to counteract expansion and shrinkage stresses and produce a flat butt weld. The advantage of 'U' and 'J' preparations is that less weld metal is deposited, with a reduction in the amount of distortion produced. This technique can also be used on T-fillet welds, where 'stitch welds' may be sufficient in some cases.

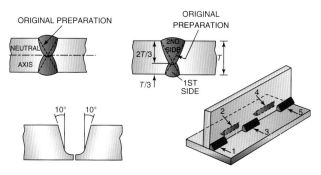

FIGURE 10.22 **Alternative joint design**

Back to back assembly is used where the shape of the fabrication will makes it possible to clamp or tack up two similar fabrications and weld alternately. The effect is once again to balance the distortion or stress on the parts being welded. Figure 10.23 shows the assembled components and a weld sequence for completion.

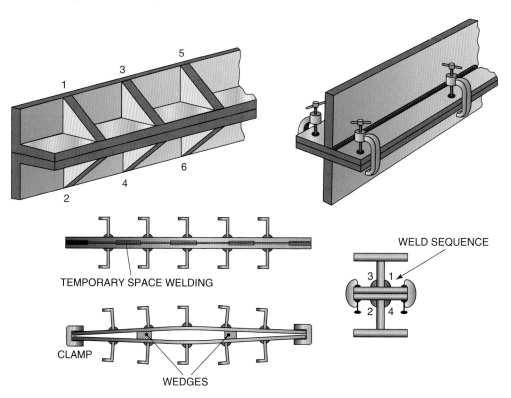

FIGURE 10.23 **Stages of a planned welding sequence**

When a number of plates are to be welded together as shown in Figure 10.24, the recommended sequence is to complete the transverse welds first, followed by the longitudinal welds.

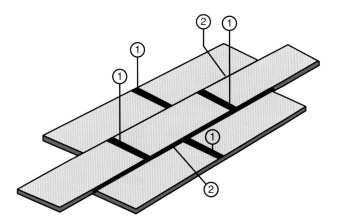

FIGURE 10.24 Sequence for welding plates

Backing bars (chills) and backing strips are used extensively to try to control distortion. These terms are sometimes confused. A 'backing bar' is used to control the amount of penetration in a butt weld and is not intended to become part of the final welded joint. It is removed after completion of the weld and should be made from copper or some other material which will not alloy with the weld metal. When large pieces of copper are not available, a backing bar may be made from steel with a copper or other metal insert to control the profile at the root of the weld. A backing bar which has a fairly large cross section controls the root penetration and also acts as a 'heat sink', helping to absorb the heat that would normally be present in the parent metal and reducing distortion. Some backing bars incorporate water cooling to great effect.

A 'backing strip' is a piece of metal that is fitted to the underside of a butt joint to support the weld metal at the root of the joint, so controlling the amount of penetration, Figure 10.25. The backing strip may itself be penetrated by the weld metal and may remain as part of the completed joint or may be removed by machining. Because the backing strip is penetrated by the weld metal, it is usually made from the same material as the parent metal.

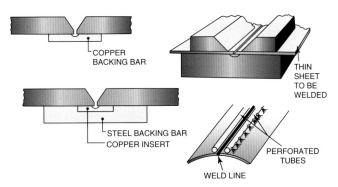

FIGURE 10.25 Backing strip is penetrated by the weld metal

Jigs and fixtures may be used to restrain or control movements of the components during welding, Figure 10.26. If close control over accuracy is required, the work piece may be restrained with the use of wedges, strongbacks, chains, clamps and stays, which remain in position during cooling. As these methods prevent most of the movement, there is a progressive build-up of stress in the fabrication. Where necessary, 'stress relief' is carried out with the stays, etc. in position. These jigs may also incorporate 'back-purging' facilities for stainless steel fabrications.

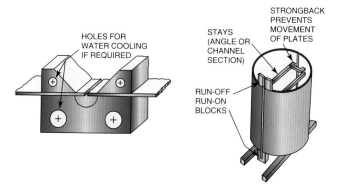

FIGURE 10.26 Fixtures to restrain or control movements of the components during welding

Peening, or lightly hammering, the weld is one way to counteract the shrinkage forces of a weld bead as it cools. Essentially, peening the bead stretches it and makes it thinner, thus relieving (by plastic deformation) the stresses induced by contraction as the metal cools, Figure 10.27. This method must be used with care, and a root bead should never be peened, because of the danger of either concealing a crack or causing one. Generally, peening is not permitted on the final run, because of the possibility of covering a crack, interfering with inspection, and because of the undesirable work hardening effect. The usefulness of this technique is limited, even though there have been instances where inter-pass peening proved to be the only solution for distortion or a cracking problem. Peening is often used on cast iron for this purpose. Before peening is used on a job, engineering approval should be obtained.

Pre-heating involves heating the area of the joint before welding is started. It is usually carried out by oxy-propane or oxy-acetylene flame, or alternatively by electrical resistance or inductance heating. The temperature for pre-heating will vary with the type and thickness of material being welded. The use of pre-heating lessens the amount of weld shrinkage and stress, which reduces distortion in thinner materials and internal stress in thicker materials. Pre-heating will also slow down the rate of cooling of the weld, and this is particularly important when welding carbon steels in which the carbon content is 0.35 per cent and above. A slow cooling rate will prevent the formation of a hard, non-ductile structure which can cause cracking.

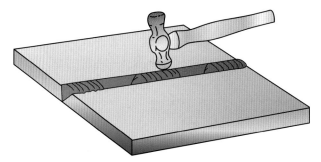

FIGURE 10.27 Peening and skip welding method

Post-heating is the heating of the weld area or the whole work piece AFTER after welding has been completed. It involves controlled heating and cooling in order to promote desirable mechanical properties.

'Stress relieving' is one such heat treatment. Large internal stresses may have been set up in the structure of the material by the 'thermal cycle' of the welding process. If these stresses are not reduced then the material may crack, with disastrous consequences. The normal procedure for stress relieving plain carbon steels and carbon manganese steels is to heat them up to 650°C (just below recrystallization temperature) and hold at this temperature for one hour for each 25mm of thickness, then slowly cool in the furnace/induction blanket to 300°C after which they may be cooled to ambient temperature in air. This treatment will enable the stresses in the welded joint to be reduced to an acceptable level.

When considering post welding techniques, it must be remembered that it costs money to remove distortion after welding. Care should be taken **NOT** to introduce too much additional stress, especially when pulling and twisting.

Mechanical Methods

A certain amount of distortion may be rectified by mechanical means such as pressing, jacking, hammering, bending rollers, or crane lifting, Figure 10.28. Care needs to be taken not to put undue stress on the welds which could cause cracking, and local heating may be necessary to allow easier movement.

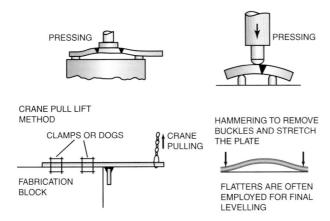

FIGURE 10.28 Rectifiying distortion using mechanical methods

Thermal Methods

A component which is buckled may be straightened without mechanical means by heating in 'local zones' on the convex side, starting at the centre of the bulge and working progressively outwards. Wedge-shaped zones should be heated to a dull red colour and unrestrained contraction allowed to take place as shown in the diagrams. The following points should be observed:

- The proposed heating zone should be marked out and heating started at the base working towards the apex. A dull red heat should not be exceeded, approximately 700ºC for steel.
- The wide part or base of the wedge should be at the outer edge and should have a width approximately one third of the length of the wedge. The apex or point of the wedge should reach the neutral axis of the assembly.
- An intense flame should be used to heat the zone rapidly before the heat disperses into the surrounding area. Oxy-acetylene is often the preferred choice.

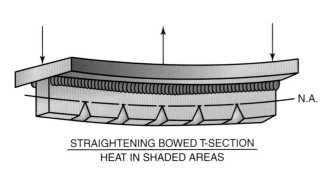

FIGURE 10.29 Straightening bowed T-section using heat wedge sequence

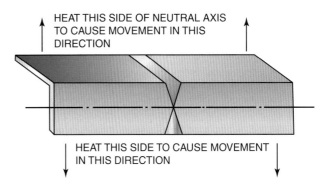

FIGURE 10.30 Using heat sequence to correct distortion, heat being applied on the opposite side to the distortion will bring alignment about a neutral axis

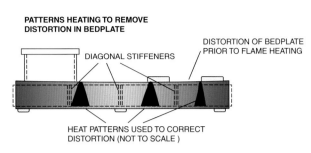

PATTERNS HEATING TO REMOVE
DISTORTION IN BEDPLATE

DIAGONAL STIFFENERS

DISTORTION OF BEDPLATE
PRIOR TO FLAME HEATING

HEAT PATTERNS USED TO CORRECT
DISTORTION (NOT TO SCALE)

FIGURE 10.31 Heat sequence heating cycles

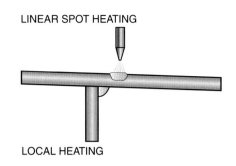

LINEAR SPOT HEATING

LOCAL HEATING

FIGURE 10.32 Localized spot heating

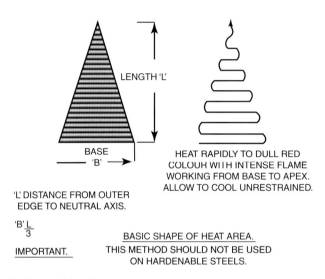

LENGTH 'L'

BASE
'B'

'L' DISTANCE FROM OUTER
EDGE TO NEUTRAL AXIS.

'B' $\frac{L}{3}$

IMPORTANT.

HEAT RAPIDLY TO DULL RED
COLOUR WITH INTENSE FLAME
WORKING FROM BASE TO APEX.
ALLOW TO COOL UNRESTRAINED.

BASIC SHAPE OF HEAT AREA.
THIS METHOD SHOULD NOT BE USED
ON HARDENABLE STEELS.

FIGURE 10.33 Heat 'wedge' calculation and heating sequence

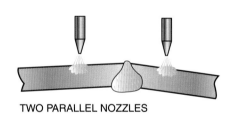

TWO PARALLEL NOZZLES

FIGURE 10.34 Dual heating cycle

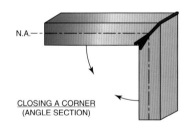

N.A.—

CLOSING A CORNER
(ANGLE SECTION)

FIGURE 10.35 Using heat on the inner side
of mitre to close joint

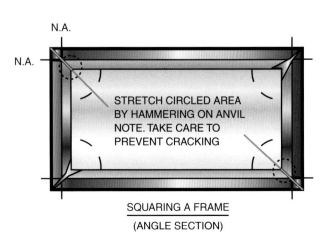

N.A.

N.A.

STRETCH CIRCLED AREA
BY HAMMERING ON ANVIL
NOTE. TAKE CARE TO
PREVENT CRACKING

SQUARING A FRAME
(ANGLE SECTION)

FIGURE 10.36 'Dressing' (hammer) technique of weld
joint to correct distortion

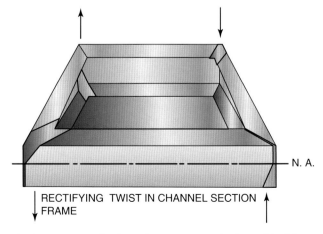

N. A.

RECTIFYING TWIST IN CHANNEL SECTION
FRAME

FIGURE 10.37 Method used to correct twist in welded frame

SUMMARY

Welding is a multi-faceted subject which embraces physics, chemistry, mathematics, drawing and metallurgy. A sound knowledge of heat input and the effects it can have on structures can save a lot of time and trouble trying to correct a problem that, with a little thought and planning, could have been predicted. Armed with the strategies discussed in this chapter, you can move forward with the knowledge that the majority of difficulties can be foreseen and avoided.

ACTIVITY AND REVIEW

1   State how the application of heat can affect a butt weld.

2   Name the three main forms of distortion.

3   List four factors which can have an effect on the amount of distortion that occurs.

4   State what effect the joint design has on distortion.

5   Explain what the term 'neutral plane' means.

6   State the advantage of using pre-setting.

7   Give a definition for 'skip' welding.

8   Draw a diagram to indicate the 'back-step' method.

9   Identify the alternative joint designs to control distortion.

10   Identify what the term 'strongback' refers to.

11   Indicate what the limitations of peening a weld are.

12   State how pre-heating can be used for distortion control.

13   Define what 'residual' stresses are.

14   State what a 'yield point' is and what bearing it has on the structure.

15   Identify how you would control distortion in a low thermal conductivity material such as stainless steel.

16   Identify three mechanical methods for rectifying distortion.

17   State the difference between a backing bar and backing strip.

18   State the purpose of post-heating.

19   State the advantages of back to back assemblies.

20   Give a brief description of taper spacing.

21   State how the wedge-shape heating cycle is used.

22   Indicate the differences between symmetrical and asymmetrical sequences of welding.

23   Describe the meaning of the term 'stress relieving'.

24   State how the root face affects heat input.

25   Identify what auxiliary equipment can be used to control distortion.

Quality Control

LEARNING OBJECTIVES

After completing this chapter, you should be able to:

- Indicate four items that would be appear on a mill certificate
- understand how a Gantt sheet is used for production
- distinguish between the symbols used in flow charts
- distinguish between qualification and certification
- list six major components listed on a weld procedure sheet (WPS)
- understand the difference between a PQR and a WPS
- identifies three common codes used in welding
- identify the names used to define the various parts of a weld deposit
- list five common defects or imperfections associated with welding
- list three mechanical tests carried out on materials
- state how the Izod and Charpy tests differ
- relate how root, face and side bends differ
- describe three methods of non-destructive testing (NDT)
- explain how macro-examination is carried out and what it reveals.

UNIT REFERENCES

NVQ units:

Working Efficiently and Effectively in Engineering.
Complying with Statutory Regulations and Organizational Safety Requirements.

Joining Materials by the Manual MIG/MAG and Other Continuous Wire Processes.
Using and Interpreting Engineering Data and Documentation.
Joining Materials by the Manual Metal Arc Welding Process.

VRQ units:

Engineering Environment Awareness.
Engineering Techniques.
Engineering Principles.
Fabrication and Engineering Principles.

KEY TERMS

brinell hardness tester a tool that characterizes the indentation hardness of materials through the scale of penetration of an indenter, loaded on a material test-piece.

delamination an imperfection such as oxide inclusion as a result of hot rolling of the material.

etching the process of using acids, bases or other chemicals to dissolve and reveal any unwanted matter on the surface of the material.

high-frequency alternating current a low amperage current that is superimposed over the welding current to assist in arc initiation in DC and to stabilize AC TIG welding operations.

quality control a system that ensures products or services are designed and produced to meet or exceed customer requirements.

radiographic inspection (RT) a non-destructive testing (NDT) method of inspecting materials for hidden flaws by using the ability of short wavelength electromagnetic radiation (high energy photons) to penetrate various materials.

rockwell hardness a hardness scale based on the indentation hardness of a material. The Rockwell test determines the hardness by measuring the depth of penetration of an indenter under a large load compared to the penetration made by a preload.

shear strength as applied to a soldered or brazed joint, it is the ability of the joint to withstand a force applied parallel to the joint.

ultrasonic inspection (UT) very short ultrasonic pulse-waves launched into materials to detect internal flaws or to characterize materials somewhat similar to sonar detection.

INTRODUCTION

Quality control describes a series of measures designed to ensure that a consistent approach is taken to all engineering operations in order to guarantee the standard of materials, manufacturing techniques and inspection procedures. Various legislation and directives from Europe and International Organization for Standardization are to be found in BS EN ISO 9000 Series.

The process starts with the material requirements and monitoring of consistency of supply, through a raft of quality checks on manufacturing processes. Quality is defined as fit for purpose. The completed components must comply with the design specification and must also have a well structured back-up service to support customers. It is important for anyone starting a manufacturing career to understand the underlying principles of quality control and to keep these principles in mind. This will help you adopt good working practices, reducing waste and costs and saving resources used to put right any defective work.

DOCUMENTATION

Documentation covers mill sheets, consumables, design details, production schedules, weld procedures and inspection detail.

Mill Certificates

These are generally issued with large purchase orders to specified design requirements or for standard products. It is becoming common practice to request a mill certificate with small discrete orders, too. A mill certificate contains the following information:

- a unique cast or batch number;
- the composition of the material (carbon content and alloying elements);
- the mechanical properties (hardness values, yield values, etc);
- the heat treatments that have been carried out (Annealed, hardened, etc);
- forming characteristics;
- pre-heating or post-heating requirements.

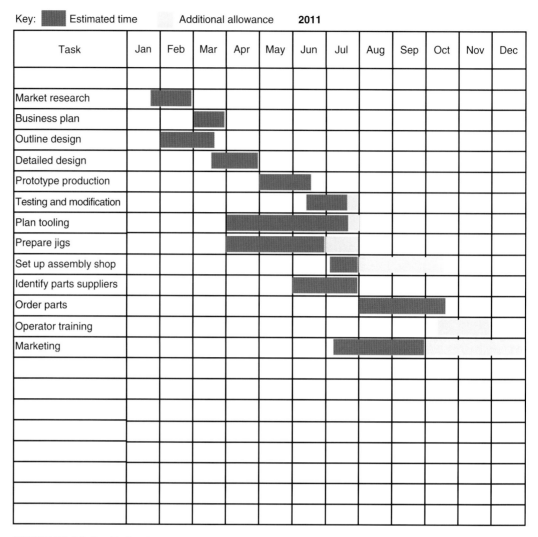

FIGURE 11.1A Gantt chart

The piece of paperwork on the following page is crucial to the design procedures, manufacturing processes and selection of appropriate welding processes to meet the desired specification. When the plate is delivered to the fabricator, an inspector checks the plate number against the mill certificate to confirm that the steel conforms to the specification to which the steel was ordered. When the plate is cut up into parts to be fabricated, the plate number, or some other identifying number, is transferred to each part considered to be of sufficient importance and this is in turn recorded on the component drawing. By doing this, traceability is incorporated into the manufacturing process and if any defect is detected it is possible to identify the areas in which that plate was used.

Consumable specification sheets are similar to the mill sheets and designate the composition of the consumable, typical application range, mechanical properties and a brief descriptor for typical applications. All of these points will comply with some form of standardization, such as British Standard (BS), European Norm (EN), International Organization for Standardization (ISO). This gives the welder a form of insurance that the specified consumables conform to the weld procedure.

fundia	IINTYG CERTIFICATE ZEUGNIS 1552/98

CERTIFICATE 3.1C EN 10204

Beställare/Customer/Besteller
STEEL PLATES & SECTIONS LTD

Stålsort/Steel grade/Stahlsorte	Beställarens order nr/Customer's order No./Auftrags-Nr. des Bestellers	Vår order/Our order No./Unsere Auftrags-Nr.
S355J2G3 EN 10 025	0034477	3008/31151

HÅLLFASTHETSDATA	MECHANICAL PROPERTIES	FESTIGKEITSEIGENSCHAFTEN

Charge Cast Schmelze	Dimension Dimension Abmessung	Sträckgräns Yield stress Streckgrenze	Brottgräns Tensile strength Zugfestigkeit	Förlängning Elongation Dehnung		Hårdhet Hardness Härte	Slagseghet Impact value Kerbschlagzähigkeit	
Nr No.	mm	R_{eH} N/mm^2	R_m N/mm^2	A_5 %	%	H B °C	ISO KV	Joule
4-6881	10, 00 4815 kg	480	598	32				

The material in question has been tested in the presence of the Society's representative with satisfactory results.

CHARGEANALYS %	CAST ANALYSIS %								SCHMELZANALYSE %

Charge nr Cast No. Schmelze Nr	C	Si	Mn	P	S	V	N	AL		CEV
4–6881	, 18	, 25	1, 07	, 011	, 010	, 05	, 009	, 020		.40

Each bundle is marked LR and has a metal tag showing cast No and size

SURVEYOR

BOXHOLM 98-09-03	ROLAND CARLSSON Kval.avd/QA

FIGURE 11.1AA Mill certificate

Production Documents

Production documents include Gantt charts which are usually displayed on the walls of the production departments and are used to programme staff, machines, materials, and movement of components. Gantt charts provide an instant picture of the progress of a project. They allow you to:

- assess how long a project should take and identify when you are behind schedule;
- plan the most efficient use of machines, manpower and materials;
- programme in holiday breaks;
- manage the dependencies between tasks;
- make decisions about what to do to bring the project back on line.

Gantt charts make it is easy to target milestones, (significant stages in the project) and help to determine whether the project will be completed on time and in budget. This is something that most manufacturers have to consider now that they are operating in a global market.

Design documentation includes diagrammatic representation in orthographic projection, exploded views, sectional views and isometric. Coupled with this is a bill of materials, specifying forming details, surface finish and dimensional tolerances.

TIP

Other documentation includes flow diagrams, pie charts and statistical process control charts which give an indication of defects and measures of how to correct these deviations.

WELD PROCEDURE SHEETS (WPS)

Because welding is now a critical activity which may have a bearing on whether a structure fails in service, guidelines were introduced to control the parameters of the procedure and specify particular operating characteristics. For components which have a specific operating function, welder codings or certification, have been established which identify the stringent conditions under which the operator has achieved success and is therefore covered by some form of insurance. A weld procedure sheet specifies the following:

- parent material specifications and thickness;
- welding consumables: electrodes, wire, flux, shielding gas and any requirements such as baking times for low hydrogen electrodes;
- welding process;
- joint configuration;
- weld sequence with diagrammatic representation;
- edge preparation with diagrammatic representation;
- welding position and operational range;
- pre-heat temperature and method of checking;
- interpass temperature control limits;
- method of cleaning   grinding, gouging or wire brushing;
- post-heat weld treatment;
- welding parameters: gas flow, amperage, voltage, etc.;
- inspection: visual, destructive, and non-destructive testing.

Where a manufacturer introduces a new process or a new configuration it will need to be proved by carrying out a Procedure Qualification Record (PQR). In this case, the joint configuration will match the typical production requirement and all the parameters will be monitored to meet the requirements of the standard. The weldment will then be subjected to mechanical testing of both plate and consumables, as well as some form of service duty cycle, to determine if it meets the standard and the manufacturer's requirements. It is common practice for this type of procedure to be witnessed by an independent third party, normally from an insurance or validating authority. Only when all parties are happy with the outcome will it become an established weld procedure sheet for that manufacturer.

INSPECTION

Inspection is an integral part of any manufacturing operation. It is a means of ensuring that only items made in accordance with the specification and any drawings are released for use. Inspections are made at various stages of the production process so that any imperfections are detected early to reduce wastage and reduce any repair costs. In the case of welded fabrication these inspections would take in to account the material specification, dimensional accuracy and surface quality. In respect of welding consumables, inspection looks at specification, cleanliness, storage and damage. Having checked the material specification, attention will be drawn to the edge preparation, fit-up of the joint configuration, and any pre-heat requirements and monitoring procedures, such as crayons, thermo-couples and pyrometers.

Calibration of equipment is an essential part of the process in order to comply with the standard. Manufacturers working to critical standards need to have records to indicate that this is occurring. When the weld is complete the inspector will check it visually for compliance before going to the more expensive methods of destructive and non-destructive testing.

We will now consider the defects that welding can produce.

COMPONENT FEATURES OF A WELD

In order to apply the various acceptance levels to the actual welded joint, it is vital to understand the different areas of the weld zone, edge preparation and joint configuration and the terminology used. This is also important when describing specific details or requirements of the weld or joint to an inspector, colleague or workmate.

A weld is defined by a range of features which to a large extent are measurable and therefore can be standardized. These features are illustrated below.

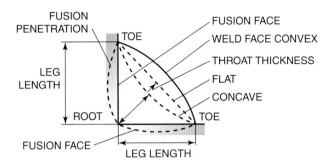

FIGURE 11.1B Features of a T-fillet weld

- *Leg length* is the distance from the intersection of the fusion faces (actual or projected) and the toe of a fillet weld, measured across the fusion face and should not be less than material thickness and is usually 1.5 x thickness of metal.
- *Toe* of a weld is the junction between the face of a weld and the parent metal.
- *Root* of a weld is the junction between the face of a weld and the parent metal.
- *Design throat thickness* is the minimum dimension of a given throat thickness for the purposes of design and strength.
- *Effective throat thickness* is the minimum distance from the root of a weld to its face less any reinforcement.
- *Fusion face* is the place where the fusion zone joins with the parent metal.
- *Fusion penetration* is the distance that fusion extends into the parent metal.
- *Throat thickness* is the shortest distance from the root to the centre of the weld face.
- *Weld face* is the total width across the face of the completed weld deposit.
- *Edge preparation* is the shaping of the edges of the metal to accommodate the required amount of weld metal, allowing for fusion and penetration.

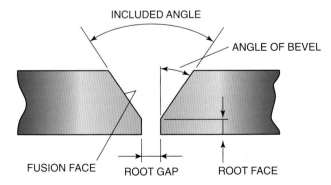

FIGURE 11.2 Edge preparation parameters for a single V-butt

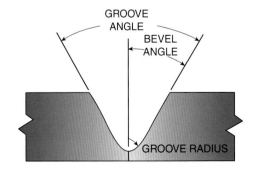

FIGURE 11.3 Edge preparation parameters for a single 'J' preparation

- *Root face* is the surface of the parent material that is filed or machined in order to act as a heat sink when depositing the root run or pass.
- *Root gap* is the gap between the parent materials to be joined in order to ensure root penetration.
- *Bevel angle* is the angle formed between the prepared edge of a member and a plane perpendicular (90°) to the surface of the member.
- *Included angle* is the total angle of the edge preparation.
- *Groove angle* is the total angle of the groove between parts to be joined in a 'J' butt weld.
- *Groove radius* is the radius used to form the shape of a 'J' or 'U' butt weld.
- *Root penetration* is the reinforcement of the weld at the side other than that from which welding was done.
- *Weld reinforcement* or excess weld metal, is the reinforcement of the weld at the side of the joint from which welding was done.
- *Heat Affected Zone (HAZ)* is the portion of the metal being welded or thermally cut that is metallurgically affected by the heat, but NOT melted.
- *Weld zone* is an area containing the weld metal and the heat-affected zone.
- *Concavity* is the maximum distance from the face of a concave fillet weld perpendicular to a line joining the toes.
- *Convexity* is the maximum distance from the face of a convex fillet weld perpendicular to a line joining the toes.
- *Incomplete fusion* is when fusion is less than complete.
- *Joint build-up sequence* is the order in which weld beads of a multiple-pass weld are deposited with respect to the cross section of the joint.
- *Complete fusion* is fusion that has occurred over the entire parent material surfaces intended for welding and between all layers and weld beads.
- *Depth of fusion* is the distance that fusion extends into the parent metal or previous pass from the surface melted during welding.
- *Capping run* is the reinforcement or final run of a welded joint. Depending on the size (width) of the weld and the type of finish required (concave/convex) and can be deposited by stringer or weaving bead action.
- *Axis of a weld* is an imaginary line drawn through a weld parallel to the root.
- *As-welded* is the condition of weld metal, welded joints, and weldments after welding but prior to any subsequent thermal, mechanical, or chemical treatments.
- *Autogenous weld* is a fusion weld made without the addition of filler metal.

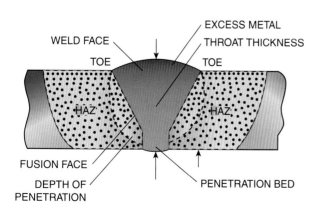

FIGURE 11.4 **Cross-section of the weld zone and parent metal**

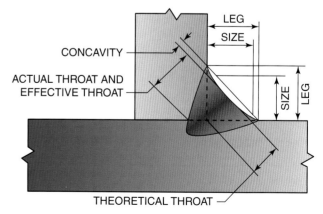

FIGURE 11.5 **Weld features of a T-fillet weld**

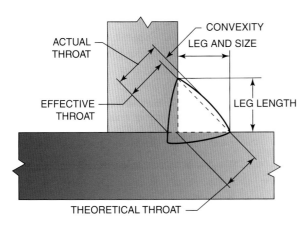

FIGURE 11.6 **Throat thickness values on T-fillet joint**

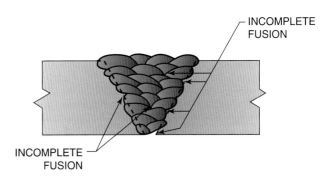

FIGURE 11.7 **Potential locations of incomplete fusion**

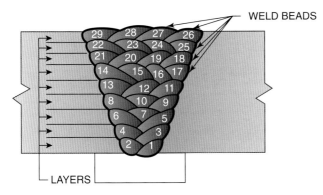

FIGURE 11.8 **Multiple weld sequence**

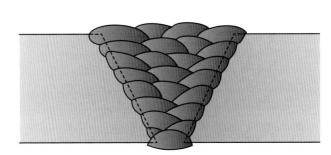

FIGURE 11.9 **Full side fusion into parent metal in multiple weld deposition**

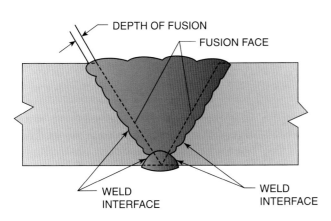

FIGURE 11.10 **Cross section of correct weld deposition**

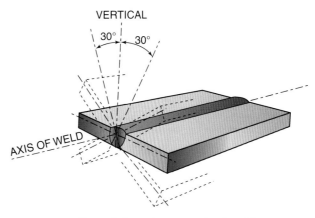

FIGURE 11.11 **Effect of distortion on axis of weld**

IMPERFECTIONS AND DEFECTS

Imperfections and flaws are interruptions in the typical structure of a weld. They may be a lack of uniformity in the mechanical, metallurgical, or physical characteristics of the material or weld. All welds have imperfections and flaws, but not all of these are necessarily defects.

A defect is an imperfection that by nature or accumulated effect renders a part or product unable to meet minimum applicable acceptance standards or specifications. The term designates rejectability.

Many acceptable products may have welds that contain imperfections, but no products may have welds that contain defects. Nevertheless, an imperfection may become so large, or there may be so many small imperfections, that the weld is not acceptable under the standards for the code for that product. Some codes are stricter than others, so the same weld might be acceptable under one code but not under another.

When a weld is evaluated, it is important to note the type of any imperfection, and its size and location. Any one or more of these factors, depending on the applicable code or standard, can change an imperfection to a defect.

The most common imperfections are:

- porosity;
- inclusions;
- inadequate joint penetration;
- incomplete fusion;
- arc strikes;
- overlap (cold lap);
- undercut;
- cracks;
- underfill;
- laminations;
- delaminations;
- lamellar tears.

Porosity

Porosity results from gas that was dissolved in the molten weld pool and was trapped as the metal cools and becomes solid. Porosity forms either spherical (ball-shaped) or cylindrical (tube- or tunnel-shaped) bubbles. Cylindrical porosity is called 'wormhole porosity'. The rounded edges tend to reduce the stresses around them; therefore, unless porosity is extensive, there is little or no loss in strength. Porosity is most often caused by improper welding techniques, contamination or an incorrect chemical balance between the filler and base metals.

Improper welding techniques may result in shielding gas not adequately protecting the molten weld pool. For example, the E7018 electrode should not be weaved wider than two and a half times the electrode diameter, because very little shielding gas is produced. If the electrode is weaved wider, parts of the weld are unprotected. Nitrogen from the air that dissolves in the weld pool can produce porosity.

The intense heat of the weld can decompose paint, dirt, or oil from machining and rust or other oxides, producing hydrogen. Hydrogen, like nitrogen, can become trapped in the solidifying weld pool, producing porosity. It can also diffuse into the heat-affected zone and produce under-bead cracking in some steels. Gas porosity can be grouped into four major types:

- Uniformly scattered porosity is most frequently caused by poor welding techniques or faulty materials, see Figure 11.12.
- Clustered porosity is most often caused by improper starting and stopping techniques, see Figure 11.13.
- Linear porosity is most frequently caused by contamination within the joint, root or interbead boundaries, see Figure 11.14.
- Piping, or wormhole, porosity is most often caused by contamination at the root, see Figure 11.15. This type of porosity forms when the gas escapes from the weld pool at the same rate as the pool is solidifying.

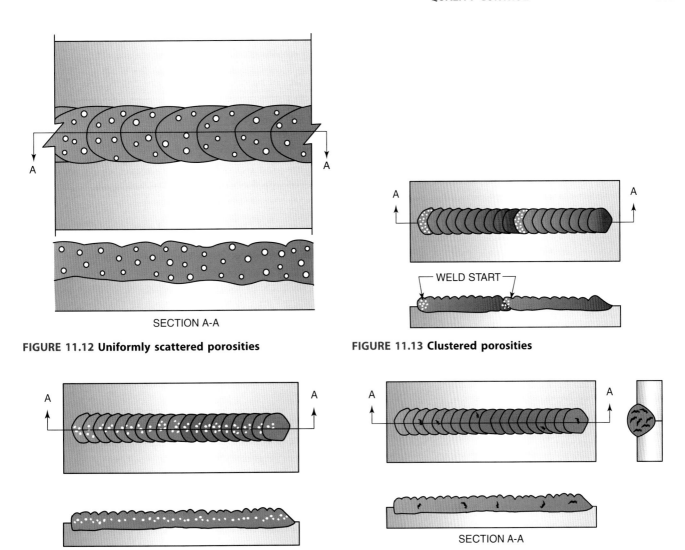

FIGURE 11.12 **Uniformly scattered porosities**

FIGURE 11.13 **Clustered porosities**

FIGURE 11.14 **Linear porosity**

FIGURE 11.15 **Piping or wormhole porosity**

Inclusions

Inclusions are non-metallic materials, such as slag and oxides, that are trapped in the weld metal, between weld beads, or between the weld and the base metal. Inclusions may be jagged and irregularly shaped. They may also form in a continuous line, causing stresses to concentrate and reducing the structural integrity of the weld.

Although not visible, inclusions can be expected if prior welds were improperly cleaned or had a poor contour. Unless care is taken in reading radiographs, the presence of slag inclusions can be misinterpreted as other defects. (Linear slag inclusions in radiographs generally contain shadow details; otherwise, they could be interpreted as lack-of-fusion defects.) These inclusions result from poor manipulation that allows the slag to flow ahead of the arc, from not removing all the slag from previous welds, or from welding highly crowned, incompletely fused welds.

Scattered inclusions can resemble porosity although they are usually not spherical. These inclusions can also result from inadequate removal of earlier slag deposits and poor manipulation of the arc. Heavy mill scale or rust can be a source of inclusions. They can also result from unfused pieces of damaged electrode coatings falling into the weld. In radiographs some detail will appear, unlike in the case of linear slag inclusions

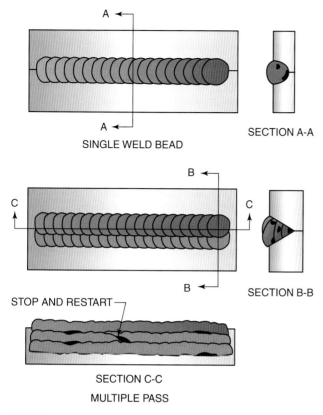

SINGLE WELD BEAD

SECTION A-A

STOP AND RESTART

SECTION C-C

MULTIPLE PASS

SECTION B-B

FIGURE 11.16 Non-metallic inclusions

Non-metallic inclusions, Figure 11.16, are caused under the following conditions:

- when slag or oxides do not have enough time to float to the surface of the molten weld pool;
- when there are sharp notches between weld beads or between the weld bead and the base metal that trap the material and prevent it floating out;
- when the joint was designed with insufficient room for the correct manipulation of the molten weld pool.

Inadequate Joint Penetration

Inadequate joint penetration occurs when the depth to which the weld penetrates the joint is less than that needed to fuse through the plate or into the preceding weld, Figure 11.17. The cross-sectional area of weld penetration in the joint may be below the minimum required depth or could become a source of stress concentration leading to fatigue failure. The criticality of such defects depends on the notch sensitivity of the metal and the factor of safety to which the weldment has been designed.

The major causes of inadequate joint penetration are:

- *Improper welding technique.* The most common cause is a misdirected arc. It may also result from a failure to use starting and run-out tabs, or a failure to back-gouge the root sufficiently, Figure 11.18.

- *Insufficient welding current.* Metals that are thick or have a high thermal conductivity are often preheated so that the weld heat is not drawn away so quickly by the metal that it cannot penetrate the joint.

- *Improper joint fit-up.* This problem results when the weld joints are not prepared or fitted accurately. Too small a root gap or too large a root face will keep the weld from penetrating adequately.

- *Improper joint design.* When joints are accessible from both sides, back-gouging is often used to ensure 100 per cent root fusion.

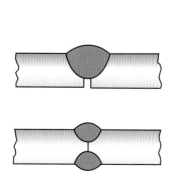

FIGURE 11.17 Inadequate joint penetration

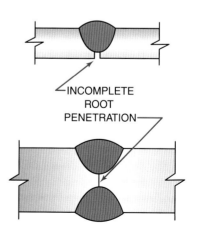

INCOMPLETE ROOT PENETRATION

FIGURE 11.18 Incomplete root penetration

Incomplete Fusion

This is the lack of fusion between the molten filler metal and previously deposited filler metal or the base metal, Figure 11.19. A lack of fusion between the filler metal and previously deposited weld metal is called 'interpass cold lap'. A lack of fusion between the weld metal and the joint face is called 'lack of sidewall fusion'. Both problems usually travel along all or most of the weld's length.

Incomplete fusion can be found in welds produced by all major welding processes. Some major causes are:

- *Inadequate agitation.* Lack of weld agitation to break up oxide layers, leaving a thin layer of oxide which prevents fusion from occurring.

- *Improper welding techniques.* Poor manipulation, such as moving too fast or using an improper electrode angle.

- *Wrong welding process.* Such as the use of dip transfer with MIG to weld plate thicker than 6mm, which may provide inadequate limited heat input to the weld.

- *Improper edge preparation.* Not removing any notches or gouges in the edge of the weld joint, such as those present on a flame-cut plate, Figure 11.20.

- *Improper joint design.* Where insufficient heat is applied to melt the base metal or the joint designer has allowed too little space for correct molten weld pool manipulation.

- *Improper joint cleaning.* Such as failure to remove oxides resulting from the use of an oxy-fuel torch, or failure to remove slag from a previous weld.

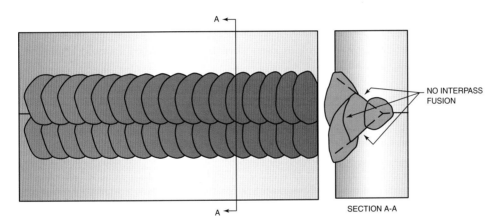

FIGURE 11.19 Incomplete fusion

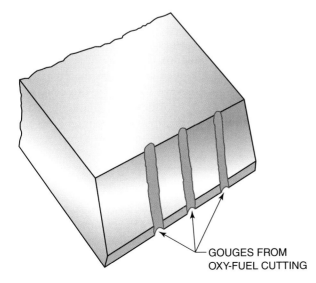

FIGURE 11.20 Gouge removal

FIGURE 11.21 Arc strikes

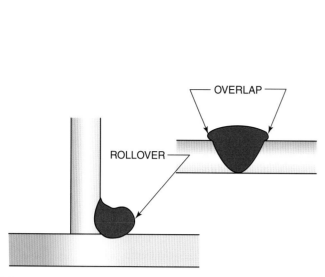

FIGURE 11.22 **Rollover or overlap**

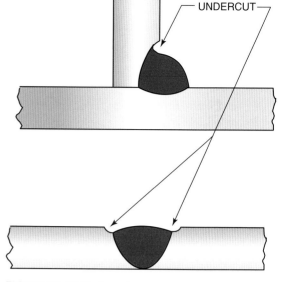

FIGURE 11.23 **Undercut**

Arc Strikes

Figure 11.21 shows arc strikes – small, localized points where surface melting occurs away from the joint. These spots may be caused by accidentally striking the arc in the wrong place or by faulty earth connections. Even though arc strikes can be ground smooth, they cannot be removed. These spots will always show if an acid etch is used. They also can be localized hardness zones or the starting point for cracking. Arc strikes, even when ground flush for a guided bend, will open up to form small cracks or holes.

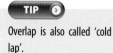

Overlap is also called 'cold lap'.

Overlap

Overlap occurs in fusion welds when weld deposits are larger than the joint is conditioned to accept. The weld metal flows over the surface of the base metal without fusing to it, along the toe of the weld bead, Figure 11.22. Overlap generally occurs on the horizontal leg of a horizontal fillet weld under extreme conditions. It can also occur on both sides of flat-positioned capping passes. With MIG welding, overlap occurs when too much electrode extension is used to deposit metal at low power. Misdirecting the arc into the vertical leg and keeping the electrode nearly vertical will also cause overlap. To prevent overlap, the fillet weld must be correctly sized to less than 10mm, and the arc must be properly manipulated.

Undercut

Undercut is the result of the arc force removing metal from a joint face that is not replaced by weld metal, along the toe of the weld bead, Figure 11.23. It can result from excessive current. It is a common problem with MIG welding when insufficient oxygen is used to stabilize the arc. Incorrect welding technique, such as incorrect electrode angle or excessive weave, can also cause undercut. To prevent undercutting, the welder can weld in the flat position using multiple passes instead of a single pass, change the shield gas and improve manipulative techniques to fill the removed base metal along the toe of the weld bead.

Crater Cracks

Crater cracks are the tiny cracks that develop in weld craters as the weld pool shrinks and solidifies, Figure 11.24. Materials with a low melting temperature gather at the crater centre during freezing and are pulled apart as the weld metal shrinks, causing cracks to form. Crater cracks can be minimized, if not prevented, by not interrupting the arc quickly at the end of a weld. This allows the crater to fill and cool more slowly. Some TIG equipment has a crater-filling control that automatically reduces the current at the end of a weld. For all other welding processes, the most effective way of preventing crater cracking is to pause and rotate the electrode or filler wire, allowing it to build up the weld bead before breaking the arc, Figure 11.25.

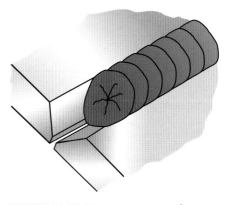

FIGURE 11.24 **Crater or star cracks**

Underfill

Underfill on a butt weld occurs when the amount of weld metal deposited is insufficient to bring the weld's surfaces to a level equal to that of the original plane or plate surface. For a fillet weld, it occurs when the weld deposit has an insufficient effective throat, Figure 11.26. This problem can usually be corrected by slowing the travel rate or making more weld passes.

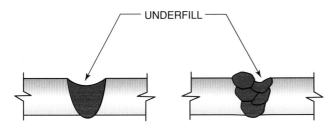

FIGURE 11.26 **Underfill**

Plate-Generated Problems

Not all welding problems are caused by the weld metal, the process, or the welder's lack of skill. Sometimes the material being fabricated can be at fault. Some problems result from internal plate defects that the welder cannot control. Others are the result of improper welding procedures that produce undesirable hard metallurgical structures in the heat-affected zone. The internal defects are the result of poor steelmaking practices. Despite the efforts of steel producers, mistakes do occur in steel production and are too frequently blamed on the welding operation.

Lamination

Located toward the centre of the plate, Figure 11.27, laminations are caused by insufficient cropping (removal of defects) of the ingots. Slag and oxidized steel are rolled out with the steel, producing the lamination. They can also occur when the ingot is rolled at too low a temperature or pressure. Laminations differ from lamellar tearing (see page 360) in that they are more extensive and involve thicker layers of non-metallic contaminants.

Delamination

When laminations intersect a joint being welded, the heat and stresses of the weld may cause some laminations to become delaminated, releasing contamination into the weld metal and leading to wormhole porosity or lack-of-fusion defects.

It is not easy to correct the problems associated with delamination. If a thick plate is installed in a compression load, it may be possible to weld over the delamination to seal it. A better solution is to replace the steel.

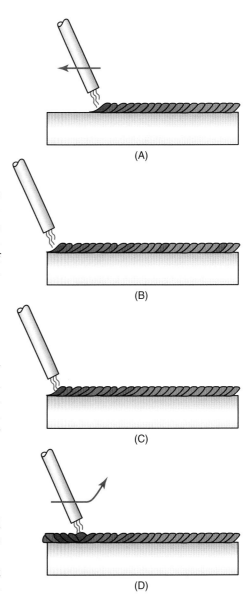

(A)

(B)

(C)

(D)

FIGURE 11.25 **Preventing crater cracking**

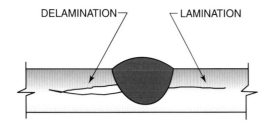

FIGURE 11.27 Lamination and delamination

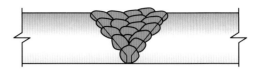

FIGURE 11.29 Using multiple welds to reduce weld stresses

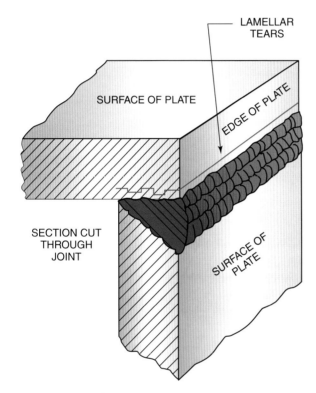

FIGURE 11.28 Lamellar tearing

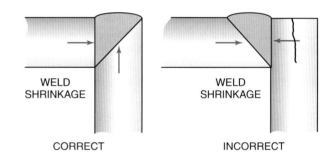

FIGURE 11.30 Correct joint design to reduce lamellar tears

Lamellar Tears

Lamellar tears appear as cracks parallel to and under the steel surface. In general, they are not in the heat-affected zone, and they have a step-like configuration. They result from thin layers of non-metallic inclusions that lie beneath the plate surface. Although barely noticeable, these inclusions separate when severely stressed, producing laminated cracks. These cracks are evident if the plate edges are exposed, Figure 11.28.

A solution to the problem is to redesign the joints to impose the lowest possible strain throughout the plate thickness. This can be accomplished by making smaller welds so that each subsequent weld pass heat-treats the previous pass, reducing the total stress in the finished weld, Figure 11.29. The joint design can be changed to reduce the stress on the weld through thickness of the plate, Figure 11.30.

DESTRUCTIVE TESTING

Destructive testing is used to determine actual values in weld metal and base metals in order to ensure that specific performance characteristics are obtained. These tests are used to establish various mechanical properties, such as tensile strength, ductility or hardness. Some tests provide values for several properties, but most are designed to determine the value for a specific characteristic of a metal.

Tensile Testing

Tensile tests are performed with specimens prepared as round bars or flat strips. Round bars are often used for testing only the weld metal, sometimes called 'all weld metal testing.' Round specimens are cut from the centre of the weld metal. This form of testing can be used on thick sections where base metal dilution into all of the weld metal is not possible. Flat bars are often used to test both the weld and the surrounding metal. They are usually cut at a 90° angle to the weld, Figure 11.31. Table 11.1 shows how a number of standard smaller-size bars can be used, depending on the thickness of the metal to be tested. Bar size also depends on the size of the tensile testing equipment available, Figure 11.32.

Two flat specimens are commonly used for testing thinner sections of metal. When welds are tested, the specimen should include the heat-affected zone and the base plate. If the weld metal is stronger than the plate, failure occurs in the plate; if the weld metal is weaker, failure occurs in the weld.

After the weld section is machined to the specified dimensions, it is placed in the tensile testing machine and pulled apart. A specimen used to determine the strength of a welded butt joint for plate is shown in Figure 11.33.

The tensile strength is obtained by dividing the maximum load required to break the specimen by the original cross-sectional area of the specimen at the middle.

The elongation is found by fitting the fractured ends of the specimen together, measuring the distance between gauge marks, and subtracting the gauge length. The percentage of elongation is found by dividing the elongation by the gauge length and multiplying by 100.

Fatigue Testing

Fatigue testing is used to determine how well a weld can resist repeated fluctuating stresses or cyclic loading. The maximum value of the stresses is less than the tensile strength of the material. Fatigue strength can be lowered by improperly made weld deposits, which may be caused by porosity, slag inclusions, lack of penetration or cracks, all of which can act as a point of stress, eventually resulting in the failure of the weld.

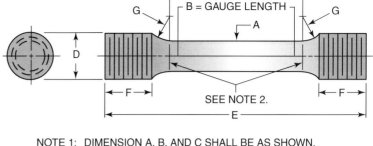

NOTE 1: DIMENSION A, B, AND C SHALL BE AS SHOWN, BUT ALTERNATE SHAPES OF ENDS MAY BE USED AS ALLOWED BY ASTM SPECIFICATION E-8.

NOTE 2: IT IS DESIRABLE TO HAVE THE DIAMETER OF THE SPECIMEN WITHIN THE GAUGE LENGTH SLIGHTLY SMALLER AT THE CENTRE THAN AT THE ENDS. THE DIFFERENCE SHALL NOT EXCEED 1% OF THE DIAMETER.

FIGURE 11.31 Tensile testing specimen

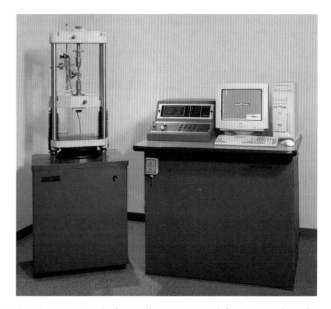

FIGURE 11.32 Typical tensile tester used for measuring the strength of welds (60,000-lb universal testing machine)

TABLE 11.1 Dimensions of tensile testing specimens

Specimen	Dimensions of Specimen						
	mm A	mm B	mm C	mm D	mm E	mm F	mm G
C-1	12.7	50.8	57.1	19.05	107.9	19.05	9.52
C-2	11.09	44.4	50.8	15.8	101.6	19.05	9.52
C-3	9.06	35.5	44.4	12.7	88.9	15.8	9.52
C-4	6.40	25.4	31.7	9.52	63.5	12.7	3.17
C-5	3.2	12.7	19.05	6.35	44.4	9.52	3.17

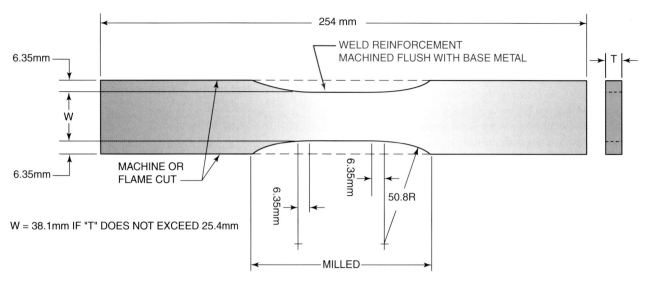

FIGURE 11.33 Tensile specimen for flat plate weld

In the fatigue test, the part is subjected to repeated changes in applied stress. Fatigue testing can be performed in several ways, depending upon the type of service the tested part must withstand. The results are usually reported as the number of stress cycles that the part will resist without failure and the total stress used.

In one type of test, the specimen is bent back and forth in a fatigue testing machine, Figure 11.34. This subjects the part to alternating compression and tension. As the machine rotates, the specimen is alternately bent twice for each revolution. Failure is usually rapid.

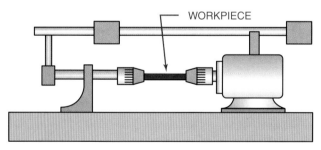

FIGURE 11.34 Fatigue testing

Shear Strength of Welds

The two forms of shear strength of welds are transverse shearing strength and longitudinal shearing strength. To test transverse shearing strength, a specimen is prepared as shown in Figure 11.35. The width of the specimen is measured in millimetres. A tensile load is applied, and the specimen is ruptured. The maximum load in kilograms is then determined.

The shearing strength of the weld, kilograms per linear millimetre, is obtained by dividing the maximum force by twice the width of the specimen:

To test longitudinal shearing strength, a specimen is prepared as shown in Figure 11.36. The length of each weld is measured in millimetres. The specimen is then ruptured under a tensile load, and the maximum force in kilograms is determined.

The shearing strength of the weld, in kilograms per linear millimetre, is obtained by dividing the maximum force by the sum of the length of welds that ruptured:

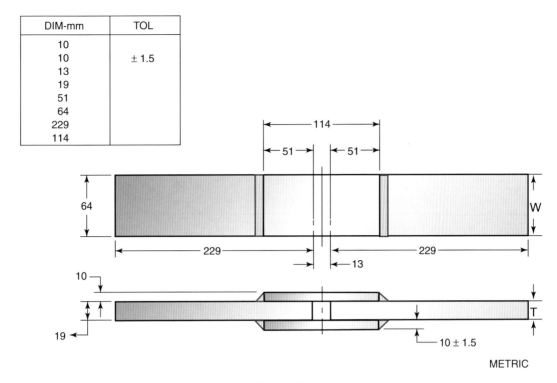

DIM-mm	TOL
10	
10	
13	± 1.5
19	
51	
64	
229	
114	

FIGURE 11.35 Transverse fillet weld shearing specimen after welding

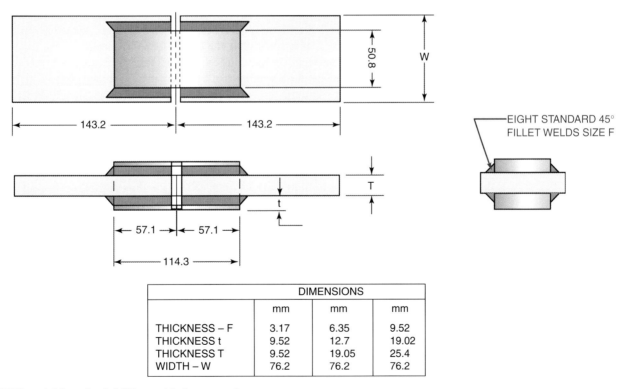

DIMENSIONS			
	mm	mm	mm
THICKNESS – F	3.17	6.35	9.52
THICKNESS t	9.52	12.7	19.02
THICKNESS T	9.52	19.05	25.4
WIDTH – W	76.2	76.2	76.2

FIGURE 11.36 Longitudal fillet weld shear specimen

Hardness Testing

Hardness is the resistance of metal to penetration and is an index of the wear resistance and strength of the metal. Hardness tests can be used to determine the relative hardness of the weld and the base metal. The two hardness testing machines in common use are the Rockwell and the Brinell.

FIGURE 11.37 Rockwell hardness tester

FIGURE 11.38 Brinell hardness tester

The Rockwell hardness tester, Figure 11.37, uses a 120° diamond cone for hard metals and a hardened steel ball of 2mm or 3mm diameter for softer metals. The method is based on measuring resistance to penetration. The depth of the impression is measured rather than the diameter. The hardness is read directly from a dial on the tester. The tester has two scales for reading hardness, known as the *B-scale* and the *C-scale*. The C-scale is used for harder metals, the B-scale for softer metals.

The Brinell hardness tester measures the resistance of material to the penetration of a steel ball under constant pressure (about 3000kg) for a minimum of approximately 30 seconds, Figure 11.38. The diameter is measured microscopically, and the Brinell number is checked on a standard chart. Brinell hardness numbers are obtained by dividing the applied load by the area of the surface indentation.

Impact Testing

A number of tests can be used to determine the impact-withstanding capability of a weld. One common test is the Izod test, Figure 11.39A, in which a notched specimen is struck by an anvil mounted on a pendulum. The energy required to break the specimen (in kg read from a scale mounted on the machine) indicates the impact resistance of the metal. The toughness of the weld metal is compared with that of the base metal.

Another similar type of impact test is the Charpy test. Whereas the Izod test specimen is gripped on one end, held vertically, and usually tested at room temperature, the Charpy test specimen is held horizontally, supported on both ends, and usually tested at a specific temperature.

A typical impact tester is shown in Figure 11.39B.

Welded Butt Joints

The three methods of testing welded butt joints are:

- the nick-break test;
- the guided bend test, and;
- the free bend test.

It is possible to use variations of these tests.

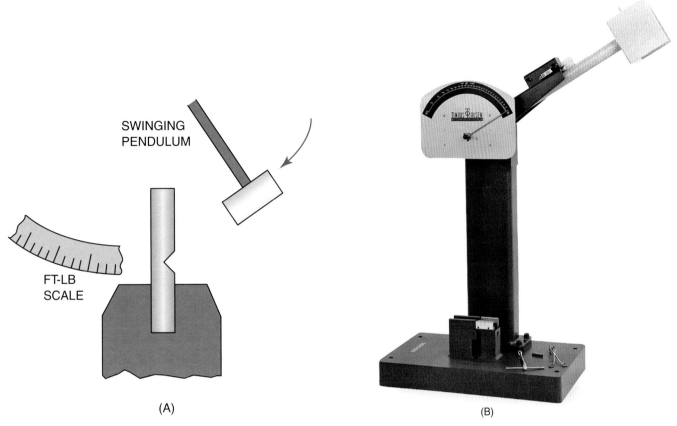

FIGURE 11.39A **Impact testing**

FIGURE 11.39B **Impact testing machine**

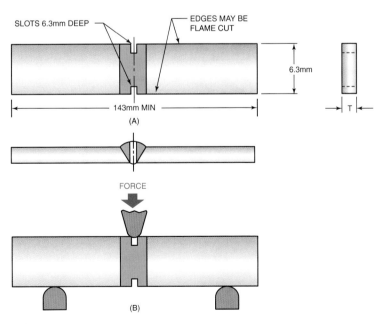

FIGURE 11.40 **Nick-break specimens**

Nick-Break Test

A specimen for this test is prepared as shown in Figure 11.40A and is supported as shown in Figure 11.40B. A force is applied, either slowly or suddenly, and the specimen is ruptured by one or more blows of a hammer. Theoretically, the rate of application can affect how the specimen breaks, especially at a critical temperature. Generally, however, there is no difference in the appearance of the fractured surface as a result of the method of applying the force. The surfaces of the fracture are checked to determine the soundness of the weld.

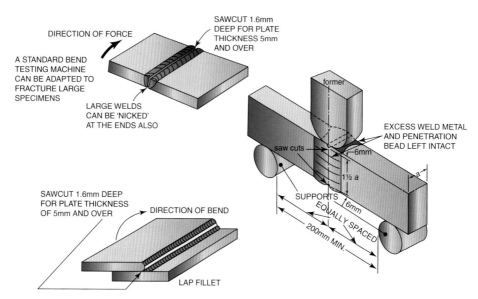

FIGURE 11.41 **Nick-break test preparation**

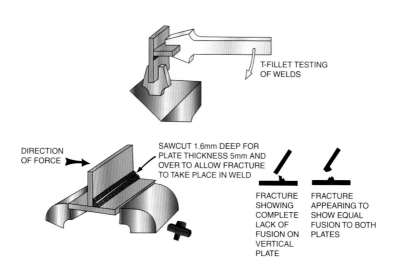

FIGURE 11.42 **T-fillet fracture testing**

Guided Bend Test

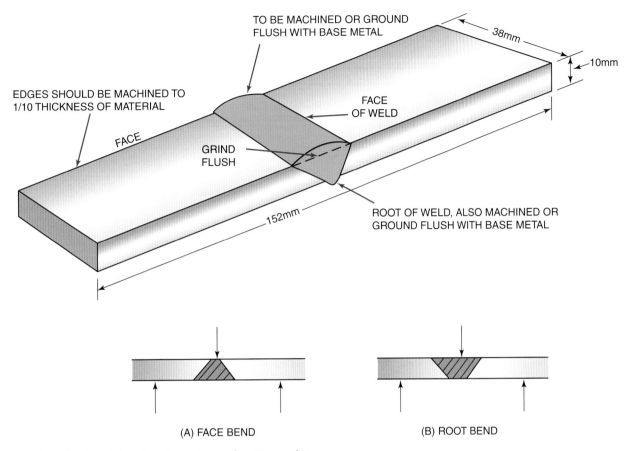

FIGURE 11.43 Root and face bend specimens for 10mm plate

To test welded, grooved butt joints on metal that is 10mm thick or less, two specimens are prepared and tested – one face bend and one root bend, Figures 11.43A and 11.43B. If the welds pass this test, the welder is qualified to make butt welds on the same plate between 5-20mm thick. These welds need to be machined as shown in Figure 11.44A. If these specimens pass, the welder will also be qualified to make fillet welds on materials of any (unlimited) thickness. For welded, V- butt joints on metal 13mm thick, two side bend specimens are prepared and tested, Figure 11.44B. If the welds pass this test, the welder is qualified to weld on metals of designated thickness.

When the specimens are prepared, care must be taken to ensure that all grinding marks run longitudinally so that they do not cause stress cracking. The edges must be rounded to reduce cracking that can radiate from sharp edges. The maximum radius of this rounded edge is 3mm.

The type of jig shown in Figure 11.45 is used to bend most specimens. Not all guided bend testers have the same bending radius. Codes specify different bending radii depending on material type and thickness, but a common radius formula used is 4 x thickness of plate. Place the specimens in the jig with the weld in the middle. Face bend specimens should be placed with the face of the weld downwards. Root bend specimens should be positioned so that the root of the weld is facing downwards. Side bend specimens are placed with either side facing up. The bend specimen must be pushed all the way through open (roller-type) bend testers and within 3mm of the bottom on fixture-type bend testers.

The specimen is removed and the convex surface is examined for cracks or other imperfections and judged according to specified criteria. Some surface cracks and openings are allowable under codes.

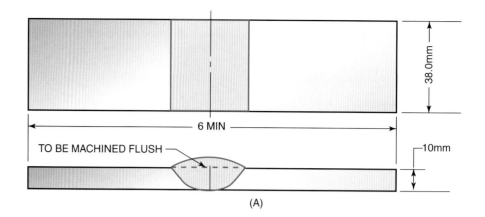

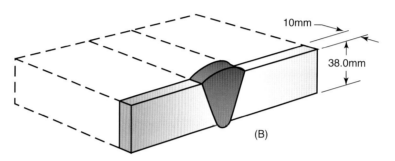

FIGURE 11.44 **Root, face and side bend specimens**

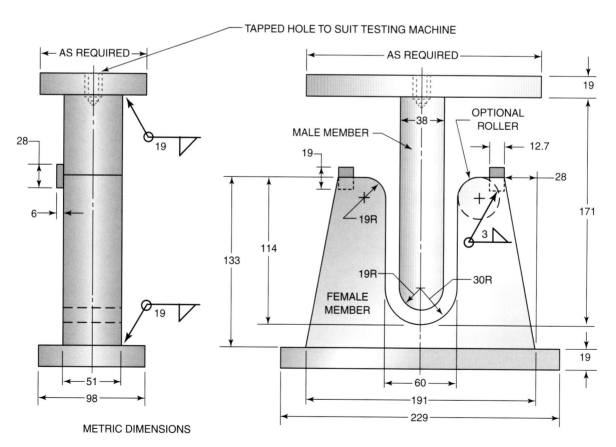

METRIC DIMENSIONS

FIGURE 11.45 **Fixture for guided bend test**

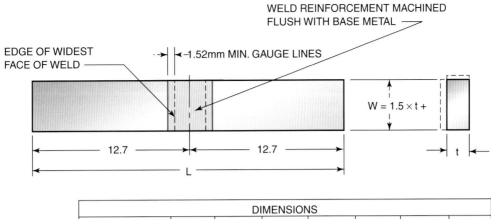

WELD REINFORCEMENT MACHINED
FLUSH WITH BASE METAL

EDGE OF WIDEST
FACE OF WELD

1.52mm MIN. GAUGE LINES

$W = 1.5 \times t +$

12.7   12.7

L   t

DIMENSIONS							
T, mm –	6.35	9.52	12.7	15.8	19.05	25.4	31.7
W, mm –	9.52	14.2	19.05	23.8	28.5	38.1	47.6
L, min, mm –	152.4	203.2	228.2	254	304.8	342.9	381
B, min, mm –	31.7	31.7	31.7	50.8	50.8	50.8	50.8

FIGURE 11.46 **Free bend test specimen**

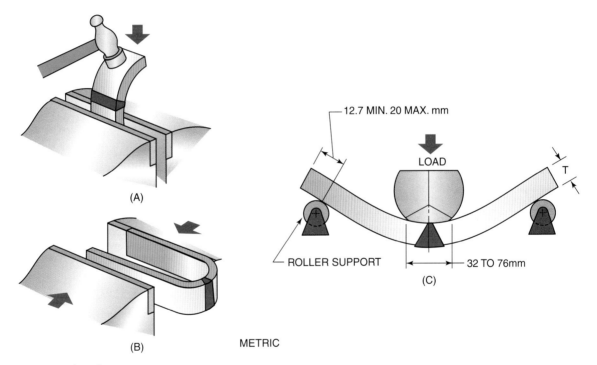

(A)

12.7 MIN. 20 MAX. mm

LOAD

T

ROLLER SUPPORT

32 TO 76mm

(C)

(B)   METRIC

FIGURE 11.47 **Free bend test**

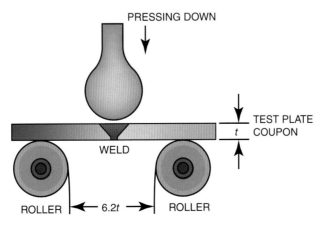

PRESSING DOWN

TEST PLATE
COUPON

t

WELD

ROLLER   6.2t ROLLER

FIGURE 11.48A **Typical setup for bend testing**

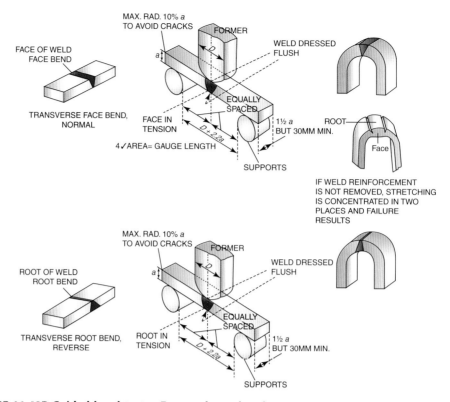

FIGURE 11.48B Guided bend tests - Face and root bend tests

Free Bend Test

The free bend test is used to test welded joints in plate. A specimen is prepared as shown in Figure 11.46. The width of the specimen is 1.5 multiplied by its thickness. Each corner lengthwise should be rounded in a radius not exceeding one-tenth the thickness of the specimen. Tool marks should run the length of the specimen.

Gauge lines are drawn on the face of the weld. The distance between the gauge lines is 3mm less than the face of the weld. The initial bend of the specimen is completed in the device illustrated in Figure 11.47. The gauge line surface should be directed toward the supports. The weld is located in the centre of the supports and loading block.

Alternate Bend

The initial bend may be made by placing the specimen in the jaws of a vice with one-third the length projecting from the jaws. The specimen is bent away from the gauge lines through an angle of 30° to 45° using a hammer. The specimen is then inserted into the jaws of a vice, and pressure is applied by tightening the vice. The pressure is continued until a crack or depression appears on the convex face of the specimen. The load is then removed.

The elongation is determined by measuring the minimum distance between the gauge lines along the convex surface of the weld to the nearest 0.25mm and subtracting the initial gauge

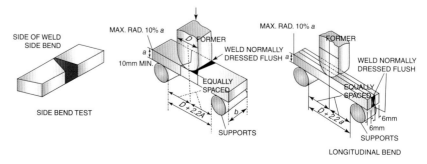

FIGURE 11.48C Side bend test for material above 10mm

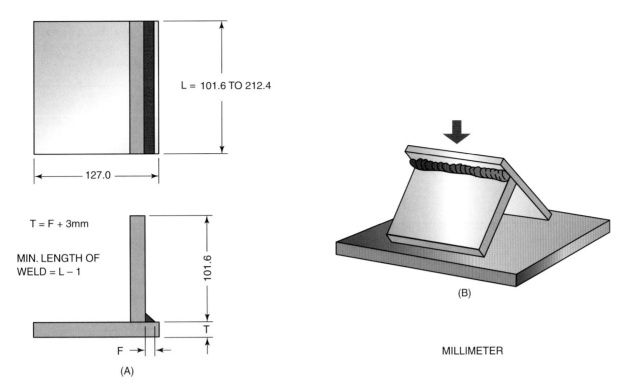

$L = 101.6$ TO 212.4

127.0

$T = F + 3mm$

MIN. LENGTH OF
WELD $= L - 1$

101.6

T

F

(A)

(B)

MILLIMETER

FIGURE 11.49 Fillet weld testing

length. The percentage of elongation is obtained by dividing the elongation by the initial gauge length and multiplying by 100.

Fillet Weld Break Test

The specimen for this test is made as shown in Figure 11.49A. In Figure 11.49B, a force is applied to the specimen until the rupture breaking of the specimen occurs. Any convenient means of applying the force may be used, such as an arbor press, a testing machine or hammer blows. The break surface should then be examined for soundness – i.e., slag inclusions, overlap, porosity, lack of fusion or other imperfections.

Macro-Examination

'Macro' comes from the Greek word meaning 'large', and macro-examination means the examination of specimens either with the naked eye or at low magnification, using a magnifying glass (up to five magnifications) or a microscope. Macro-examination is important when studying fracture faces, or welded or brazed joints. It is also useful in the detection of surface corrosion or large scale defects such as inclusions or porosity in castings.

Preparation

The section is normally cut by hand or with a suitable cut-off machine, taking care that the cutting process does not overheat the material. If the section is small enough it may be possible to mount it in an acrylic type material before it is polished. The specimen is first filed with successive grades of cut to produce a relatively smooth surface free of lines, it is then polished with various grades of wet and dry paper. Care should be exercised here when transferring from one grade of wet and dry to another, that the abrasive from the coarser paper does not impregnate the finer grade. This can be overcome by passing the specimen under a water jet prior to transfer. The specimen is then examined in this un-etched condition, and any significant features noted. Etchant can be applied at this stage. However, more detail will be revealed if the specimen is further polished on a selvyt wheel impregnated with alumina (aluminium oxide) and using a diamond lapping spray to achieve a highly polished finish. Any residue of lapping spray or from the wheel is removed by passing the specimen under running water followed by hot air drying. The specimen is now ready for etching.

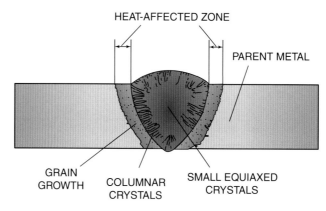

FIGURE 11.50 Metallic-arc weld in mild steel

It is essential at this stage that the specimen is handled carefully. At no time should your fingers touch the inspection face as this can affect the results (salt from your finger tips). The etchant is then applied to the surface of the section in the form of swabbing, cotton wool dipped in etchant applied by tongs as shown in the diagram.

Since most etching reagents are either acids or alkalis, care must be taken in their use. In the case of steel 'Nital' is the most common etchant reagent and is made up from 2–15 per cent (2ml–15ml) concentrated nitric acid in 100ml alcohol. The etchant shows up certain structural features such as some grain formations, large inclusions or porosity, by attacking one form of surface material more deeply than another. The actual etching time may vary from a few seconds to several hours in the case of large forgings. After being etched, the section is swilled with water, and then swabbed with methyl alcohol to promote rapid drying.

On examination the heat-affected zone reveals how much of the parent material was affected by the welding heat. This will vary from process to process even when welding the same material, MMA would have a smaller HAZ than that produced by oxy-acetylene.

If we examine a weld made by using MMA, we would be able to see the number of runs, and if the correct amount of metal has been deposited by each run. While equi-axed grains are too small to be revealed, columnar crystals are distinctly visible at this low magnification. The refinement of runs which have been previously deposited in a multi-run weld is clearly indicated. Besides the grain size there is a lot of other information available and any non-metallic inclusions (such as slag) are distinctly revealed. Porosity is shown in the form of blowholes which appear round or oval, and indicate that gas was trapped during solidification. On multi-run welds, lack of fusion to the sides or parent metal may occur and this would show up as a distinct crack on an otherwise good looking weld, which could lead to failure of the weld.

Any cracks or laminations of any sort will be revealed clearly.

Hydrochloric Acid

Equal parts by volume of concentrated hydrochloric (muriatic) acid and water are mixed by adding the acid into the water. The welds are immersed in the reagent at or near the boiling temperature. The acid usually enlarges gas pockets and dissolves slag inclusions, enlarging the resulting cavities.

Ammonium Persulfate

A solution is prepared consisting of one part of ammonium persulfate (solid) to nine parts of water by weight. The surface of the weld is rubbed with wool saturated with this reagent at room temperature.

Nitric Acid

One part of concentrated nitric acid is mixed with nine parts of water or alcohol by volume. The reagent is applied to the surface of the weld with a glass stirring rod at room temperature. Nitric acid has the capacity to etch rapidly and should be used on polished surfaces only.

TIP

The most commonly used etching solutions are hydrochloric acid, ammonium persulfate and nitric acid.

HEALTH & SAFETY

When diluting an acid, always pour the acid slowly into the water while continuously stirring the water. Carelessly handling this material or pouring water into the acid can result in burns, excessive fuming or explosion. Be sure to wear safety glasses and gloves to prevent injuries.

After etching, the weld is rinsed in clear, hot water. Excess water is removed, and the etched surface is then immersed in ethyl alcohol and dried.

NON-DESTRUCTIVE TESTING (NDT)

Non-destructive testing is a method used to test welds for surface defects such as cracks, arc strikes, undercuts and lack of penetration. Internal or subsurface defects can include slag inclusions, porosity and un-fused metal in the interior of the weld.

Visual Inspection

Visual inspection is the most frequently used non-destructive testing method and is the first step in almost every other inspection process. The majority of welds receive only visual inspection. If the weld looks good, it passes; if it looks bad, it is rejected. This procedure is often mistakenly overlooked when more sophisticated non-destructive testing methods are used.

An active visual inspection schedule can reduce the finished weld rejection rate by more than 75 per cent. Visual inspection can easily be used to check for fit-up, inter-pass acceptance, welder technique and other variables that will affect the weld quality.

Visual inspection should be used before any other non-destructive or mechanical tests are used to eliminate (reject) the obvious problem welds. Eliminating welds that have excessive surface defects that will not pass the code or standard being used saves preparation time.

TIP

Minor problems can be identified and corrected before a weld is completed. This eliminates costly repairs or rejection.

Penetrant Inspection

Penetrant inspection is used to locate minute surface cracks and porosity. Two types of penetrants are now in use: colour-contrast and fluorescent versions. Colour-contrast penetrants contain a coloured (often red) dye that shows under ordinary white light. Fluorescent penetrants contain a more effective fluorescent dye that shows under ultraviolet light.

The following steps should be followed when a penetrant is used:

1 The first step is pre-cleaning. Suspected flaws are cleaned and dried so that they are free of oil, water, or other contaminants.

2 The test surface is covered with a film of penetrant by dipping, immersing, spraying, or brushing which draws the penetrant into the flaw by capilliary attraction.

3 After a designated time period has elapsed, the test surface is then gently wiped, washed, or rinsed free of excess penetrant. It is dried with cloths or hot air.

4 A developing powder (usually white) is applied to the test surface to acts as a blotter to speed the process by which the penetrant seeps out of any flaws by reverse capilliary attraction in to the developer.

5 Depending upon the type of penetrant applied, visual inspection is made under ordinary white light or ultraviolet light, Figure 11.45. In the latter case, the penetrant fluoresces a yellow-green colour, which clearly defines the defect.

Magnetic Particle Inspection

Magnetic particle inspection uses fine ferromagnetic particles (powder) to indicate defects open to the surface or just below the surface on magnetic materials.

A magnetic field is induced in a part by passing an electric current through or around it. The magnetic field is always at right angles to the direction of current flow. Magnetic particle inspection registers an abrupt change in the resistance in the path of the magnetic field, such as would be caused by a crack lying at an angle to the direction of the magnetic poles at the crack. Finely divided ferromagnetic particles applied to the area will be attracted and outline the crack.

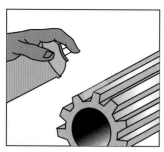

1. PRE-CLEAN INSPECTION AREA. SPRAY ON CLEANER/REMOVER – WIPE OFF WITH CLOTH.

2. APPLY PENETRANT, ALLOW SHORT PENETRATION PERIOD.

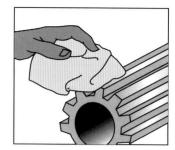

3. SPRAY CLEANER/REMOVER ON WIPING TOWEL AND WIPE SURFACE CLEAN.

4. SHAKE DEVELOPER CAN AND SPRAY ON A THICK, UNIFORM FILM OF DEVELOPER.

5. INSPECT. DEFECTS WILL SHOW AS BRIGHT RED LINES IN WHITE DEVELOPER BACKGROUND.

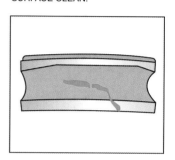

FIGURE 11.51 Penetrant testing

The flow or imperfection interrupting the magnetic field in a test part can be either longitudinal or circumferential, Figure 11.52. A different type of magnetization is used to detect defects that run down the axis, as opposed to those occurring around the girth of a part. For some applications you may need to test in both directions.

In Figure 11.53A, longitudinal magnetization allows the detection of flaws running around the circumference of a part. The part is placed inside an electrified coil to induce a magnetic field down the length of the part. In Figure 11.53B, circumferential magnetization allows the detection of flaws running down the length of a test part by sending an electric current down the length of the part to be inspected.

Radiographic Inspection

Radiographic inspection is a method for detecting flaws inside weldments. Radiography gives a picture of all imperfections that are parallel (vertical) or nearly parallel to the source. Imperfections that are perpendicular (flat) or nearly perpendicular to the source may not be seen on the X-ray film. These methods use invisible, short-wavelength rays developed by X-ray machines, radioactive isotopes (gamma rays), and variations. These rays are capable of penetrating solid materials and reveal most flaws in a weldment on an X-ray film or a fluorescent screen. After exposure and processing, flaws are revealed on films as dark or light areas against a contrasting background, Figure 11.54.

The defect images in radiographs measure differences in how the X rays are absorbed as they penetrate the weld. The weld itself absorbs most X rays. If something less dense than the weld is present, such as a pore or a lack-of-fusion defect, fewer X rays are absorbed, darkening the film. If something more dense is present, such as heavy ripples on the weld surface, more X rays will be absorbed, lightening the film. The X-ray image is a shadow of the flaw. The further the flaw is from the X-ray film, the fuzzier the image appears.

Skilled readers of radiographs who are also very knowledgeable about welding can interpret the significance of the light and dark regions by their shape and shading.

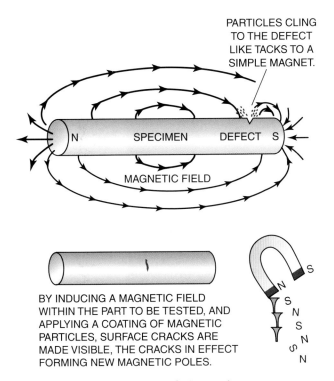

FIGURE 11.52 **Magnetic particle inspection**

PARTICLES CLING TO THE DEFECT LIKE TACKS TO A SIMPLE MAGNET.

BY INDUCING A MAGNETIC FIELD WITHIN THE PART TO BE TESTED, AND APPLYING A COATING OF MAGNETIC PARTICLES, SURFACE CRACKS ARE MADE VISIBLE, THE CRACKS IN EFFECT FORMING NEW MAGNETIC POLES.

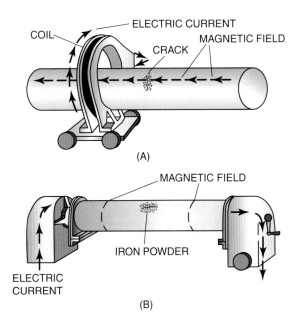

FIGURE 11.53 **Magnetic fields**

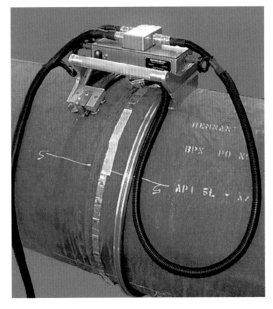

FIGURE 11.55 **New mobile X-ray equipment**

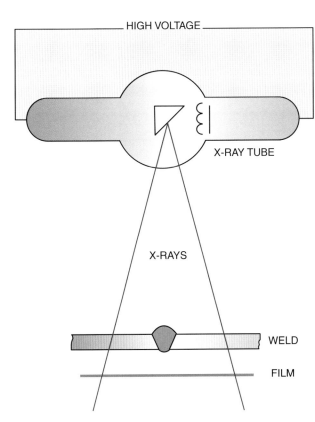

FIGURE 11.54 **Schematic of an X-ray system**

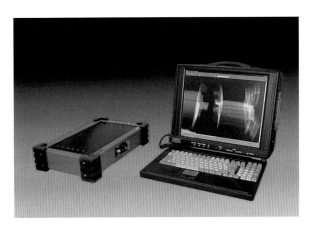

FIGURE 11.56 **Preparing to test the quality of a weld on a pipe using X-ray equipment**

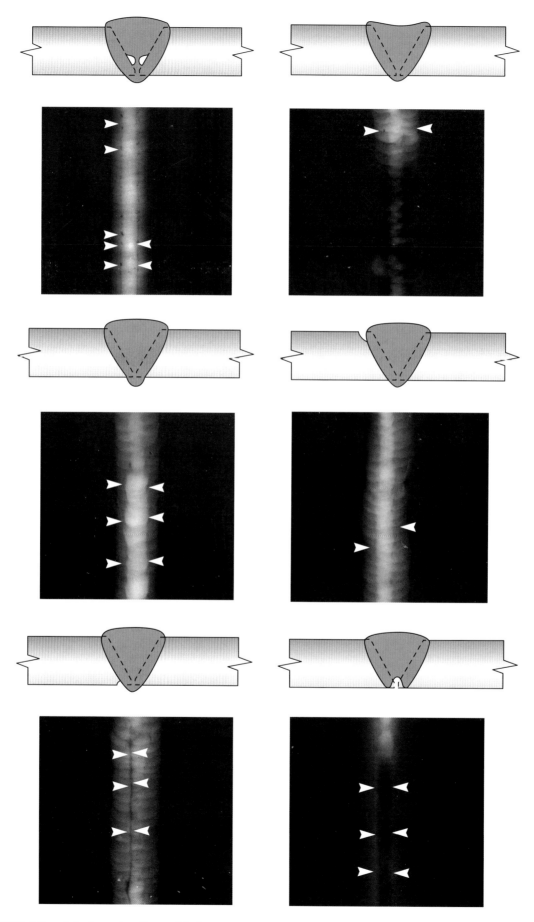

FIGURE 11.57 Welding defects and radiographic images

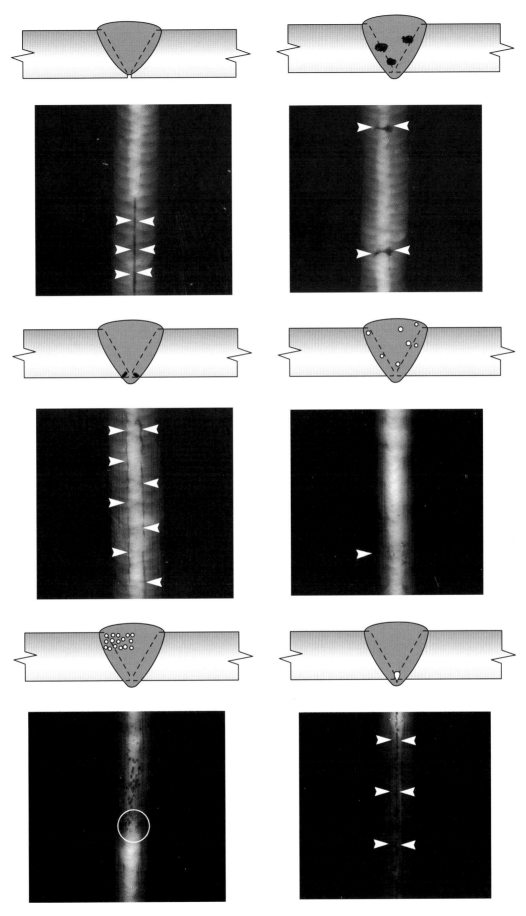

FIGURE 11.57 Welding defects and radiographic images (continued)

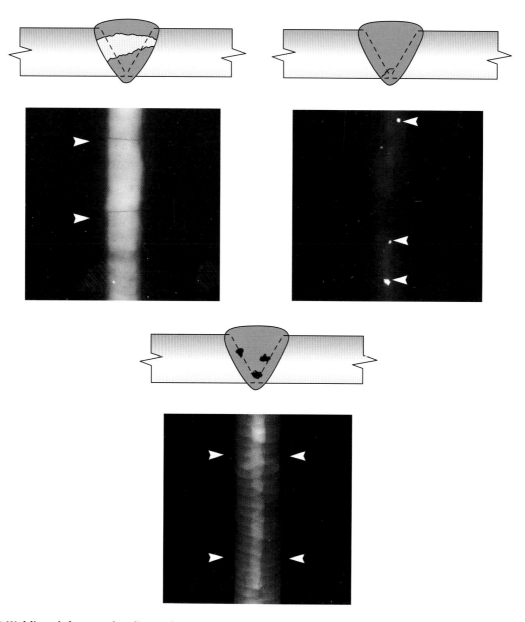

FIGURE 11.57 Welding defects and radiographic images (continued)

Figure 11.55 shows samples of common weld defects and a representative radiograph for each. Four factors affect the selection of the radiation source:

- thickness and density of the material;
- absorption characteristics;
- time available for inspection;
- location of the weld.

Portable equipment is available for examining fixed or hard-to-move objects. The selection of the correct equipment for a particular application is determined by the specific voltage required, the equipment's degree of utility, the economics of inspection and the production rates expected, Figures 11.56 and 11.57

Ultrasonic Inspection

Ultrasonic inspection is fast and uses few consumable supplies, which makes it inexpensive for colleges to use. However, because of the time required, it is sometimes not considered economic in the field as a non-destructive testing method. The ultrasonic inspection method employs electronically produced high-frequency sound waves, which penetrate metals and many other materials at speeds of several metres per second. A portable ultrasonic inspection unit is shown in Figure 11.58.

The two types of ultrasonic equipment are pulse and resonance. The pulse-echo system, most often employed in the welding field, uses sound generated in short bursts or pulses. Since the high-frequency sound used is at a relatively low power, it has little ability to travel through air, so it must be conducted from the probe into the part through a medium such as oil or water.

Sound is directed into the part from a probe held at a pre-selected angle or in a pre-selected direction so that flaws will reflect some energy back to the probe. These ultrasonic devices operate much like depth sounders, or sonar. The speed of sound through a material is a known quantity. The equipment measures the time taken for a pulse to return from a reflective surface. Internal computers calculate the distance and present the information on a display screen so that an operator can interpret the results. The signals can be monitored electronically to operate alarms, print systems or recording equipment. Sound not reflected by flaws continues into the part. If the angle is correct, the sound energy will be reflected back to the probe from the opposite side. Flaw size is determined by plotting the length, height, width and shape using trigonometric rules.

Figure 11.59 shows the path of the sound beam in butt welding testing. The operator must know the exit point of the sound beam, the exact angle of the refracted beam, and the thickness of the plate when using shear wave and compression wave forms.

FIGURE 11.58 **Portable ultrasonic inspection unit**

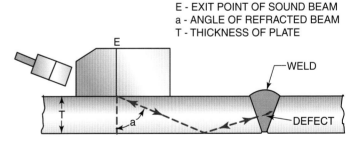

E - EXIT POINT OF SOUND BEAM
a - ANGLE OF REFRACTED BEAM
T - THICKNESS OF PLATE

FIGURE 11.59 **Ultrasonic testing**

Leak Checking

In leak checking, leaks can be found by filling a welded container with either a gas or a liquid. Additional pressure may be applied to the material in the weldment. Water is the most frequently used liquid, although sometimes a liquid with a lower viscosity is used. If gas is used, it can be detected with an instrument when it escapes through a flaw in the weld or as bubbles from an air leak.

Eddy Current Inspection

Eddy current inspection is based on a magnetic field that induces eddy currents within the material being tested. An eddy current is an induced electric current circulating wholly within a mass of metal. This method is effective in testing both non-ferrous and ferrous materials for internal and external cracks, slag inclusions, porosity, and lack of fusion on or very near the surface. Eddy current inspection cannot locate flaws that are not near the surface.

A coil carrying high-frequency alternating current is brought close to the metal to be tested. A current is produced in the metal by induction. The magnitude and phase difference of these currents are indicated by the impedance value of the pick-up coil. Careful measurement of this impedance allows the detection of defects in the weld.

SUMMARY

Quality is something that must be built into a product. The purpose of testing and inspection is to verify that the required level of quality is being maintained. The most important standard that a weld must meet is fitness for service. A weld must be able to meet the demands placed on the weldment without failure. No weld is perfect and both the welder and the welding inspector must be aware of the appropriate level of weld imperfections. Producing or inspecting welds to an excessively high standard will result in a product that is unnecessarily expensive.

Producing high-quality welds is a matter of skill and knowledge, and not luck. Knowing the causes and effects of weld defects and imperfections will help you develop as a welder. Good welders often know whether welds they produce will or will not pass inspection. Inspecting your welds as part of your training program will help you develop this skill.

ACTIVITY AND REVIEW

1 Name four components that would be listed on a mill certificate.

2 Define the term 'fit for purpose'.

3 Identify what the following abbreviations stand for:

   a BS
   b EN
   c ISO.

4 State what type of standard BS EN ISO 9000 applies to.

5 Describe how Gantt charts are used in production and what advantages can be seen from using them.

6 State what the term 'milestones' refers to.

7 Explain what a Statistical Process Control chart indicates.

8 A Weld Procedure Sheet (WPS) supplies a welder with specific information, name SIX sources of information supplied.

9 State how a Procedure Qualification Record (PQR) differs from a Weld Procedure Sheet (WPS).

10 Explain the term 'witnessed' and what it refers to.

11 Describe why calibration of equipment is important to critical welding applications.

12 State what changes an imperfection into a defect.

13 Name six features of a weld.

14 Explain what the abbreviation HAZ stand for.

15 Identify what the term 'as welded' means.

16 List six imperfections.

17 Describe what an autogenous weld is.

18 What is a lamination and how is it created?

19 Identify what gases are associated with porosity.

20 Explain how a delamination differs from a lamination.

21 Indicate how 'wormhole porosity' appears to the naked eye.

22 Name three sources of non-metallic inclusions.

23 State three types of porosity.

24 Give a brief description of three type of mechanical testing.

25 Describe the differences between the Brinell and Rockwell tests.

26 State the difference between root, face and side bend tests.

27 Explain how the Izod and Charpy impact tests vary.

28 State the procedure for carrying out a Macro-examination.

29 List three common etching solutions.

30 Explain how Nital is produced, and what material you would use it for.

31 Name the four most commonly used methods of Non-destructive testing.

32 Indicate what property must a penetrant have to be effective.

33 Describe how ultrasonic testing work.

34 List the commonly used NDT processes that are effective for surface or near surface defects.

35 Name the two methods of radiographic inspection.

Appendix

Brazing and filler wires for non-ferrous materials

Filler Rod	Diameter (mm)	Flux	Applications	Standard	Notes
Copper Silver Alloy	1.5, 3.2, 4.5, 6.4	Silver solder flux	Welding of arsenical, commercial, and de-oxidised copper	EN 1044 CP 102 BS: 1845 CP 1	Suitable for welding copper components, including electrical parts
Bronze Welding (Silicone)	1.5, 2.3, 3.2, 5.5, 6.4	Bronze Weld Flux	Bronze welding copper and Brass, sheet, tubes, also ferrous metals	EN 1044 CU 302 BS: 1453 C2 BS: 1845 CZ6	Also available as a ready fluxed rod
Bronze Welding (Nickel Bronze)	1.5, 3.2, 4.5	Bronze Weld Flux	Bronze welding steel or malleable iron	EN 1044 CU 305 BS 1453 C5 BS: 1845 CZ8	Building up worn surfaces, welding copper/zinc/nickel alloys of similar composition
Bronze Welding (Manganese)	3.2, 4.5, 6.4	Bronze Weld Flux	Bronze welding cast or malleable iron	BS: 1453 C4 BS: 1845 CZ7	Building up worn surfaces or fractured castings
Copper Phosphorous Brazing Alloy	2.3, 3.2, 4.5	Self-fluxing no need for additional fluxing	Brazing copper, brass and bronze parts for electrical switchgear, motors, cables, etc.	EN 1044 CP 201 BS: 1845 CP3	Not to be used directly on steel or cast-iron
Strip Spelter (Brazing)	2.3 & 6.4	Borax based flux	Brazing copper, copper alloys and ferrous metals	BS: 1845 CZ1	
Aluminium Bronze	3.2	Aluminium Bronze Flux	Welding aluminium bronze where shock, fatigue and dilute acid conditions exist		
Aluminium (Pure)	1.5, 2.3, 3.2, 4.5	Aluminium Welding	Used in the welding of pure aluminium sheet, tube and extrusions	BS: 1453 G1B	
Aluminium Alloy (5% Copper)	1.5, 3.2, 4.5, 6.4	Aluminium Welding	Welding aluminium castings especially those containing 5% copper		
Aluminium Alloy (5% Silicon)	1.5, 2.3, 3.2, 4.5, 6.4	Aluminium Welding	Welding pure aluminium and aluminium castings not containing zinc	BS: 1453 NG 21 BS: 1845 AL 4	
Aluminium Alloy (10–13% Silicon)	1.5, 2.3, 3.2, 4.5	Aluminium Welding or Brazing Flux	Welding high silicon aluminium alloys. Brazing pure aluminium & aluminium/ magnesium alloys up to 2% magnesium	EN ISO 18273 S AL BS: 2901 4047A (NG2)	
Aluminium Alloy (10% Silicon–4% Copper)	1.5 & 3.2	Aluminium Brazing Flux	Brazing aluminium and its alloys	BS: 1845 AL1	Difficult to apply on aluminium alloys containing more than 2% magnesium. Heat treatable deposit
Aluminium Alloy (5% Copper–2% Silicon)	6.4	Aluminium Welding Flux	Welding crankcases, gear boxes, or castings of similar composition		
Aluminium Alloy (5% Magnesium)	1.5, 2.3, 3.2, 4.5	Aluminium Welding Flux	Welding alloy parts in MG5 for automobiles and aeronautical work	BS: 1453 NG 6	Can also be used on corrosion-resisting tanks

Brazing table non-ferrous materials (continued)

Filler Rod	Diameter (mm)	Flux	Applications	Standard	Notes
Magnesium Alloy (1–2% Manganese)	3.2	Magnesium Flux	Welding alloys similar to DTD 118A	BS: 1453 D2	These alloys are better welded by the TIG process
Magnesium Alloy (10% Aluminium)	3.2 & 4.5	Magnesium Flux	Welding alloys containing 10% aluminium	BS: 1453 D1	
Zinc-base Die Casting Metal	3.2	Use aluminium Flux if necessary	Welding castings such as carburettors, and washing machine components	BS: 1004 A	
Non-ferrous Hardfacing Alloys	3.2, & 6.4	Set carburizing flame correctly. Use Cast Iron flux for Cast Iron surface	Primarily to withstand abrasion impact at various temperatures, high corrosion resistance and retains hardness at 'red' heat		A range of hardfacing alloys is available and careful selection needs to be used to ensure the right alloy is found to cover most requirements

Welding wires/filler for ferrous materials

Filler Rod	Diameter (mm)	Flux	Applications	Standard	Notes
Low Carbon Mild steel	1.5, 2.0, 2.4, 3.2, 4.5, 6.4	None	Welding of mild steel and wrought iron	BS: 1453 A1 EN 12536: 01	Often copper coated for protection
Silicon Manganese Steel	1.5, 2.3, 3.2, 4.5	None	Welding steel where mechanical properties are required	BS: 1453 A2 EN 12536:011	Used for welding steel pipework
High Tensile Silicon Manganese Steel	1.5, 2.3, 3.2, 4.5, 6.4	None	For high-tensile Steels of similar composition	BS: 1453 A3	Material with tensile strength 26–32 tons/in^2
3% Nickel Steel	1.5, 3.2	None	Repair work and building up nickel steels. Can be hardened and tempered	BS: 1453 A4	Useful for building up of worn camshafts, shafts and gears
Wear-resisting Alloy Steel	3.2, 4.5, 6.4	None	Restoring worn components-rock drills. Hardenable in oil	BS: 1453 A5	Can be used on any steel or cast-iron surface subject to shock and abrasion
Carbon Molybdenum Alloy	1.5, 3.2, 4.5	None	Used to weld boiler and super heater tubes or components subject to heat and stresss	BS: 1453 A6	Creep-resisting steel
Stainless Steel (Niobium Bearing)	1.5, 2.3 3.2, 4.5	Stainless Steel Flux	For welding stainless steel tubes, sheets of austenitic quality	BS: 1453 A8, Nb	
Heat-resisting Steel	1.5, & 3.2	Stainless Steel Flux	Welding heat resisting steels	BS: 1453 A11, Nb	
Stainless Steel (Molybdenum Bearing)	1.5, & 3.2	Stainless Steel Flux	Welding molybdenum bearing stainless	BS: 1453 A12	
High-Carbon & Low Alloy Steels	1.5, & 3.2	Use a flux to weld Cast Iron	For building up or repair work		Not recommended for metal working tools, hardenable in oil or water
Cast Iron	4.5, 6.4, 8.0, 9.5	Cast Iron or High Carbon Steel	Welding high grade castings, machine-able deposit	BS: 1453 B1	Suitable for cylinder blocks, brackets, lathe beds, etc.

Student welding report

STUDENT WELDING REPORT

Student Name: _____   Date: _____

Instructor: _____   Class: _____

Experiment or Practice #: _____   Process: _____

Briefly Describe Task: _____

_____

_____

_____

INSPECTION REPORT

Inspection	Pass/Fail	Inspector's Name	Date
Safety:			
Equip. Setup:			
Equip. Operation:			
Welding	Pass/Fail	Inspector's Name	Date
Accuracy:			
Appearance:			
Overall Rating:			

Comments:

Student Grade: _____   Instructor Initials: _____   Date: _____

Glossary

acetone a fragrant (garlic smelling) liquid chemical used in acetylene cylinders. The cylinder is filled with a porous material (kapok or prepared charcoal) and acetone is then absorbed by this material to stabilize the gas. Acetylene is then added and absorbed by the acetone, which can absorb up to twenty five times its own volume of the gas.

American Society of Mechanical Engineers (ASME) a validating body which determines and overseas welding of pressurized systems such as pressure vessels or pipe work.

American Welding Society (AWS) a multi-faceted, non-profit organization with a goal to advance the science, technology and application of welding and related joining disciplines.

ampere a unit of electrical current.

arc cutting a group of thermal cutting processes that severs or removes metal by melting with the heat of an arc between an electrode and the work piece.

arc plasma a state of matter found in the region of an electrical discharge (arc). See also plasma.

autogenous weld a weld in which all of the weld metal has come from the parent material only and no filler material is added.

automated operation operations are performed repetitively by a robot or other machine that is programmed flexibly to do a variety of processes.

automatic operation operations are performed repetitively by a machine that has been programmed to do an entire operation without the intervention of the operator.

back gouging a process of cutting a groove in the back side of a joint that has been welded.

back mark a measurement taken to ensure accuracy when marking out holes on structural steel sections.

blowback occurs when the nozzle end is partially blocked, which results in a build-up of pressure to clear the obstruction and a large bang. This condition can be remedied by cleaning the nozzle orifice.

brazing a process that uses heat from a fuel-gas flame or electrical induction with a low melting point filler material to flow between a joint by capillary attraction.

brinell hardness tester a tool that characterizes the indentation hardness of materials through the scale of penetration of an indenter, loaded on a material test-piece.

certification approval a widely respected document certifying that a welder has passed a performance qualification test at an accredited test facility.

combination welding symbol a welding symbol which indicates multiple operations.

confined spaces a space with limited or restricted means for entry or exit, and it is not designed for continuous employee occupancy. Confined spaces include, but are not limited to, underground vaults, tanks, storage bins, manholes, pits, silos, process vessels and pipelines.

coupling distance the distance to be maintained between the inner cones of the cutting flame or plasma cutter and the surface of the metal being cut.

cutting tip the part of an oxygen cutting torch from which the gases issue.

defect an imperfection that is unable to meet minimum acceptance standards or specifications.

delamination an imperfection such as oxide inclusion as a result of hot rolling of the material.

dexterity the ability to manipulate one or more objects.

drag (thermal cutting) the off-set distance between the actual and straight line exit points of the gas stream or cutting beam measured on the exit surface of the base metal.

drag lines high-pressure oxygen flow during cutting forms lines on the cut faces. A correctly made cut has up and down drag lines (zero drag); any deviation from the pattern indicates a change in one of the variables affecting the cutting process; with experience the welder can interpret the drag lines to determine how to correct the cut by adjusting one or more variables.

dross a mass of solid impurities (iron oxide) attached on the underside of a cut.

earmuffs a type of hearing protection that covers the entire ear.

earplugs a type of hearing protection that is fitted into the ear.

electric shock an electric shock can occur upon contact of a human's body with any source of voltage high enough to cause sufficient current through the body.

electrical ground a common return path for electric current (earth return or ground return), or a direct physical connection to the Earth.

electrical resistance a ratio of the degree to which an object opposes an electric current through it, measured in Ohms.

electrode setback the distance the electrode is recessed behind the constricting orifice of the plasma arc torch,measured from the outer face of the nozzle.

electrode tip the end of a welding electrode that is closest to the work.

electrolytic a reaction that takes place between the anode and cathode of an electrical circuit in the presence of some electrically conducting fluid.

empathy a feeling for the subject.

endothermic reaction a reaction in which heat is created internally by the combination of substances, in thermit welding all of the heat produced comes from the chemical reaction of elements.

etching the process of using acids, bases or other chemicals to dissolve and reveal any unwanted matter on the surface of the material.

exhaust pickup a component of a forced ventilation system that has sufficient suction to pick up fumes, ozone and smoke from the welding area and carry the fumes, etc., outside of the area.

exothermic gases combustable gasses providing heat as a by-product.

exothermic reaction a reaction in which heat is given off as in oxy-fuel cutting.

explosimeter a piece of equipment which can indicate the parts per million of explosive compounds present within a structure or air.

fillet weld a weld of approximately triangular cross section joining two surfaces approximately at right angles to each other in a lap joint, T-joint, or corner joint.

fitting the mechanical or hand machining of components or assemblies.

flash burn a burn caused by Ultra Violet (UV) light produced by a welding arc.

flash glasses eye protection specifically designed to filter out UV light.

flashback a serious occurrence in which the flame travels back up the torch and possibly the hoses by back pressure towards the gas bottles. This occurrence results in a loud squealing noise with sparks emitting from the end of the nozzle.

flux cored arc welding (FCAW) an arc welding process that uses an arc between a continuous filler metal electrode and the weld pool. The process is used with shielding gas from a flux contained within the tubular electrode, with or without additional shielding from an externally supplied gas, and without the application of pressure.

forced ventilation to remove excessive fumes, ozone or smoke from a welding area, a ventilation system may be required to supplement natural ventilation.

forge welding a solid state welding process that produces a weld by heating the workpieces to welding temperature and applying blows sufficient to cause permanent deformation at the hot surfaces.

french chalk available in rectangular form commonly used for marking metal.

frustrum is a cone with its top cut off parallel to the base of the cone.

full face shield protective equipment designed to cover the entire face.

fusion the flowing together or deposition into one body of the materials being welded.

fusion welding any welding process or method that uses fusion (joining metals to form one) to complete the Metal Inert Gas Shielded (MIG) and Metal Active Gas Shielded (MAGS). Arc welding processes that use an arc between a continuous filler metal electrode and the weld pool. The process is used with shielding from an externally supplied gas and without the application of pressure. In the case of MAGS the shielding gas is mixed with another gas to make it 'active' which gives improved mechanical properties such as increased surface deposition.

goggles special eye protection designed to seal around each eye.

gouging removal of metal by oxy-fuel or the plasma arc processes.

graphite a mineral form of carbon.

groove an opening or a channel in the surface of a part or between two components, that provides space to contain a weld.

hard dross a form of dross caused by oxy-fuel cutting that is difficult to remove.

hard facing the surface deposition of a wear resistant compound on an inferior carrier material.

heat-affected zone the area of base material which has had its microstructure and properties altered by welding or heat intensive cutting operations.

high-frequency alternating current a low amperage current that is superimposed over the welding current to assist in arc initiation in DC and to stabilize AC TIG welding operations.

high-speed cutting tip a special cutting tip usually constructed in two pieces that is designed for mechanized oxy-fuel cutting operations.

hot work permit a document required to be completed before beginning hot work operations in areas not specifically designated for welding or cutting.

housekeeping performance of duties to keep a welding shop or job site clean and free of hazards.

ignition point the temperature at which metals will start to oxidize in oxy-fuel cutting.

imperfection an interruption of the typical structure of a material, such as a lack of bonding in its mechanical, metallurgical or physical characteristics. An imperfection is not necessarily a defect.

infrared a form of electromagnetic radiation whose wavelength is longer than that of visible light, but shorter than that of microwaves. Infrared radiation is heat that can be felt at a distance.

ionized gas see Plasma.

joint configuration the type of joint produced by the joint dimensions. The details set out in prints or plans to note the shape required.

joint type a weld joint classification based on the five basic arrangements of the component parts such as a butt joint, corner joint, edge joint, lap joint and T-joint.

joules the SI unit of energy.

kerf the width of a saw, oxy-fuel or plasma cut.

layout putting work pieces in place in preparation for joining, with or without the assistance of clamps or fixtures.

machine mounted cutting torch cutting with equipment that requires manual adjustment of the equipment controls in response to visual observation of the cutting operation, with the torch held by and controlled by a mechanical device such as a profile mechanism.

machine operation welding operations are performed automatically under the observation and control of the operator.

mandatory the law to conform with statutory regulations.

manual metal arc is an arc welding process in which an arc is created between a covered electrode and the weld metal. The process is used with shielding from the decomposition of the electrode covering, without the application of pressure, and with filler metal from the electrode.

manual operation the entire welding process is manipulated by the welding operator.

Material Safety Data Sheet (MSDS) hazards and properties form containing data regarding the properties of a particular substance.

measuring the process of estimating the magnitude of some attribute of an object, such as its length or weight, relative to some standard (unit of measurement), such as a metre or a kilogram.

mechanical testing tests carried out to the specimen's failure in order to understand its structural performance or material behaviour under different loads. These tests are generally much easier to carry out, give more information and are easier to interpret than non-destructive testing.

metallurgy the science of materials and structures formed.

MPS gases a group of flammable gasses containing Methylacetylene Propadiene, stabilized.

natural ventilation the process of supplying and removing air through an indoor space by natural means.

non-destructive testing (NDT) testing that does not destroy the test object; also called non-destructive examination (NDE) and non-destructive inspection (NDI).

nozzle the end piece of a welding torch or gun that directs the flow of gas.

nozzle insulator a non-conductive piece that separates the nozzle from the contact tube.

nozzle tip the end of a nozzle, sometimes replicable on heavy duty MIG, MAG or FCAW equipment.

ohm a unit of electrical resistance.

orifice a term used to describe the hole at the end of a tube or nozzle.

oxy-acetylene hand cutting torch hand held apparatus used for making oxy-fuel cuts.

oxy-fuel gas cutting a group of oxygen cutting processes that use heat from an oxy-fuel gas flame.

oxy-fuel gas welding a group of welding processes that produces fusion of the work pieces by heating them with an oxy-fuel gas flame. The processes are used with or without filler metal.

part dimensions the size, shape and contour of a part represented on a drawing, sketch or blueprint.

peening using the ball part of a ball-peen hammer to counteract contraction on weld deposits in certain materials.

pilot arc a low-current arc between the electrode and the constricting nozzle of the plasma arc torch to ionize the gas and facilitate the start of the welding arc.

plasma a gas that has been heated to an at least partially ionized condition, enabling it to conduct an electric current.

plasma arc cutting (PAC) an arc cutting process that uses a constricted arc and removes the molten metal with a highvelocity jet of ionized gas issuing from the constricting orifice.

plasma arc gouging a thermal gouging process that removes metal by melting with the heat of a plasma arc torch.

plug weld a joint commonly used in motor vehicle bodywork in which one of the components to be joined has a preprepared section removed to allow access to the bottom panel. The weld is completed by filling the section and the bottom panel together.

pre-heat flame brings the temperature of the metal to be cut above its ignition point, after which the high-pressure oxygen stream causes rapid oxidation of the metal to perform the cutting.

pre-heat holes the cutting tip has a central hole through which the oxygen flows. Surrounding this central hole are a number of other holes called pre-heat holes. The differences in the type or number of pre-heat holes determine the type of fuel gas to be used in the tip.

projection drawings are sketches, drawings or computer models presented in one of several view types, such as isometric or orthographic projections.

prototype a one-off product designed to prove specification and become the model for future production.

qualification see preferred terms welder performance qualification and procedure qualification.

quality control a system that ensures products or services are designed and produced to meet or exceed customer requirements.

radiographic inspection (RT) a non-destructive testing (NDT) method of inspecting materials for hidden flaws by using the ability of short wavelength electromagnetic radiation (high energy photons) to penetrate various materials.

rockwell hardness a hardness scale based on the indentation hardness of a material. The Rockwell test determines the hardness by measuring the depth of penetration of an indenter under a large load compared to the penetration made by a preload.

safety glasses eye protection worn on the face to protect the eyes from impact, sparks or dust.

semi-automatic operation during the welding process, the filler metal is added automatically, and all other manipulation is performed manually by the operator.

shear strength as applied to a soldered or brazed joint, it is the ability of the joint to withstand a force applied parallel to the joint.

slag a non-metallic product resulting from the mutual breakdown of flux and non-metallic impurities in some welding processes.

snifting the rapid opening and closing of a gas bottle to remove any residue that may have accumulated during storage of the gas cylinders. This procedure should *never* be carried out with hydrogen as it is possible to produce an explosion by the sudden release of hydrogen in the atmosphere.

soft dross the most porous form of dross which is most easily removed.

spatter a fine deposition of filler material as the result of too high a current range or too long an arc length.

specification a specification is an explicit set of requirements to be satisfied by a material, product or service.

stack cutting thermal cutting of stacked metal plates arranged so that all the plates are severed by a single cut.

stand-off distance the distance between a nozzle and the workpiece.

standard an established norm or requirement. A technical standard is usually a formal document that establishes uniform engineering or technical criteria, methods, processes and practices.

strongback a structure used to support the underside of a welded joint and to maintain shape by controlling distortion and act as a 'heat sink'.

synchronized wave form an alternating current (AC) type where the positive and negative sides of the AC cycle may be manipulated.

tack weld a weld made to hold the parts of a weldment in proper alignment until the final welds are made.

tip cleaners tools used to clean the orifices of oxy-fuel torch tips made of abraded steel wires of specific sizes.

tolerance the allowable deviation in accuracy or precision between the measurement specified and the part as laid out or produced.

tufting a surface deposition of flux on the end of the filler material when brazing or silver soldering. The 'tuft' is produced by heating the end of the filler wire and dipping into the flux which causes adhesion of the flux.

tungsten inert gas shielded (TIG) and tungsten active gas shielded (TAGS) are arc welding processes that use an arc between a tungsten electrode (non-consumable) and the weld pool. The process is used with shielding gas and without the application of pressure.

ultrasonic inspection (UT) very short ultrasonic pulse-waves launched into materials to detect internal flaws or to characterize materials somewhat similar to sonar detection.

ultraviolet light is a form of electromagnetic radiation with a "wavelength shorter than that of visible light, but longer than X-rays."

valve protection cap a protective cover which fits on a compressed gas cylinder.

ventilation the intentional movement of air from outside a building to the inside.

venturi a condition in which gas or fluid can be introduced or accelerated. This effect can be induced in welding operations by reducing the angle of the nozzle which reduces gas coverage and increases potential for atmospheric gases to be drawn into the welding zone.

venturi effect which acts like a slipstream to draw other gases into the weld.

visible light the visible spectrum (sometimes called the optical spectrum) is the portion of the electromagnetic spectrum that is visible to (can be detected by) the human eye.

volt a unit of electrical pressure.

warning label a form of hazard communication that attaches directly to an object.

water table a special table designed for plasma arc cutting operations, where the torch head is submerged under water in order to reduce smoke and noise.

weld a localized joining of metals or non-metals produced either by heating the materials to suitable temperatures, with or without the application of pressure, or by the application of pressure alone and with or without the use of the filler material.

weld joint the junction of members or the edges of members that are to be joined or have been joined by welding.

weld test a welding performance test to a specific code or standard.

welder certification a widely respected document certifying that a welder has passed a performance qualification test at an accredited test facility.

welder performance qualification the demonstration of a welder's or welding operator's ability to produce welds meeting prescribed standards.

welding code a document or specification governing aspects of process and procedures for the joining of materials by welding.

welding helmet safety equipment designed to protect the welder's face and head from radiation, sparks, spatter and fumes associated with welding and cutting operations.

welding position the relationship between the weld pool, joint, joint members and the welding heat source during welding.

welding procedure qualification record (WPQR) a record of welding variables used to produce an acceptable test weldment and the results of tests conducted on the weldment to qualify a welding procedure specification.

welding procedure specification (WPS) a document providing in detail the required variables for specific application to assure repeatability by properly trained welders and welding operators.

welding schedule a written statement, usually in tabular form, specifying values of parameters and the welding sequence for performing a welding operation.

welding symbol a graphical representation of a weld.

weldment a component in which one or more joints are joined by a range of welding processes.

witness mark dot punch indentations used to define a line to which a material is to be filed, machined or cut to.

workmanship standards the quality requirements for a part to be produced.

Index